FOREST PATTERN AND ECOLOGICAL PROCESS

Dedications

To the great and extraordinary montane ash forest – the true source of learning, inspiration and insight

To my field ecology team – it would not be possible to ask for a better and more valuable contribution

To all my many and greatly valued research colleagues and their wonderful contributions

To the statistical acumen of Ross Cunningham, Christine Donnelly, Jeff Wood, Alan Welsh and Emma Knight

To the memory of Steve Craig

To the memory of the people and the town of Marysville

FOREST PATTERN AND ECOLOGICAL PROCESS

A Synthesis of 25 Years of Research

DAVID LINDENMAYER

CSIRO PUBLISHING

© David Lindenmayer 2009

All rights reserved. Except under the conditions described in the *Australian Copyright Act* 1968 and subsequent amendments, no part of this publication may be reproduced, stored in a retrieval system or transmitted in any form or by any means, electronic, mechanical, photocopying, recording, duplicating or otherwise, without the prior permission of the copyright owner. Contact **CSIRO** PUBLISHING for all permission requests.

National Library of Australia Cataloguing-in-Publication entry

Lindenmayer, David.

Forest pattern and ecological process : a synthesis of 25 years of research / David Lindenmayer.

9780643096608 (pbk.)

Includes index.
Bibliography.

Eucalyptus regnans – Ecology – Victoria.
Forest ecology – Victoria.
Forest conservation – Victoria.
Biodiversity conservation – Victoria.

583.7661709945

Published by

CSIRO PUBLISHING
150 Oxford Street (PO Box 1139)
Collingwood VIC 3066
Australia

Telephone: +61 3 9662 7666
Local call: 1300 788 000 (Australia only)
Fax: +61 3 9662 7555
Email: publishing.sales@csiro.au
Web site: www.publish.csiro.au

Front cover photo by Esther Beaton.
Back cover photo by David Lindenmayer.

Set in 10/13 Adobe Minion and ITC Stone Sans
Edited by Adrienne de Kretser, Righting Writing
Cover and text design by James Kelly
Typeset by Desktop Concepts Pty Ltd, Melbourne
Index by Russell Brooks
Printed in China by 1010 Printing International Ltd

The paper this book is printed on is certified by the Forest Stewardship Council (FSC) © 1996 FSC A.C. The FSC promotes environmentally responsible, socially beneficial and economically viable management of the world's forests.

CSIRO PUBLISHING publishes and distributes scientific, technical and health science books, magazines and journals from Australia to a worldwide audience and conducts these activities autonomously from the research activities of the Commonwealth Scientific and Industrial Research Organisation (CSIRO).
The views expressed in this publication are those of the author(s) and do not necessarily represent those of, and should not be attributed to, the publisher or CSIRO.

Preamble

One wonderful thing about having children is reading them stories. In children's stories, forests are often portrayed as dark and foreboding places where it's all too easy to become lost and threatened by terrifying animals. This imagery has been repeated for centuries, from the fables of Little Red Riding Hood, Snow White and the Seven Dwarfs, and Hansel and Gretel, to modern classics such as the Harry Potter series and the Lord of the Rings.

Many of these perceptions flow from fiction into reality. Life imitating art in this way partly results from a lack of understanding of forests and how they work, and the relationships between plants and animals that underpin the way forests function.

This book outlines much of what has been learned from my 25 years of work in the extraordinary montane ash forests of the Central Highlands of Victoria. This learning highlights why forests are to be marvelled at. Just like learning to appreciate the art of literature, its patterns and process, there is a certain magic in learning to understand the patterns which characterise a natural forest and the ecological processes which give rise to those patterns.

David Lindenmayer
March 2009

Preface

Although this book is about how a forest works, it also tells the story of much of my life. It's a book that has taken 25 years to write – and perhaps may never be truly finished, even in another 25 years. My first 'real job' in ecological science was in Victoria's montane ash forests. That was in 1983. That initial work focused on understanding the distribution, abundance and conservation of one species – the nationally endangered arboreal marsupial Leadbeater's Possum. This iconic animal is one of Victoria's two faunal emblems and its distribution coincides almost entirely with the wet ash-type eucalypt forests of the Central Highlands of Victoria. More than 25 years after I began, we are still working on Leadbeater's Possum, but, as this book shows, the scope of the research has been greatly broadened. The work now encompasses many species, including other arboreal marsupials and many taxa from other groups such as birds and terrestrial mammals. Understanding interrelationships between wildlife and their habitats has demanded a better understanding of the forest. As a result, the research program has expanded to include the structure and composition of stands and landscapes. Forests are dynamic entities and our research now includes work on natural disturbances and human disturbances, and the similarities and differences between them. Of course, work on disturbance has entailed examining the impacts of logging on forest structure, forest composition and forest biodiversity. A key thread throughout the 25 years of research has been learning how to mitigate logging effects.

The body of research summarised in this book first began in 1983. In late 2007 my close friend and colleague, Professor Jerry Franklin, insisted that it was time to draw together the published material and general insights from the 25 years of research. Another friend, Professor Gene Likens, reinforced the importance of ecological synthesis and further encouraged me to undertake the task of writing this book. That idea was supported by Gordon Hickey, formerly of the Victorian Department of Sustainability and Environment, and John Manger of CSIRO Publishing. The task proved to be far from simple – the work in the Central Highlands of Victoria has generated approximately 165 scientific papers and four previous books. It has spanned several broad themes: from single species work to species assemblages research, spatial patterns (e.g. forest cover and composition) to work on ecological processes (e.g. fire and tree fall), empirical work and simulation modelling, observational studies to true experiments, and ecological theory versus applied ecological research.

An appreciation of how a forest works has developed not only from a deep understanding of particular topics but also from linking those topics – through synthesis and identifying synergies, and through understanding the patterns which characterise forest landscapes and the ecological processes that give rise to those patterns. This book aims to capture some of the thinking on how the montane ash forests of Victoria work, not only through a summary of the extensive research from the past 25 years but also by connecting the common ground among broad research themes.

This book has been written for a broad audience, including other researchers, resource managers, policy-makers, naturalists and readers with a general interest in forests.

David Lindenmayer
March 2009

Postscript

On the morning of Friday 6 February 2009, I completed a full draft of this book and sent it to CSIRO Publishing in Melbourne. Less than 48 hours later – on Black Saturday – the Central Highlands began burning. This means that much of what was written has formed the backdrop for learning about post-fire ecological and biodiversity recovery. This is another learning journey about forest pattern and ecological process – a journey that will take at least another 25 years. Perhaps, if I am lucky enough to still be working in the forests of the Central Highlands of Victoria in 2034, there might be a second edition of this book that will document further exciting discoveries and insights about the patterns and processes which make montane ash forests the extraordinary places they are.

Acknowledgements

This book was written as part of completing milestones for a monitoring grant from the Victorian Department of Sustainability and Environment. Ian Miles, Gordon Hickey, Kylie White, Wil Blackburn and Lee Miezis have been the champions for this project.

This book spans many research and management themes. The quality of work in the themes was maintained through close collaboration with leading professional and accredited statisticians. In particular, Ross Cunningham, Alan Welsh, Jeff Wood, Christine Donnelly and Emma Knight have played critical roles in the design, implementation, analysis, interpretation and writing of much of the work that is summarised in this book.

I have been fortunate to work with many other outstanding scientists over the past 25 years. It is not possible to name them all without forgetting (and offending!) someone. However, several people deserve special mention. They include Sam Banks, Mark Burgman, Andrew Claridge, Don Driscoll, Jean Dubach, Joern Fischer, Jerry Franklin, Malcolm Gill, Mike Hutchinson, Bob Lacy, Brendan Mackey, Mike McCarthy, Henry Nix, Hugh Possingham, Kim Ritman and Mick Tanton.

I am most grateful to Andrew Smith for first giving me an opportunity to work in the montane ash forests.

Over the past decade, the work in the Central Highlands of Victoria has been co-ordinated on the ground by a series of experienced Senior Research Officers, each of whom has done an outstanding job of running complex projects in difficult terrain. I am indebted to the outstanding work of Mason Crane, Ryan Incoll, Chris MacGregor, Lachie McBurney, Damian Michael, Rebecca Montague-Drake and Matthew Pope.

In 1999 and 2000 it was a privilege to work closely with the outstanding nature photographer Esther Beaton. She has taken many wonderful photographs of the forests of the Central Highlands of Victoria. Some of her photographs appear in this book and in a previous one (*Life in the Tall Eucalypt Forests*). My best photos fall far short of her discards!

The work in the Central Highlands of Victoria has progressed because of the support of many organisations and funding bodies. These include:

- Victorian Department of Sustainability and Environment;
- Parks Victoria;
- Earthwatch Institute;
- Victorian Department of Primary Industries;
- Federal Department of Agriculture, Fisheries and Forestry;
- Federal Department of the Environment, Water, Heritage and the Arts;
- Forest and Wood Products Australia;
- Australian Greenhouse Office;
- Myer Foundation;
- Herman Slade Foundation;
- Poola Foundation;
- Kendall Foundation;
- Land and Water Australia.

Two good friends, Jim Atkinson and Di Stockbridge, have been very generous private donors to my work for many years.

There are literally thousands of volunteer observers who were vital in completing much of the work reported in this book. In particular, assistance from the Earthwatch Institute, the Field Naturalists' Club of Victoria and the Friends of Leadbeater's Possum has been essential for completing night-time surveys of arboreal marsupials.

Rachel Muntz assisted in the collection of important reference material and subediting for this book. John Stein completed the spatial analyses that facilitated the maps in Chapter 2. Steve Holliday provided picture research. Clive Hilliker's expertise in graphical design assisted in the redrawing of many of the figures in this book.

Comments from Mark Antos, John Manger, Rachel Muntz and an anonymous reviewer improved an earlier version of the text.

I am deeply grateful to John Manger for encouraging me to sit down and write this book. He is undoubtedly the best publisher in Australia.

Finally, my sincerest thanks go to my family and my parents, whose critical support allowed me to complete this book – an exercise that proved to be far more difficult and protracted than initially anticipated.

Contents

Preamble	v
Preface	vii
Acknowledgements	ix

Part I Introduction and background — 1

1. Introduction — 3
2. Background — 9
3. Field survey methods — 29

Part II Forest cover and composition of the forest — 45

4. The ash-type eucalypt forest — 47
5. The rainforest — 61

Part III The structure of the forest — 71

6. Key structural features: overstorey trees with hollows — 73

   Colour Plates — 85

7. Key structural features: understorey trees and the shrub layer — 113
8. Key structural features: logs — 125

Part IV Animal occurrence — 131

9. Distribution and abundance of individual species — 133
10. Viability of populations of individual species — 151
11. Composition of animal communities — 167

Part V Disturbance regimes — 179

12. Natural disturbance regimes: fire — 181
13. Human disturbance: logging — 193
14. Salvage logging effects — 205

Part VI Forest management and biodiversity conservation — 215

15. Reserves — 217
16. Mitigating logging impacts — 229
17. Monitoring — 245

Part VII Conclusions and future directions — 255

18. Conclusions and future directions — 257

Bibliography	271
Index	303

PART I

INTRODUCTION AND BACKGROUND

There are three chapters in Part I. Together they contain introductory material important for better understanding the context for, and the results of, the research described in subsequent chapters.

Chapter 1 outlines the rationale behind writing the book, the breadth of material in this volume, how the book is structured and the reasons for that structure. It also discusses some of the omissions that had to be made – what was left out and why.

Details of the study area in the Central Highlands of Victoria are set out in Chapter 2. These include climate, patterns of human settlement, land tenure, geology, past natural disturbances and past and ongoing human disturbances. Chapter 2 also summarises some of the biology of the key vertebrate and plant species in the study area.

Chapter 3 describes the range of field methods used to study forest animals, particularly arboreal marsupials, birds and small terrestrial mammals, which have been a core focus of our work for the past two-and-a-half decades. It also outlines the field techniques used to characterise vegetation cover data which were subsequently part of other studies in the Central Highlands of Victoria.

1 Introduction

The wet montane ash forests of the Central Highlands of Victoria have been the broad location for many decades of field research. Indeed, one of the first published papers in the discipline of ecology was from the Central Highlands of Victoria (Jarrett and Petrie 1929). The forests in the region are truly spectacular; the tallest flowering plants in the world are from this region (Ashton 1975a; Beale 2007). Some of the world's most enigmatic and iconic animal species are found in the Central Highlands of Victoria. Virtually the entire known distribution of the nationally endangered Leadbeater's Possum (*Gymnobelideus leadbeateri*) occurs in the region (Lindenmayer 2000). The ash-type forests are also important for the production of water (O'Shaughnessy and Jayasuriya 1991; Vertessey and Watson 2001; Watson *et al.* 1999) and the production of timber and pulpwood (Gooday *et al.* 1997; Govt of Victoria 1986; Lindenmayer 2007). The inherent tensions between forestry, water production and biodiversity conservation in the montane ash forests of the Central Highlands of Victoria (Ajani 2007) also occur in many other forests worldwide (Fries *et al.* 1997; Haynes *et al.* 2006; Lindenmayer and Franklin 2002; Spies *et al.* 2007). Therefore, it is hoped that the body of work in this book might prove to be a model research and management program for other places, particularly through illustrating the many kinds of ways in which different scientific ideas evolve and are interlinked.

The body of work from the Central Highlands of Victoria is perhaps different from many other projects around the world in that many different questions have been posed and many kinds of work have been undertaken in the one region. The continuously expanding depth and breadth of work over the past 25 years has been a remarkable personal journey. This made it hard to know when to write this book, as the work has no clear endpoint. New insights lead to new questions and continue to revitalise the research – a characteristic of what some authors consider to be successful long-term research and monitoring (Strayer *et al.* 1986). Indeed, the events of February 2009 in which large areas of montane ash forest were intensively disturbed by wildfires mean that many years of work are needed to document the process of ecological recovery and biotic response to that major conflagration.

Initially, the primary focus of the work in the Central Highlands of Victoria was on where Leadbeater's Possum occurred within montane ash forests. However, an early (and obvious) insight was that the species did not exist in isolation from other species. It

Figure 1.1: The broad range of topics tackled over the past 25 years in the Central Highlands of Victoria. (a) Single species research – Leadbeater's Possum. (b) Sooty Owl – a predator of Leadbeater's Possum (illustrating multi-species interactions). (c) Landscape pattern – forest cover. (d) Ecological processes – logging. (Photos by David Lindenmayer/Esther Beaton).

was clear that information on other species of arboreal marsupials might help better identify where Leadbeater's Possum did and did not occur, and the underlying reasons for those patterns of occurrence. Hence, studies of a single species rapidly expanded to encompass other species and ultimately the entire assemblage of possums and gliders.

The composition and structure of the forests strongly influence where these animals live. It therefore became critical to better understand temporal and spatial variation in the structure and composition of stands and landscapes. Habitat suitability is strongly shaped by the ecological process of disturbance, making it essential to quantify the impacts of fires and logging on stand structure and landscape composition. Learning how to mitigate the impacts of human disturbance on forest biodiversity is therefore informed by quantifying where species occur, why they occur where they do, and how disturbance dynamics influence the key resources required by the target species. Of particular concern is how logging changes key attributes of stand structure and landscape composition and, in turn, how logging practices might be altered to reduce such impacts.

WHY THIS BOOK WAS WRITTEN

The overarching aim of this book is to describe some (certainly not all!) of the different kinds of work our team at The Australian National University has under-

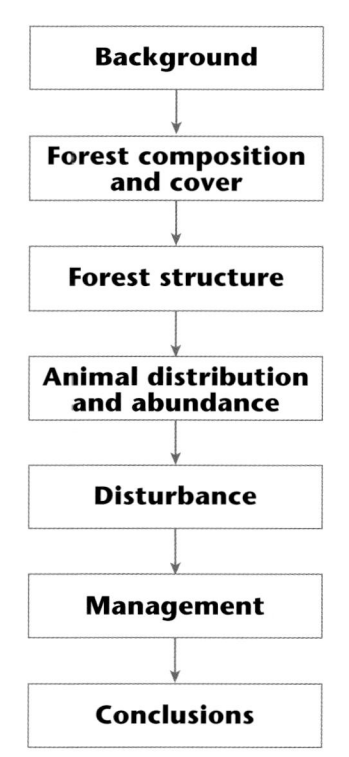

Figure 1.2: Linked research themes tackled in the ash forests of the Central Highlands of Victoria.

taken in the Central Highlands of Victoria, and highlight how they are interlinked. The approach is not to clumsily rehash thousands of pages of scientific publications. Rather, the focus is on a summary of information, with a particular emphasis on the synthesis of material across themes, the important lessons learned and the new insights gained. Such syntheses have not featured in previous books on the montane ash forests of Victoria. Those have focused on particular topics such as the biology of Leadbeater's Possum (Lindenmayer 1996; Lindenmayer and Possingham 1995c) and disturbance regimes in these forests (Mackey et al. 2002).

The hope is that a synthesis of the many reports scattered in different journals and books might be more than the sum of the parts, with such a body of work being useful for others with interests in studying the impacts of landscape change on biodiversity and/or with interests in establishing large-scale multi-faceted field research programs. This, in turn, may help speed progress on ways to better conserve forest biodiversity, including in landscapes used for the production of such commodities as timber, pulpwood and woodchips and landscapes that have been heavily disturbed by major wildfires such as those that occurred in February 2009.

STRUCTURE OF THIS BOOK

This book is a series of seven logically linked parts. Part I contains two chapters in addition to this one. Chapter 2 provides some background to the research in the Central Highlands of Victoria, particularly the characteristics of the study region, and its flora and fauna. Chapter 3 outlines the various field methods that have been used over the past 25 years.

Part II is about patterns of forest cover and composition. The first chapter in Part II concerns the pattern and composition of wet eucalypt forest (Chapter 4). The second (Chapter 5) focuses on cool temperate rainforest, which can be common in parts of the Central Highlands of Victoria.

Stand structure is the core theme of Part III. The three chapters in this part of the book encompass large overstorey trees with hollows (Chapter 6), understorey trees and tall shrubs (Chapter 7) and the ground layer, including logs and coarse woody debris (Chapter 8).

Part IV is about animal occurrence and comprises three chapters. Single species work features in Chapter 9, particularly research on the factors influencing the occurrence of individual species of arboreal marsupials, birds and small terrestrial mammals. The persistence of species in a given area obviously has a large influence on patterns of distribution and abundance. Therefore, Chapter 10 summarises a body of work on population viability analysis. Chapter 11 focuses on assemblages of animals, again with the emphasis on mammals and birds.

Part V is about the key ecological process of disturbance in the Central Highlands ash forests. Chapter 12 examines wildfire and how it varies in severity and impacts within and between landscapes. Chapters 13 and 14 focus on two kinds of human disturbance. These are traditional logging practices, and salvage logging after natural disturbance (wildfire).

Forest management is the unifying theme of Part VI. Chapter 15 is about reserves and their management. Off-reserve management aimed at mitigating the impacts of logging in areas broadly designated for timber harvesting is the key topic of Chapter 16. Improving management and monitoring the effectiveness of management actions is discussed in Chapter 17.

The final part of book (Part VII) contains a single chapter (Chapter 18). This is an overview of the body of the research to date and some comments about future directions, emerging challenges and key questions to be addressed.

STRUCTURE OF CHAPTERS

The 18 chapters vary markedly in length. Some, like this one, are very short. Others, such as Chapters 6, 9 and 16, are far longer and more detailed. This simply reflects the different emphases of research topics and programs over the past decades.

Most of these chapters follow a particular format. A short introduction is followed by a summary table of studies on the chapter topic in the montane ash forests over the past 25 years. The background datasets used in those studies are also described. The main part of each chapter is an overview of the key research work and the findings from that work. Almost all things are connected to other things in ecology – an important way forward in the discipline is to break apart the links so that scientifically tractable questions can be posed and defined pieces of work can be completed to address those questions. Otherwise, a researcher can be 'snowed by a blizzard of ecological details', hampering scientific progress and thwarting improved understanding. The main section of each chapter therefore concentrates on the findings of defined and contained pieces of ecological research. Because the main section is often long and multi-faceted it is followed by a summary, as well as a section on lessons learned. Despite the extensive work in the montane ash forests over the past 25 years, there is still much that is not known, or known only poorly. The penultimate section of each chapter is therefore one on knowledge gaps. This is not a shopping list of additional research projects, but key areas of work that need to be addressed.

While ecological science is often most successfully undertaken by the reduction of a problem into tractable questions, it is also important to comprehend how different bodies of work are linked. Therein lies the art of ecological synthesis. Therefore, the final section in each chapter (other than the background chapters, 1–3) focuses on synthesis, in particular the interrelationships between forest patterns and ecological processes and, in turn, the key linkages between different topics and bodies of work. This section underscores an important overarching theme of the book, as reflected in the title – developing a better understanding of how the montane ash forest ecosystems actually function.

Ecological processes and forest patterns

Although often short, the sections on forest pattern and ecological process proved very difficult to write. There are many key ecological processes in montane ash forests, as in all forests, and they influence a wide range of forest patterns. However, there is often not a simple linear relationship between ecological processes and forest patterns. Indeed, there are many cases where a pattern (e.g. the cover of rainforest) might influence an ecological process (e.g. the intensity and spatial contagion of a natural disturbance). In other cases, a given forest pattern (e.g. the spatial arrangement of hollow trees) can affect another pattern (e.g. the occurrence of cavity-dependent animals). There are also cases where a given ecological process is a fundamental driver of other ecological processes. Natural disturbance is a classic example; it can alter carbon budgets and nutrient cycling, hydrological regimes and subsequent natural disturbance events. These nuances mean that it is not straightforward to define precisely what is an ecological process and what is a forest pattern. In fact, the same entity might be a pattern in one context but a process in another. Therefore, no attempt has been made in this book to set unyielding rules about which entities are processes and which are patterns.

The discussion on ecological processes and forest patterns serves to illustrate the strong linkages between the various topics in the different chapters. The section on ecological processes and forest patterns in each chapter therefore often contains what has been termed a topic interaction diagram to illustrate these linkages – a set of conceptual models of how different components of montane ash forest function via connections between processes and patterns as well as between different patterns. These interaction diagrams are often presented in a highly simplified manner to communicate links or sometimes relationships between process and pattern. The emphasis is on 'simplified' because, clearly, these process–pattern and pattern–pattern interrelationships are often not simple.

CAVEATS

This book contains many references to my own publications arising from collaborative work in the Central Highlands of Victoria. This will frustrate some readers and an apology is issued to those who find it annoying. Self-citation was somewhat unavoidable, but the work of others has been referenced where appropriate.

There are strong links between many of the research topics that have been completed over the 25 years. This has resulted in some unavoidable overlap between chapters. This has arisen, in part, because data on particular groups (e.g. arboreal marsupials and birds) were used in several kinds of studies and for different questions. It also is due to the fact that many studies are relevant to several themes discussed in the book. As an example, our studies of nest boxes are relevant not only to the discussion in Chapter 9 on tree hollows and stand structure but also to forest management and mitigating the impacts of logging (Chapter 16).

Omissions

The particular focus of this book is the biology and ecology of montane ash forests. Extensive work on the biology, life history and genetics of individual species, such as Leadbeater's Possum and the Mountain Brushtail Possum, has been left out, either because it has been included in earlier books (Lindenmayer 1996; Lindenmayer and Possingham 1995) or because it will be part of forthcoming books (e.g. on the Mountain Brushtail Possum). This book does not include a detailed exploration of silvicultural systems and logging methods, other than the impacts of traditional kinds of clearfelling and post-wildfire salvage logging on forest structure and vertebrate biota (Chapters 13, 14 and 16). This will be considered an oversight by some traditional foresters, but large bodies of work on silvicultural systems already exist for these forests (e.g. Campbell 1997; Flint and Fagg 2007). It was a deliberate decision not to repeat what others have done.

Finally, despite the best efforts of the team of statisticians and ecologists involved in the work in the Central Highlands of Victoria, the studies completed to date are far from perfect. Perfect ecological studies do not exist – even in the minds of the strictest of theoreticians! Some limitations of the work are obvious, but others will have undoubtedly been overlooked. Criticisms of the work are welcomed so that future research in the Central Highlands of Victoria (and possibly a future edition of this book) might be improved.

SUMMARY

Many broad themes of work have been completed in the Central Highlands of Victoria over the past 25 years. These include:

- quantifying patterns of distribution and abundance of individual species of mammals and birds and improving the understanding of the factors influencing why animals occur where they do;
- quantifying the composition of mammal and bird assemblages and how they change in response to stand age, disturbance history and other factors;
- identifying the spatial and temporal factors shaping the composition and structure of the forest, including the development of cool temperate rainforest;
- improving the understanding of natural disturbance regimes such as wildfires and how they can vary in intensity and frequency across landscapes;
- documenting the impacts of human disturbance regimes, particularly clearfell logging, on forest structure, forest composition and forest biodiversity;
- developing improved approaches to forest management, especially more environmentally sensitive logging methods and cost-effective approaches to long-term monitoring of forest structure and biodiversity.

The broad aims of this book are to describe some of the key findings of these studies and identify the key links between research themes.

2 Background

INTRODUCTION
This chapter provides background and context for the more detailed discussions of the research and other work that follows in subsequent chapters. Some of the key features of the Central Highlands region are summarised, types and patterns of vegetation cover are outlined and attributes of some of the flora and fauna are described, especially those targeted for detailed research in montane ash forests.

LOCATION
The Central Highlands of Victoria lies about 120 km north-east of the city of Melbourne. The region covers approximately half a degree of latitude and 1 degree of longitude (37°20'–37°55'S and 145°30'–146°20'E), an area of about 400 000 ha. Major towns in the area include Healesville, Warburton, Marysville, Alexandra, Powelltown and Noojee (Figure 2.1).

The montane ash forests of the Central Highlands are dominated by largely monotypic stands of Mountain Ash (*Eucalyptus regnans*), Alpine Ash (*Eucalyptus delegatensis*) or Shining Gum (*Eucalyptus nitens*). Mountain Ash typically occurs at altitudes between 200 m and 1100 m, Alpine Ash between 900 m and 1450 m and Shining Gum between 600 m and 1500 m (Boland *et al.* 2006; Costermans 1994; Lindenmayer *et al.* 1996b).

There is approximately 170 000 ha of montane ash forest in the Central Highlands of Victoria (Macfarlane and Seebeck 1991) of which ~121 000 ha is Mountain Ash (Lindenmayer 2007). Most montane ash forest is in public ownership (Commonwealth of Australia and Dept of Natural Resources and Environment 1997; Land Conservation Council 1994). Approximately 30% of the montane ash on public land occurs within the Yarra Ranges National Park and other conservation reserves, where it is exempt from logging. Much of the Yarra Ranges National Park encompasses the Maroondah, Upper Yarra and O'Shannassy catchments, administered by Melbourne Water. The remaining ~70% is broadly designated for wood production, although streamside zones, steep and rocky terrain and special protection zones for biodiversity protection will not be logged (Commonwealth of Australia and Dept of Natural Resources and Environment 1997; Dept of Natural Resources and Environment 1996).

The Central Highlands of Victoria is characterised by an elevation gradient of 400–1500 m above sea level (Colour Plate 1). The terrain varies from mountainous and steep to undulating. Permanent

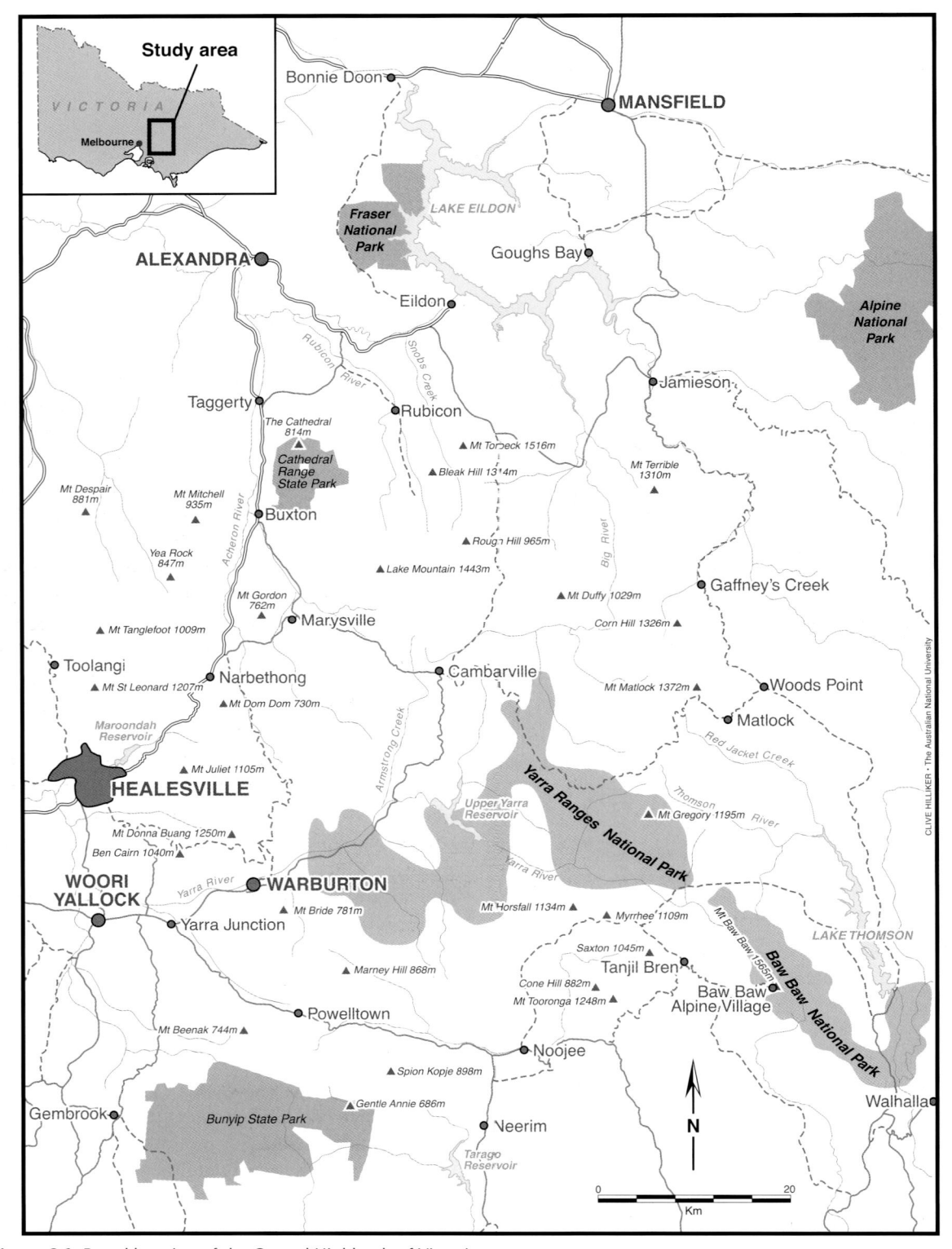

Figure 2.1: Broad location of the Central Highlands of Victoria.

Figure 2.2: Watts Creek Falls, part of the closed catchment system that generates water for Melbourne. (Photo by David Lindenmayer).

streams flow through the valleys, and there are semi-permanent wetlands and soaks.

CLIMATE

The Central Highlands of Victoria experiences mild humid winters with occasional periods of snow. Summers are generally cool. The area is classified as Cfb under the Köppen system of climatic classification (Dick 1975). Climatic analyses of the study area using the computer program BIOCLIM (Nix 1986) gave a range of 7.8° to 13.4°C in the mean annual temperature. Mean annual precipitation varied from 975 mm to 1700 mm (Lindenmayer *et al*. 1996a, 1991f) (Colour Plate 2).

GEOLOGY AND SOILS

Geology directly influences key characteristics of soils such as fertility and water-holding capacity. These (and other) attributes of soils influence the types of plants that can grow there and the growth rates of those plants (Florence 1996). Animal distribution and abundance is strongly affected not only by vegetation communities but also by other directly or indirectly related factors such as nutrient levels of tree foliage (Braithwaite 1984).

The basement rocks of the Central Highlands of Victoria are Cambrian submarine volcanics and sediments (Colour Plate 3). These basement rocks are overlaid by marine sediments including sandstones, siltstones, mudstones, shales and slates, most of which were deposited during the Ordovician and Middle Devonian periods. In addition, there are a number of Late Devonian plutons such as the Mt Baw Baw Batholith, which are coarse-grained igneous rocks. A period of orogeny or mountain-building occurred in the Middle Devonian, which created metamorphosed rocks. Late Devonian acid volcanics overlay folded sediments in parts of the Central Highlands region. Further tectonic activity in the Carboniferous period resulted in additional faulting and folding. In the Cainozoic, there was extrusion of basaltic lava, particularly in the southern parts of the region.

The soils in the Central Highlands of Victoria are strongly influenced by geology, the length of weathering, topography and climate. They include Ordovician metasediments, Devonian sediments and extrusives, granitic soils and alluvium. Some of the most fertile soils are deep red earths which overlay igneous felsic intrusive parent material. These have a high soil water-holding capacity and nutrient availability compared with most forest soils in Australia.

FLORA

A range of forest types occur in the Central Highlands of Victoria (Commonwealth of Australia and Dept of

Table 2.1: Paleozoic periods

Cambrian	Ordovician	Silurian	Devonian	Carboniferous	Permian

--------------------500--400--------------------------------300 million years ago

Figure 2.3: Leaves, fruits and bark of Mountain Ash. (Photo by Paul Gullan/Viridans Images).

Natural Resources and Environment 1997; Lumsden *et al.* 1991) (Colour Plate 4). These include dry mixed-species stands dominated by tree species such as Messmate (*Eucalyptus obliqua*), Red Stringybark (*Eucalyptus macrorhyncha*) and Silver-top Ash (*Eucalyptus sieberi*), tall forests dominated by Manna Gum (*Eucalyptus viminalis*), Narrow-leaved Peppermint (*Eucalyptus radiata*) and Mountain Grey Gum (*Eucalyptus cypellocarpa*), and high-elevation stands of Snow Gum (*Eucalyptus pauciflora*) (Costermans 1994; Lumsden *et al.* 1991). Other tree species in the region include (among others) Brown Stringybark (*Eucalyptus baxteri*) and Candlebark (*Eucalyptus rubida*) (Boland *et al.* 2006). The particular focus of our research over the past 25 years has been on montane ash forest, hence it is the focus of this book.

Montane ash forests are dominated by Mountain Ash, Alpine Ash or Shining Gum. Of these species, old-growth stands of Mountain Ash can include mature and old trees with heights approaching 100 m, making these trees the tallest flowering plants in the world (Beale 2007). Mountain Ash and Alpine Ash are in the group of eucalypts known as *Monocalyptus*; Shining Gum is in the Southern Blue Gum assemblage of trees belonging to the *Symphyomyrtus* (Boland *et al.* 2006). The three species are readily distinguished by differences in leaves, fruits and bark, and the persistence of the bark on the trunk.

Montane ash forests are typically monotypic; that is, stands are dominated by a single species of overstorey eucalypt. However, stands comprising a mixture of Mountain Ash and Alpine Ash occur at the elevational boundaries of the two taxa (Lindenmayer *et al.* 1993b).

Figure 2.4: Leaves, fruits and bark of Alpine Ash. (Photo by Paul Gullan/Viridans Images).

Mountain Ash and Alpine Ash may also occur in mixed stands at the ecological limits of other eucalypt species (e.g. Messmate and Mountain Grey Gum) (Campbell 1984). Shining Gum often occurs as relatively small patches in the Central Highlands of Victoria, characterised by low values for the mean minimum temperature of the coldest month (Lindenmayer et al. 1996b) (see Chapter 4). Hence, mixtures of Shining Gum and Alpine Ash and Shining Gum and Mountain Ash are not uncommon. Finally, ash-type eucalypts are known to hybridise with a number of other eucalypts. For example, hybrids between Red Stringybark and Mountain Ash have been well studied (Ashton and Sandiford 1988). Mountain Ash and Messmate also form hybrids in several places in the Central Highlands of Victoria (Ashton 1981a).

The understorey tree and shrub layers of montane ash forests are well developed and support a range of plant species (Lindenmayer and Ough 2006; Mueck 1990; Ough 2002). Understorey trees in montane ash forests include Myrtle Beech (*Nothofagus cunninghamii*), Southern Sassafras (*Atherosperma moschatum*), Silver Wattle (*Acacia dealbata*), Blackwood (*Acacia melanoxylon*), Forest Wattle (*Acacia frigiscens*) and Mountain Hickory Wattle (*Acacia obliquinervia*). These plants can be relatively tall, exceeding 20 m in height (Adams and Attiwill 1984).

The shrub layer in montane ash forest varies from 2 m to 15 m in height and supports a diverse array of plants. These include Soft Tree Fern (*Dicksonia antarctica*), Rough Tree Fern (*Cyathea australis*), Musk Daisy Bush (*Olearia argophylla*), Blanket Leaf (*Bedfordia arborescens*), Dusty Daisy Bush (*Olearia phlogopappa*), Hazel Pomaderris (*Pomaderris aspera*), Mountain Correa (*Correa lawrenciana*), Dogwood (*Cassinia aculeata*), Tree Geebung (*Persoonia arborea*),

Figure 2.5: Leaves, fruits and bark of Shining Gum. (Photo by Paul Gullan/Viridans Images).

Stinkwood (*Zieria arborescens*), Victorian Christmas Bush (*Prostanthera lasianthos*) and Austral Mulberry (*Hedycarya angustifolia*) (Costermans 1994; Mueck 1990; Ough and Ross 1992).

FAUNA

The following section outlines some of the fauna that occurs in the montane ash forests of the Central Highlands of Victoria. A particular emphasis is on the species which have been the focus of detailed research over the past 25 years – arboreal marsupials, small terrestrial mammals and birds.

Arboreal marsupials

The ash-type eucalypt forests of the Central Highlands of Victoria support four species of marsupial gliders – the Greater Glider (*Petauroides volans*), Yellow-bellied Glider (*Petaurus australis*), Sugar Glider (*Petaurus breviceps*) and Feathertail Glider (*Acrobates pygmaeus*) – and four species of possums. Possum species are Leadbeater's Possum (*Gymnobelideus leadbeateri*), Common Ringtail Possum (*Pseudocheirus peregrinus*), Mountain Brushtail Possum (*Trichosurus cunninghami*) and Eastern Pygmy Possum (*Cercartetus nanus*). Leadbeater's Possum is virtually confined to the Central Highlands region (Lindenmayer 2000), but the other seven species are distributed throughout relatively large areas of Australia (Goldingay and Jackson 2004; Strahan 1995).

The eight species of arboreal marsupials are from four families – the Petauridae, Phalangeridae, Burramyidae and Pseudocheiridae. Each of these (except Phalangeridae) is represented by at least one gliding and one non-volant member. Two of the volant species, the Yellow-bellied Glider and the Greater Glider, have

Figure 2.6: A subset of understorey plants. (a) Myrtle Beech. (b) Rough Tree Fern. (c) Silver Wattle. (d) Hazel Pomaderris. (Photos by Paul Gullan/Viridans Images).

highly developed gliding abilities and can volplane distances of 120–140 m and 100 m, respectively (Henry and Craig 1984; Lindenmayer 2002; McKay 1983). The smaller Feathertail Glider is a proficient glider and can volplane distances of more than 20 m. These species vary substantially in body size (10–4000 g), breeding system (monogamous to polygamous), diet (specialist exudivores, specialist folivores and omnivores) and home range (1–60 ha) (Lindenmayer 1997). These differences are highlighted in Table 2.2 and summarised briefly in the following paragraphs.

The Eastern Pygmy Possum weighs 15–40 g and eats a diet comprising arthropods, nectar, pollen, seeds and fruits (Huang *et al.* 1987; Turner 1983). It can make nests in tree hollows (Lindenmayer *et al.* 1991c), under the bark of eucalypts, in abandoned bird nests and in burrows (Ward 1990b).

Leadbeater's Possum is a small (120–140 g) and cryptic species of non-volant, nocturnal arboreal marsupial that is colonial (group sizes typically comprising 3–12 animals) and has a monogamous matriarchal mating system (Harley 2004; Smith 1980, 1984a). Its diet includes arthropods, gums produced from wound sites on *Acacia* spp. trees, manna (a white, carbohydrate-rich crystalline substance that occurs on *Eucalyptus* leaves) and honeydew (a sugar-rich secretion produced by nymphal stages of phloem-feeding lerp-forming insects of the Order Hemiptera) (Smith 1984b).

The Common Ringtail Possum is a folivore that consumes not only the leaves of *Eucalyptus* trees, but also foliage from a range of understorey and shrub species (Pahl 1984, 1987; Wayne 2005). Its diet may include fruits and flowers (Pahl 1984). The home

Table 2.2: Life history attributes of arboreal marsupials recorded in the montane ash forests of the Central Highlands of Victoria

Common name	Home range (ha)	Mean body mass (g)	Social organisation	Mating system[a]	Diet
Feathertail Glider	0.4–2.1	10–14	Colonial (up to 40 individuals); possibly not territorial; up to 29 individuals observed to share a den site	**Promiscuous**, polygynous, monogamous	Arthropods, insect and plant exudates (pollen, honeydew, nectar and seeds)
Sugar Glider	0.5–7.1	115–160 (male), 95–135 (female)	Colonial (up to 12 individuals)	**Polygynous**	Arthropods, insect and plant exudates (wattle gum, eucalypt sap, nectar, pollen, manna, honeydew)
Yellow-bellied Glider	20–85	450–700	Colonial (up to 11 individuals); up to five individuals observed to share a den tree	**Monogamous**, polygynous, polygamous	Arthropods, insect and plant exudates (honeydew, eucalypt sap, nectar and pollen)
Greater Glider	0.7–3	900–1700	Solitary; males' larger home ranges overlap with those of several females; co-occupancy of den trees rare except between matched pairs in the breeding season	**Polygamous**, monogamous, polygynous	Eucalypt foliage specialist – various species of eucalypt taken
Leadbeater's Possum	1–2	120–140	Colonial (up to 12 individuals)	**Monogamous**	*Acacia* spp. sap, manna, honeydew, invertebrates
Common Ringtail Possum	0.07–2.6	660–900	Pairs and small groups of animals	**Polygamous**	Foliage of a wide range of plants, flowers, fruits
Common Brushtail Possum	0.7–11	1500–3500	Overlapping male–female pairs, small groups of animals	**Monogamous**, polygamous	Foliage of a wide range of plants, flowers, fruits, occasionally bird nestlings
Mountain Brushtail Possum	1–7	2500–4500	Overlapping male–female pairs, small groups of animals	**Monogamous**, polygamous	Foliage of a wide range of plants, seeds, above-ground fungi, truffles
Eastern Pygmy Possum	0.2–1.7	15–40	Small groups of animals	**Polygamous**	Nectar, pollen, fruit and arthropods

a The mating system of some (perhaps many) species of arboreal marsupials may change with availability and quality of food resources. The system likely to be most common in the Central Highlands of Victoria is shown in bold.

range of the species is typically 1 ha or smaller (Lindenmayer et al. 2008c; Munks et al. 2004; Smith et al. 2003). The average weight of adults is approximately 750 g. The mating system of the Common Ringtail Possum is poorly known but animals appear to be colonial, with group sizes varying from two to five (Kerle 2001; Munks et al. 2004).

The Mountain Brushtail Possum is a relatively large non-volant animal with adults weighing up to 4 kg (Viggers and Lindenmayer 2000). Its diet comprises leaves of several understorey trees and shrubs as well as various types of fungi (Claridge and Lindenmayer 1993, 1998; Seebeck et al. 1984). Adults in some populations appear to live in pairs and occupy a territory of 2–6 ha (How 1972). The Mountain Brushtail Possum can be long-lived and sedentary and occupy the same area for up to 15 years (Banks et al. 2008; Viggers and Lindenmayer 2004). The species is considered to maintain a monogamous mating system (How 1972), although this may vary between populations in different parts of the species' distribution (Viggers and Lindenmayer 2004).

The Feathertail Glider is the smallest arboreal marsupial in the montane ash forests and typically weighs 10–15 g. Its diet is very similar to that of the Eastern Pygmy Possum (Ward 1990a). The species is colonial; group size can vary from two to 29 (Ward 1990a). The mating system appears to be flexible but is most often promiscuous.

The biology and ecology of the Sugar Glider is very similar to that of Leadbeater's Possum. The species weighs 90–170 g and lives in social groups of up to 10 animals that have a home range of ~0.5–7 ha (Lindenmayer 2002). Its diet is similar to that of Leadbeater's Possum, but includes eucalypt sap, nectar and pollen (Goldingay et al. 1991; Smith 1982; Suckling 1982).

The Yellow-bellied Glider is an intermediate-sized animal, with adults weighing approximately 650 g (Henry and Craig 1984). The home range of the Yellow-bellied Glider is 30–60 ha (Craig 1985; Goldingay and Kavanagh 1993), substantially larger than any of the other species of arboreal marsupials which occur in the study area. The Yellow-bellied Glider has a diverse diet which encompasses invertebrates, nectar and sap from eucalypt trees, pollen and honeydew (Goldingay 1986). The breeding system appears to be variable, changing in response to factors such as resource availability (Lindenmayer 2002).

Adults of the Greater Glider weigh up to 1300 g and are entirely folivorous, feeding exclusively on the leaves of *Eucalyptus* spp. trees (Kavanagh and Lambert 1990). The home range of the Greater Glider is about 1–2 ha (Henry 1984; Pope et al. 2004) and the species is usually solitary except during the breeding season, when pairs of animals are often recorded (Henry 1984). The mating system may vary between monogamy and polygamy depending on resource availability (Norton 1988).

Terrestrial native mammals

Montane ash forests support a rich array of terrestrial native mammals (Lumsden et al. 1991). These include two native murid rodents – the Bush Rat (*Rattus fuscipes*) and Broad-toothed Rat (*Mastacomys fuscus*). The Bush Rat is an omnivorous mammal that is widely distributed in eastern Australia (Lunney 1983; Robinson 1987). The species' habitat includes upland and lowland rainforest, wet sclerophyll forest, subalpine scrub or heath, wet scrub or heath, and sedge. Males weigh up to 250 g and are larger than females (up to 150 g). Breeding is usually restricted to a period spanning November to January, during which females produce one or two litters (Lunney 1983). Post-breeding mortality among adults appears to be common (Lee and Cockburn 1985). Juveniles of both sexes disperse from their natal territory in May (Robinson 1987). Home ranges vary between 0.1–1.2 ha and females in particular are highly territorial (Peakall et al. 2003). Population densities of the Bush Rat are among some of the highest recorded anywhere in mainland eastern Australia (Cunningham et al. 2005).

The Broad-toothed Rat is a strictly herbivorous small mammal that feeds largely on grasses (Carron et al. 1990). It occurs in high-rainfall areas throughout upland parts of eastern Australia from Tasmania to Barrington Tops in coastal central New South Wales (Green and Osborne 2003; Menkhorst and Knight 2001). Animals weigh 95–145 g. Females produce a litter of two to three young in summer. The Broad-toothed Rat was regularly captured in trapping studies 15–35 years ago within montane ash forests in parts of

Figure 2.7: Arboreal marsupials found in the montane ash forests of the Central Highlands of Victoria: (a) Feathertail Glider. (Photo by Ann Jelinek). (b) Sugar Glider. (Photo by Mike Greer). (c) Yellow-bellied Glider. (Photo by David Lindenmayer). (d) Greater Glider. (Photo by David Lindenmayer). (e) Eastern Pygmy Possum. (Photo by David Lindenmayer). (f) Leadbeater's Possum. (Photo by David Lindenmayer). (g) Common Ringtail Possum. (Photo by Mike Greer). (h) Mountain Brushtail Possum. (Photo by David Lindenmayer).

Table 2.3: Life history attributes of terrestrial mammals recorded in the montane ash forests of the Central Highlands of Victoria

Common name	Home range (ha)	Mean body mass (g)[a, b]	Social organisation[a, c]	Diet[a, c]
Bush Rat	0.1–1.2[c]	125 (50–225)	Solitary	Arthropods, seeds, fruits, fungi
Swamp Rat	0.2–4.0[a, c]	122 (55–160)	Solitary	Mainly sedges, grasses
Agile Antechinus	0.4–5.3[c]	16–44	Seasonally communal	Invertebrates, small vertebrates
Dusky Antechinus	0.4–1.5[d]	38–170	Solitary	Invertebrates, small vertebrates
Long-nosed Bandicoot	1.5–5.2[e]	850–1100	Solitary	Invertebrates, seeds, leaves, fungi
Spotted-tailed Quoll	88–2560[f]	1500–7000	Solitary	Small to medium-sized mammals, birds, carrion, invertebrates
Eastern Grey Kangaroo	8–430[c, g]	35 000–66 000	Group	Grasses, forbs
Red-necked Wallaby	12–32[c]	11 000–27 000	Adult males solitary, females and subadults in small groups	Grasses, forbs
Swamp Wallaby	10–27[h]	10 000–20 500	Mainly solitary	Wide variety of plants including fungi, shrubs, grasses, sedges, bracken
Common Wombat	5–23[c]	20 000–39 000 (26 000)	Solitary	Mainly grasses, rushes; also sedges, roots, tubers, mosses, fungi
Echidna	40–70[c]	2000–7000	Solitary	Mainly ants and termites, some other invertebrates

a Strahan (1995). b Menkhorst and Knight (2001). c Menkhorst (1995). d Sanecki et al. (2006). e Scott et al. (1999). f Claridge et al. (2005). g Dawson (1995). h Troy and Coulson (1993).

the Central Highlands of Victoria (Seebeck 1971; Beezobs and Sanson 1997). However, more recent surveys have rarely resulted in captures of the species, including in areas where substantial populations used to occur, such as Bellell Creek, near Lake Mountain and Marysville.

Two species of small carnivorous marsupials – Agile Antechinus (*Antechinus agilis*) and Dusky Antechinus (*Antechinus swainsonii*) – are particularly common in the montane ash forests of the Central Highlands of Victoria. The Agile Antechinus is a small scansorial marsupial that occurs in all forest types and heathland habitats to about 1200 m above sea level, from southern Victoria to south-eastern New South Wales (Dickman 1995a; Dickman et al. 1998). Males weigh up to 40 g and can be twice the size of females (up to 24 g) (Dickman 1995a; Kraaijeveld-Smit et al. 2003). All the males die after a two-week mating season in August–September. Females give birth to a litter of 6–10 young (depending on population-specific teat number) approximately one month after mating. Litters are usually sired by more than one father (Kraaijeveld-Smit et al. 2002a). These young stay attached to the teat for five weeks, followed by a period of about two months in the nest. Most juvenile males disperse away from the natal territory in January–February (Cockburn et al. 1985). Home ranges of males are around 3 ha, those of females 1.5 ha and there is overlap between and within sexes (Lazenby-Cohen and Cockburn 1991).

Figure 2.8: Small terrestrial mammals found in the montane ash forests of the Central Highlands of Victoria. (a) Bush Rat. (Photo by Esther Beaton). (b) Agile Antechinus. (Photo by Esther Beaton). (c) Dusky Antechinus. (Photo by Tony Robinson/Viridans Images).

The Dusky Antechinus is a ground-dwelling marsupial found in Tasmania and the eastern part of mainland Australia (Dickman 1995b). It occupies a range of habitat types from subalpine heath and woodland, to wet sclerophyll forest and upland and lowland rainforest (Dickman 1995b). This species is much larger than the Agile Antechinus, with males reaching 120 g and females 70 g (Dickman 1995b). The life history of this species is similar to that of the Agile Antechinus, particularly in relation to its synchronised annual breeding season followed by male die-off (Cockburn and Lee 1988). Litter size varies from six to eight and juvenile males disperse. Most juvenile females remain in their natal territory or disperse relatively short distances (Kraaijeveld-Smit *et al.* 2007). Little is known about home range sizes and territoriality (Dickman 1995b).

Three species of macropod marsupials commonly observed in the Central Highlands region are the Black Wallaby (*Wallabia bicolor*), Eastern Grey Kangaroo (*Macropus giganteus*) and Red-necked Wallaby (*Macropus rufogriseus*). Two other native mammals frequently recorded in montane ash forests are the Common Wombat (*Vombatus ursinus*) – the world's largest burrowing mammal – and the Echidna (*Tachyglossus aculeatus*), one of the world's three species of living monotremes. Three species of native terrestrial mammals are rare in the region – the Spotted-tailed Quoll (*Dasyurus maculatus*), the Long-nosed Bandicoot (*Parameles nasuta*) and the Swamp Rat *(Rattus lutreolus)*.

Bats

The montane ash forests of the Central Highlands of Victoria provide habitat for a range of species of microchiropteran bats (Brown *et al.* 1989, 1997; Brown and Howley 1990; Churchill 2008; Lumsden *et al.* 1991). All the species in montane ash forests are relatively small, ranging from ~5 g to ~35 g in adult body mass. They include the White-striped Freetail Bat (*Austronomus australis*), Gould's Wattled Bat (*Chalinolobus gouldii*), Chocolate Wattled Bat (*Chalinolobus morio*), Large Forest Bat (*Vespadelus darlingtoni*), Southern Forest Bat (*Vespadelus regulus*), Little Forest Bat (*Vespadelus vulturnus*), Lesser Long-eared Bat (*Nyctophilus geoffroyi*), Gould's Long-eared Bat (*Nyctophilus gouldii*), Eastern Broad-nosed Bat (*Scotorepens orion*) and Eastern Falsistrelle (*Falsistrellus tasmaniensis*).

The limited work completed to date on bats within montane ash forests suggests that foraging activity increases with increasing levels of vertical heterogeneity of stands (Brown *et al.* 1997), particularly in old-growth areas.

Terrestrial feral animals

A number of species of terrestrial feral mammals have been recorded in the montane ash forests of the Central Highlands of Victoria. Six are observed relatively commonly. They are the European Rabbit (*Oryctolagus cuniculus*), Red Fox (*Vulpes vulpes*), Feral Cat (*Felis catus*), Dingo and/or wild dog (*Canis familiaris dingo*) and Sambar Deer (*Cervus unicolor*). The Dingo is classified as an introduced species because it arrived in Australia approximately 5000 years ago (Savolainen *et al.* 2004), initially as a commensal with indigenous people. Four other species have been occasionally observed but are generally rare – Black Rat (*Rattus rattus*), House Mouse (*Mus musculus*), Feral Pig (*Sus scrofa*) and Goat (*Capra hircus*).

Birds

Birds are by far the most species-rich vertebrate assemblage in the montane ash forests of the Central Highlands of Victoria (Loyn 1985, 1998). These forests support more than 70 species of birds. It is a forest-dominated assemblage and includes such well-known and highly charismatic taxa as the Superb Lyrebird (*Menura novaehollandiae*) and several psittacid and cacatuid parrots – the Crimson Rosella (*Platycercus elegans*), King Parrot (*Alisterus scapularis*), Gang-gang Cockatoo (*Callopcephalon fimbriatum*) and Yellow-tailed Black Cockatoo (*Calyptorhynchus funereus*).

A number of species in montane ash forests are listed as threatened or are thought to be declining significantly elsewhere in Australia. For example, the Olive Whistler (*Pachycephala olivacea*) and Red-browed Treecreeper (*Climacteris erythrops*) are uncommon species of conservation concern in parts of their ranges. Notably, a number of the bird species which vary from uncommon to moderately common in montane ash forests are known to be declining significantly in the woodlands of south-eastern Australia (Barrett *et al.* 2003). These include the Crested Shrike-tit (*Falcunculus frontatus*), Flame Robin (*Petroica phoenicea*) and Eastern Yellow Robin

Figure 2.9: Birds found in the montane ash forests of the Central Highlands of Victoria. (a) Crested Shrike-tit. (Photo by Graeme Chapman). (b) Red-browed Treecreeper. (Photo by Julian Robinson). (c) Flame Robin. (Photo by David Cook/David Cook Wildlife Photography).

Table 2.4: Life history attributes of bats recorded in the montane ash forests of the Central Highlands of Victoria

Common name	Scientific name	Mean weight (g)	Diet	Foraging	Roosting	Breeding
White-striped Freetail bat	*Austronomus australis*	36.3–39	Mainly moths and beetles	Primarily above canopy or cleared areas, often 50 m or more above ground	In trees, singly or small groups	Single young, usually Dec.–Jan., weaned by May
Gould's Wattled Bat	*Chalinolobus gouldii*	13.8	Wide range of flying insects. Moths and bugs predominant	Just below or within tree canopy and along forest edges, fast-flying	Tree hollows, females in colonies up to 80. Males often solitary	Usually twins, born Oct.–Nov., independent in six weeks
Chocolate Wattled Bat	*Chalinolobus morio*	8.1	Opportunistic, mainly moths in Victorian studies	Mostly between top of the understorey and the canopy	Tree hollows, other cavities, females in colonies up to 70, sometimes larger. Males usually solitary	One, sometimes two, young, born in November, independent by February
Large Forest Bat	*Vespadelus darlingtoni*	7.2	Small insects, can forage on mild winter nights when other species are hibernating	Mainly within spaces among trees and between the canopy and understorey	In tree hollows and buildings, usually small groups of females or solitary males	Single young born Nov.–Dec., free-flying by late Jan.–early Feb.
Southern Forest Bat	*Vespadelus regulus*	5.2	Aerial insects, mainly moths, flies	Less than half canopy height, often very close to vegetation, very agile	Colonies of up to 100 in tree hollows, males usually separate except in breeding season	Single young born Nov.–Dec., weaned six weeks later
Little Forest Bat	*Vespadelus vulturnus*	4.3	Aerial insects	Fly below forest canopy, very agile, often forage close to vegetation	Usually in small colonies in tree hollows or roofs of buildings	Single young born Nov.–Dec., free-flying by mid January

Common name	Scientific name	Mean weight (g)	Diet	Foraging	Roosting	Breeding
Lesser Long-eared Bat	*Nyctophilus geoffroyi*	8.2	Moths are the most common prey but will take a wide variety including wingless insects from plants or ground	Forage close to vegetation and into understorey, slow-flying and highly manoeuvrable	In crevices such as tree hollows, in buildings, under bark. Usually singly or in small groups	Usually twins, born late spring–early summer, young independent by early February
Gould's Long-eared Bat	*Nyctophilus gouldii*	12.3	Mainly moths in Victoria. Variety of prey capture techniques, will take insects on wing, or from vegetation or ground	Forage within vegetation and spaces among trees, below canopy height	In tree hollows and under bark, females in groups of 20 or more, males singly or small groups	One or two young born Oct.–Nov., young independent by January
Eastern Broad-nosed Bat	*Scotorepens orion*	11	Unknown	Unknown	Tree hollows	Single young Nov.–Dec.
Eastern Falsistrelle	*Falsistrellus tasmaniensis*	20.5	Moths and other aerial insects	Within or just below canopy, flight swift and direct	Hollow trunks of eucalypts in colonies, usually single sex	Single young born in December

Data from Churchill (1998).

Figure 2.10: Reptiles found in the montane ash forests of the Central Highlands of Victoria. (a) Tiger Snake. (Photo by David Lindenmayer). (b) Highlands Copperhead. (Photo by Esther Beaton). (c) Spencer's Skink. (Photo by Damian Michael). (d) Forest Skink. (Photo by Damian Michael).

(*Eopsaltria australis*) (Cunningham *et al.* 2008; Ford *et al.* 2001). The montane ash forests provide habitat for threatened species of owls including the Sooty Owl (*Tyto tenebricosa*), Powerful Owl (*Ninox strenua*) and Masked Owl (*Tyto novaehollandiae*).

As would be expected from such a species-rich assemblage, the birds in montane ash forests vary substantially in a wide range of life history attributes. These include body weight, group type (solitary, pairs or flock), social system (monogamous, polygamous etc.), type of nest (hollow, cup, mud bowl etc.), nest placement (horizontal fork, ground etc.), nesting height, number of eggs laid in a clutch, broods per year and movement behaviour (resident versus migrant, latitudinal or altitudinal migrant). They also vary in foraging strategy or guild (*sensu* Mac-Nally 1994).

Reptiles

Reptiles are Australia's most diverse vertebrate group (Wilson and Swan 2007). However, the reptile assemblage found in the montane ash forests of the Central Highlands of Victoria is somewhat depauperate (Brown and Nelson 1993) relative to many other Australian environments. These forests support no goannas, geckoes, legless lizards or pythons and only front-fanged (elapid) snakes (see below). Low levels of species richness are likely to

Table 2.5: Life history attributes of reptiles recorded in the montane ash forests of the Central Highlands of Victoria

Common name	Scientific name	Snout-vent	Body	Life-form	Common shelter site	Activity pattern	Mode of thermoregulation	Mode of reproduction	Clutch/off-spring size	Foraging mode	Diet
Black Rock Skink	Egernia saxatilis intermedia	110	210	Saxicolous	Crevices	Diurnal	Heliotherm	Viviparous	4	Sit and wait	Insects
Southern Water Skink	Eulamprus tympanum	75	230	Terrestrial	Debris	Diurnal	Heliotherm	Viviparous	4	Active	Arthropods
Highlands Forest Skink	Anepischtos maccoyi	45	120	Fossorial	Leaf litter	Nocturnal	Thigmotherm	Oviparous	4	Active	Insects
Coventry's Skink	Niveoscincus coventryi	45	120	Terrestrial	Debris	Diurnal	Heliotherm	Viviparous	5	Active	Arthropods
Spencer's Skink	Pseuemoia spenceri	50	130	Terrestrial	Crevices	Diurnal	Heliotherm	Viviparous	3	Active	Insects
White's Skink	Liopholis whitii	75	180	Terrestrial	Rocks	Diurnal	Heliotherm	Viviparous	4	Active	Arthropods
Tussock Skink	Pseudomoia entrecasteauxii	55	135	Terrestrial	Debris	Diurnal	Heliotherm	Viviparous	2-6	Active	Insects
Cunningham's Skink	Egernia cunninghami	230	430	Saxicolous	Crevices	Diurnal	Heliotherm	Viviparous	3-8	Active	Plant material, invertebrates
Tiger Snake	Notechis scutatus	890	1050	Terrestrial	Debris	Diurnal	Heliotherm	Viviparous	20+	Active	Frogs, rodents
Highland Copperhead	Austrelaps ramsayi	760	910	Terrestrial	Debris	Diurnal	Heliotherm	Viviparous	15	Active	Reptiles, frogs
Red-bellied Black Snake	Pseudechis porphyriacus	1050	1230	Terrestrial	Debris	Diurnal	Heliotherm	Viviparous	12	Active	Frogs, lizards, small mammals
Small-eyed Snake	Cryptophis nigrescens	365	425	Terrestrial	Debris	Nocturnal	Heliotherm	Viviparous	4	Active	Skinks

Data from Cogger (2000); Jenkins and Bartell (1980); Shine (1991); Wilson and Swan (2008).

result from the predominance of cool and wet environmental conditions in the study area.

Skinks are the most species-rich reptile group in montane ash forests (Brown and Nelson 1993; Lumsden *et al.* 1991). Skinks found include the Southern Water Skink (*Eulamprus tympanum*), Highlands Forest Skink (*Anepischtos maccoyi*), Coventry's Skink (*Niveoscincus conventryi*), Spencer's Skink (*Pseudomoia spenceri*), Tussock Skink (*Pseudomoia entrecasteauxii*), Black Rock Skink (*Egernia saxatilis*), White's Skink (*Liopholis whitii*) and Cunningham's Skink (*Egernia cunninghami*). The last three, with the Eastern Small-eyed Snake (*Cryptophyis nigrescens*), are often associated with granite outcrops and other kinds of rocky areas.

Elapid snakes are the next most common reptile group encountered in montane ash forests. They include the Highland Copperhead (*Austrelaps ramsayi*), Tiger Snake (*Notechis scutatus*), Red-bellied Black Snake (*Pseudechis porphyriacus*) and Eastern Small-eyed Snake. The first two are reasonably common and all are highly venomous (Wilson and Swan 2007). None are regarded as vulnerable or of conservation concern.

Frogs

As in the case of reptiles, the amphibian fauna of the montane ash forests of the Central Highlands of Victoria is relatively species-poor. Species recorded from the relatively limited studies include Common Eastern Froglet (*Crinia signifera*), Pobblebonk (*Limnodynastes dumerilii*), Peron's Tree Frog (*Littoria peronii*), Whistling Tree Frog (*Littoria ewingii*), Southern Toadlet (*Pseudophryne semiornata*) and Baw Baw Frog (*Philoria frosti*) (Brown *et al.* 1989; Lumsden *et al.* 1991).

Of these species, the Baw Baw Frog is of major conservation concern because it is one of Australia's most critically endangered amphibians (Hollis 2004). It is a high-elevation specialist, endemic to the Baw Baw Plateau and one of the most range-restricted species of amphibians in Australia (Hollis 2004). Its distribution encompasses a range of vegetation types on the Baw Baw Plateau. In the mid 1990s, field surveys extended the known distribution of the Baw Baw Frog to lower-elevation montane ash forests off the plateau escarpment. Populations of the Baw Baw Frog have declined several orders of magnitude over the past 15–20 years

Figure 2.11: Frogs found in the montane ash forests of the Central Highlands of Victoria. (a) Baw Baw Frog. (Photo by Greg Hollis). (b) Peron's Tree Frog. (c) Pobblebonk. (Photos by David Cook/David Cook Wildlife Photography).

and this once common species (~10 000–15 000 individuals) is now 1–2% of what it was 25 years ago (Hollis 2004). Recent monitoring data suggest that population declines are occurring throughout the species' range, although the rate of decline appears to be faster among populations at higher elevations (Greg Hollis, *pers. comm.*).

Fish

A number of species of fish have been recorded from aquatic environments in montane ash forests,

Table 2.6: Life history attributes of frogs recorded in the montane ash forests of the Central Highlands of Victoria

Common name	Scientific name	Body size (mm)	Habitat and breeding	No. of eggs	Tadpole type	Tadpole description
Common Eastern Froglet	Crinia signifera	18–25 (female), 19–28 (male)	Extremely widespread and highly variable species found in a variety of habitats including those heavily modified by humans. Eggs are laid in clumps attached to sticks, grass or leaf litter in shallow water	100–150	Lentic	As in the case of adults, tadpoles can be extremely variable
Pobblebonk	Limnodynastes dumerilii	52–70 (female), 52–83 (male)	Breeds in a wide variety of habitats including permanent and temporary waterholes. Floating foam nests made in aquatic areas	3900	Lentic	Dark-grey to black
Peron's Tree Frog	Litoria peronii	44–53 (female), 46–55 (male)	Breeds in a wide variety of habitats including permanent and temporary waterholes	1303–2391	Lentic	Pale golden-yellow, up to 44 mm
Whistling Tree Frog	Litoria ewingii	20–40 (female), 32–46 (male)	Breeds in a wide variety of habitats including permanent and temporary waterholes Eggs attached to clumps of submerged vegetation	17–33	Lentic	Transparent, up to 50 mm
Southern Toadlet	Pseudophryne semimarmorata	30	Found in a variety of damp situations in sclerophyll forests. Mainly autumn breeder. Eggs laid in nest burrows in seasonally flooded areas	8–194	Lentic	Dark-grey with some iridescence, up to 34 mm
Baw Baw Frog	Philoria frosti	45	Found beneath logs or rocks adjacent to streams or in tunnels in sphagnum bogs. Breeds in late spring. Eggs laid in foam nests in sphagnum moss, tadpoles usually complete development in nest	50–185	Usually terrestrial	Brown to black, up to 24 mm

Data from Anstis (2002); Barker et al. (1995); Cogger (2000); Tyler (1994).

although relatively limited work has been completed on this group (Doeg and Joehn 1990). For example, native freshwater fish species known from the Acheron River in the Central Highlands of Victoria include the Two-spined Blackfish (*Gadopsis bispinosis*), River Blackfish (*Gadopsis marmoratus*), Mountain Galaxias (*Galaxias olidus*) (Tunbridge and Glenane 1983) and the Short-finned Eel (*Arguilla australis*). The Barred Galaxias (*Galaxias fuscus*) is known to occur on Lake Mountain and in the nearby Taggerty River and is listed as critically endangered in Victoria (Parks Victoria 2002).

The Acheron River and many other rivers in montane ash forests also contain populations of introduced fish species including Rainbow Trout (*Salmo gairdneri*), Brown Trout (*Salmo trutta*) and Tench (*Tinca tinca*) (Brown et al. 1989).

Invertebrates

There have been relatively few studies of the invertebrate biota of montane ash forest, although there has been a bias in work on commercially important invertebrate pest species. Outbreaks of phasmid stick insects (Neumann *et al.* 1977) and psyllids (Coy and Burgess 1994) have been reported in different decades over the past 50 years (Flint and Fagg 2007). Other insects that can damage stands of montane ash forest include a range of termite species and the larvae of wood-degrading moths (Elliott *et al.* 1998).

Ashton (1979) found extremely large populations of several species of ants in Mountain Ash forests. He suggested that harvesting by ants of seed dropped by the overstorey trees was one of the key reasons for low rates of germination despite high rates of seed fall. Studies conducted elsewhere in south-eastern Australia (O'Dowd and Gill 1984) have shown that seed theft by ants is an important ecological process in the vegetation dynamics of Alpine Ash forest.

Neumann (1991, 1992) examined the arthropod fauna of Mountain Ash forests and compared it with that of sites subject to clearfell logging. He found large differences in the composition of communities of ants and litter-dwelling arthropods before and after disturbance, and documented high levels of invertebrate diversity. This is consistent with the findings of Milne and Short (1999), who found over 40 morphospecies from 14 orders associated with three *Dicranoloma* spp. mosses.

A wide range of other species of invertebrates occur in montane ash forests, including carnivorous gastropods, isopods and earthworms (*Megascolex stelli* and *Megascolex dorsalis*), of which there may be a staggering 1.6 million individuals per ha. These and other invertebrates are important food items for other animals such as the Superb Lyrebird (Ashton 1975c).

SUMMARY

The montane ash forests of the Central Highlands of Victoria support a rich and diverse biota, although it is somewhat less rich than other environments. For example, bird species richness in the woodlands of southern Australia is significantly higher (~150 species) than that recorded in montane ash forests (~70 species). Similarly, Australia's desert environments are far more reptile species-rich (Wilson and Swan 2007) than the wet ash forest.

Of the array of animal species found in montane ash forests, very few (e.g. Leadbeater's Possum and the Baw Baw Frog) are almost exclusively associated with those forests. Leadbeater's Possum has been found in other kinds of environments in recent years (Harley 2004; Jelinek *et al.* 1995) and the Baw Baw Frog also occupies a range of habitats at elevations higher than those where montane ash forests typically occur (Hollis 2004).

3 Field survey methods

INTRODUCTION
Good field information about the management of natural resources, including biodiversity, underpins any robust ecological study and thus well-informed decisions (Sutherland 1996). A wide range of field methods have been used in studies in montane ash forests over the past 25 years. Some of these are briefly outlined in this chapter. Each method has inherent strengths and limitations; some are useful for particular taxa but highly problematic for others. As an example, trapping can be useful for studying the Sugar Glider, but the Feathertail Glider and the Greater Glider are difficult to capture using traditional trapping techniques. Therefore, the results of field studies are often strongly influenced by the survey methods used, making it important to be aware of the potential advantages and deficiencies of each (Sutherland 1996).

The first section describes some of the field methods used to study arboreal marsupials. Subsequent sections outline field survey methods used in work on small terrestrial mammals and birds. A major aspect of the work in the Central Highlands of Victoria has been the structure and composition of the vegetation. Therefore, the final section is dedicated to describing the field methods used to capture data on forest structure and composition. This chapter does not include any information on the statistical methods employed to analyse different datasets over the past 25 years. This information is available in the many publications co-authored by professional statisticians (Table 3.1).

ARBOREAL MARSUPIALS
Six major field methods have been employed in studies of arboreal marsupials – stagwatching, spotlighting, trapping, hairtubing, radio-tracking and the installation of nest boxes. These methods are briefly outlined in the remainder of this section. Trapping and nest box studies have enabled blood and tissue samples to be gathered from two species – Leadbeater's Possum and the Mountain Brushtail Possum. These samples have allowed a range of genetic studies to be conducted, as briefly outlined in following sections.

Stagwatching
A 'stag' is an old living or dead tree with many hollows in its branches or trunk. Stagwatching involves the careful observation of large trees with hollows that provide potential nest sites for possums and gliders (Lindenmayer *et al.* 1991a; Seebeck *et al.* 1983). A selected stag, or tree with hollows, is watched from an hour before dusk until an hour after dusk. This is

Table 3.1: Studies directly or indirectly examining field survey methods for various animal groups in the montane ash forests of the Central Highlands of Victoria

Study description	Reference
Stagwatching methods for arboreal marsupials	Lindenmayer et al. (1991a)
Comparison of stagwatching, spotlighting and trapping for arboreal marsupials	Smith et al. (1985); Smith et al. (1989a)
Stagwatching detections of radio-collared individuals of the Mountain Brushtail Possum	Lindenmayer et al. (1996c)
Efficacy of different kinds of hairtubes for terrestrial mammals and arboreal marsupials	Lindenmayer et al. (1994d); Lindenmayer et al. (1999d)
Comparison of hairtubing and scat surveys for small terrestrial mammals	Lindenmayer et al. (1996e)
Effectiveness of nest boxes for detecting cavity-using vertebrates	Lindenmayer et al. (2003c); Lindenmayer et al. (2009c)
Trap–recapture methods for Mountain Brushtail Possum	Lindenmayer et al. (1998)
Effectiveness of spotlighting for detecting arboreal marsupials[a]	Lindenmayer et al. (2001a)
Trap–recapture methods for small terrestrial mammals	Cunningham et al. (2005)
Application of radio-tracking methods to den use in arboreal marsupials	Lindenmayer and Meggs (1996); Lindenmayer et al. (1996c)
Evaluation of bird survey methods[a]	Cunningham et al. (1999); Lindenmayer et al. (2009a)

[a] These studies were conducted in areas outside the Central Highlands of Victoria, but results were used to inform field protocols.

when arboreal marsupials leave their den site to begin foraging (Lindenmayer et al. 1991c). The species that is using the tree as a nest site is identified from its size and shape in silhouette as it moves out of its hollow and into the surrounding forest to commence foraging. Spotlights are not used because they deter an animal from moving out of the tree (Lindenmayer 1996). Teams of volunteer observers are needed to simultaneously stagwatch all the trees with hollows on a site (usually a 1–3-ha patch of forest). It is important to survey all trees simultaneously because an individual arboreal marsupial will often change between different den sites, sometimes on successive nights (Gibbons and Lindenmayer 2002; Lindenmayer et al. 1996c). Therefore, surveying only parts of a given site will lead to inaccurate counts of the number of animals.

Although the stagwatching technique is extremely labour-intensive, it has been used successfully to count populations of most species of arboreal marsupials in the montane ash forests of the Central Highlands of Victoria. It was found to be more effective than other methods such as spotlighting and hairtubing for detecting most species of arboreal marsupials in the Central Highlands (Smith et al. 1989a). Two kinds of studies have been undertaken to examine the effectiveness of the stagwatching technique. First, numbers of animals observed by trapping, spotlighting and stagwatching on 32 sites in ash-type forests were compared. This showed that trapping and spotlighting typically resulted in lower numbers of detections than did stagwatching. Second, 16 radio-collared Mountain Brushtail Possum individuals were radio-tracked during the day to their nest sites in large trees with hollows on a 35-ha site at Cambarville in the Central Highlands (Lindenmayer et al. 1996c). Repeated stagwatching surveys were then conducted using volunteer observers of stands of trees, some of which contained collared animals. On all occasions, observers recorded animals emerging from the trees confirmed earlier by radio-tracking to contain collared individuals (Lindenmayer et al. 1996c).

All field survey techniques, including the stagwatching method, have limitations. First, like spotlighting, stagwatching has had only limited success in detecting the Feathertail Glider and Eastern Pygmy

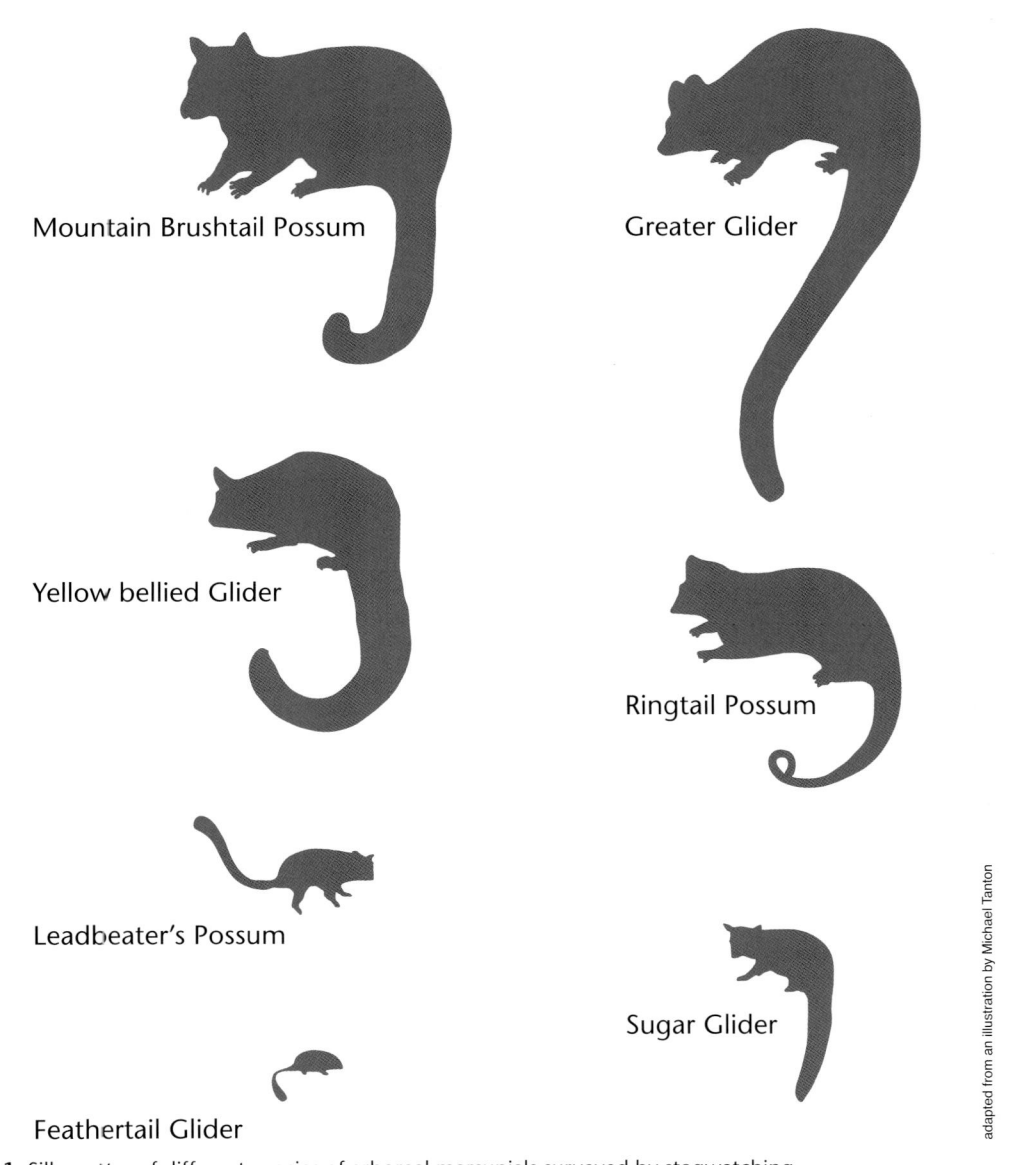

Figure 3.1: Silhouettes of different species of arboreal marsupials surveyed by stagwatching.

Possum, possibly because these species are so small that they are readily overlooked. Second, in some landscapes, there can be so many trees with hollows that it is logistically impossible for all of them to be stagwatched even if a survey site is only 1 ha in size. Third, volunteer observers require training to distinguish the different species of arboreal marsupials. Fourth, despite the success of stagwatching, animals may still be overlooked because they emerge high in a tree without being seen by observers. A fifth problem is that there is potential for confounding between the amount of survey effort (number of observers on a site) and the number of trees with hollows on that site. To test problems that might be created by such confounding, stagwatching surveys were completed on a number of sites that did not support any trees with hollows. Observers were stationed at the corners of 25 × 25 m grid squares set out on each site. No arboreal marsupials were observed in any of these surveys (Lindenmayer *et al.* 1991d).

Figure 3.2: Perspective from the base of a large tree with hollows (stag) during a stagwatch. (Photo by David Lindenmayer).

Spotlighting

Spotlighting is one of the most widely used methods to survey arboreal marsupials (Figure 3.3). It involves using a large hand-held spotlight. Rods and cones in the retina of an animal's eye reflect the light from the observer's spotlight, giving what is known as 'eye-shine' (Lindenmayer and Press 1989). The colour of eyeshine varies between species of possums and gliders and is a key diagnostic tool. For example, the dull orange eyeshine of the Sugar Glider and Leadbeater's Possum contrasts markedly with the very bright white eyeshine characteristic of most populations of Greater Glider (Smith and Winter 1984). This feature makes the Greater Glider one of the species most readily detected using spotlighting. Once an animal is detected with a spotlight, it is important to confirm its identity using binoculars. This is because the colour of eyeshine can sometimes be influenced by the angle at which the light beam contacts an animal's eyes.

The effectiveness of spotlighting can be increased by including short silent 'lights-off' periods when it is possible to hear animals moving in the canopy or understorey (e.g. the Greater Glider can often be heard as it lands in a tree). Their precise location can then be determined by switching the spotlight beam back on and directing it to the source of the noise.

Although spotlighting is often used as a field survey technique, it has significant limitations and

Table 3.2: Numbers and frequency of occurrence of possums and gliders detected by stagwatching and spotlighting at 32 survey sites

	Stagwatch		Spotlight		No. stg/No. spt[a]	Sts Stg/sts spt[b]	Sts spt only (%)[c]	Sts stg only (%)[c]
	No.	Sites	No.	Sites				
Mountain Brushtail Possum	65	23	17	12	3:8	1:9	15	52
Greater Glider	11	6	9	7	1:2	0:9	50	42
Common Ringtail Possum	8	5	11	6	0:7	0:8	50	42
Yellow-bellied Glider	12	7	5	5	2:4	1:4	13	25
Sugar Glider	8	5	3	3	2:7	1:7	29	57
Leadbeater's Possum	87	17	9	6	9:7	2:8	0	65

a No. stg/No. spt is the ratio between numbers detected by stagwatching and numbers detected by spotlighting.
b Sts stg/sts spt is the ratio between the number of sites at which each species was detected by stagwatching and the number of sites at which each species was detected by spotlighting.
c Sts spt and sts stg are the percentage of the total number of sites at which the species was detected by either stagwatching or spotlighting.
Redrawn from Smith et al. (1989a).

Figure 3.3: Spotlighting an arboreal marsupial. (Photo by Alex Green).

many animals can be overlooked. A study testing the ability of an experienced spotlight observer to detect a radio-collared population of Greater Glider found that 10–60% of animals remained undetected (Lindenmayer et al. 2001). Therefore, spotlighting surveys can substantially underestimate the numbers of arboreal marsupials in an area – even in the case of those believed to be relatively easily detected, such as the Greater Glider. Species that are smaller and more active than the Greater Glider and that have comparatively dull eyeshine (e.g. Leadbeater's Possum and Sugar Glider) are even more likely to be overlooked with spotlighting (Lindenmayer 2002).

Spotlighting has been found to be particularly problematic in ash-type eucalypt forests because the dense understorey of trees obscures a clear vision of animals, particularly those high in the canopy. Even for large arboreal marsupials, like the Mountain Brushtail Possum, detections by spotlighting can be comparatively uncommon, including in places where population densities are extremely high (more than one individual per hectare). Other methods, such as stagwatching, have been found to be more productive for detecting arboreal marsupials in montane ash forests (Smith et al. 1989a). Finally, the effectiveness of spotlighting can be influenced by some environmental factors. Not surprisingly, animals are less likely to be observed in poor weather or when it is windy (Davey 1989).

Trapping

Trapping is required in many types of studies of arboreal marsupials. For example, it is essential, if the aim is to gather body measurements and growth rate data, to determine patterns of reproduction, longevity and mortality, or to radio-track animals (see below). There are many different kinds of traps and baits, that can have a marked influence on the kinds of animals that are trapped. They can also influence the capture rates of individual species and therefore affect measures, such as population estimates, that might be calculated from trapping data (Lebreton et al. 1992).

Traps set high in trees have been used routinely in trapping studies of Leadbeater's Possum (Smith 1984a), sometimes at heights of 8–15 m or more (Lindenmayer and Meggs 1996). The bait is typically a mix of water and honey and/or golden syrup, mimicking some of the kinds of natural foods consumed by the species. Trails of honey thinned with water are splashed around the branches leading to a trap to further entice animals to be captured (Meggs et al. 1991; Smith 1980). Large cage traps set on the ground have been used for over 17 years at Cambarville as part of a long-term trap–recapture study of the Mountain Brushtail Possum (Banks et al. 2008; Lindenmayer et al. 1998). The bait in these traps is green apple, which is more effective than other varieties of apple.

For the small traps used in capturing Leadbeater's Possum and the large cage traps used in studies of the Mountain Brushtail Possum, a plastic sleeve is placed over a trap. This prevents captured animals from becoming wet if it rains or snows. For traps set high in trees, non-absorbent cotton wool is placed inside the trap to allow an animal to make a

Figure 3.4: Elliott traps set high above the forest floor for Leadbeater's Possum. (Photo by David Lindenmayer).

Figure 3.5: Large wire cage trap used in trap–recapture studies of the Mountain Brushtail Possum. (Photo by David Lindenmayer).

temporary nest. These simple procedures reduce the chance of an animal becoming cold and wet, risking death from hypothermia.

A system of animal marking has been an important part of the trapping programs in the Central Highlands of Victoria. This has ensured that individuals can be distinguished, which is critical for estimates of population size and determining the gender and age composition of groups or colonies. In recent years, microchipping has replaced the procedure of applying a tattooed number to the ear of captured animals. A tiny chip, the size of a grain of rice, with a unique combination of letters and numbers, is inserted under the loose skin between the shoulder blades. Each time the animal is recaptured, the microchip number can be read (like a barcode) using a portable scanning machine. The marking procedures are quick and harmless, but animals are nevertheless sedated or anaesthetised to ensure minimal pain and stress.

An important consideration in the interpretation of trapping investigations in the montane ash forests is that the studies have been biased towards areas with high population densities where it has been possible to make many successful captures and complete a statistically rigorous investigation. Insights from these studies on life history attributes, such as home range size, mating systems and fecundity, may not apply to places where population densities are significantly lower (Lindenmayer 2002).

Hairtubing

The hairtubing method is used to determine the presence of various species of mammals by attracting them to an open-ended plastic tube that contains a bait. A sticky surface such as a double-sided tape is fixed to the inside of a hairtube. Fur from animals entering a tube may adhere to the tape and later be analysed to identify them. Identification is based on an assessment of the gross morphology of a hair sample, as well as an examination of sections of individual hairs under a microscope (Figure 3.6). Assessing hair samples requires considerable expertise and experience.

Different types of hairtubes can be deployed in field surveys. Hairtubes with different dimensions detect different species – small animals such as the Feathertail Glider can enter and leave large tubes without leaving fur samples, and are more frequently recorded in small hairtubes. Hairtubes can be placed on the ground or in trees, and their location clearly influences the types of species detected. Not surprisingly, tree-dwelling arboreal marsupials are rarely recorded in hairtubes placed on the ground. The bait used in hairtubes can also affect the range of mammals detected. Hairtubes baited with honey have been effective in attracting the Sugar Glider and Feathertail Glider; mixtures of peanut butter, rolled oats and honey have resulted in much lower rates of detection of these species.

Surveys with hairtubes can have several advantages. Hairtubing is not a particularly labour-intensive method and large numbers of hairtubes can be deployed and left for several weeks, without any checking or maintenance, before being retrieved. In addition, a range of different species of mammals can be detected in any given hairtube.

Like all survey techniques, hairtubing has limitations. While hair samples enable researchers to identify particular species that visited a hairtube, they provide no information on a range of other parameters such as:

- the age, sex and identity of individual animals;
- the number of times a given animal visited a particular hairtube;
- the range of tubes and therefore locations visited by a given individual.

As a result, hairtubing provides presence-only data and not abundance information on the species detected during the sampling period.

Finally, some species such as the Greater Glider are almost never recorded by hairtubing. For example, we have made no detections of the species from more than 5000 hairtube plots in the ash-type forests of the Central Highlands of Victoria (Lindenmayer *et al.* 1994c, 1994d), despite the fact that the Greater Glider can be quite common in some parts of the region (Lindenmayer 2002).

Nest boxes

Nest boxes have been used in many studies of cavity-dependent animals, including arboreal marsupials (Beyer and Goldingay 2006). They have been used in two major experimental studies in the montane ash forests of the Central Highlands of Victoria (Lindenmayer *et al.* 2003c, 2009c). A total of 196 nest boxes have been installed, varying in:

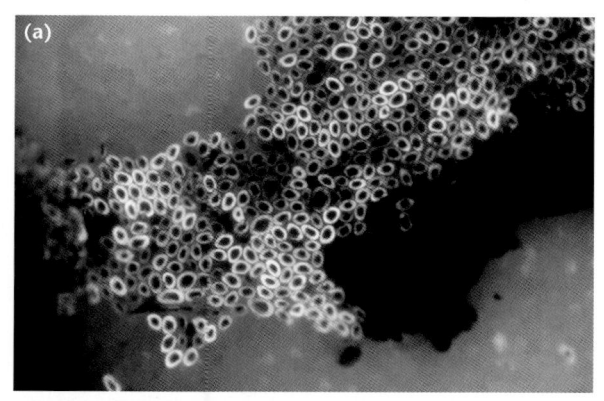

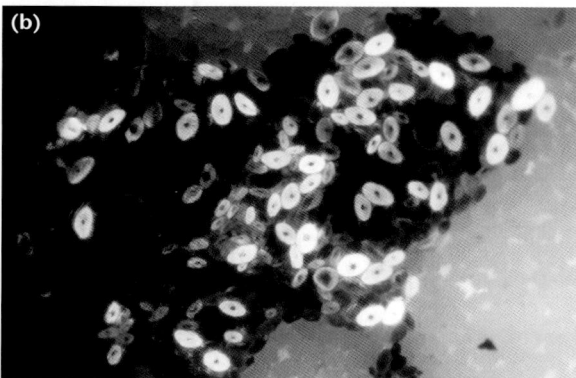

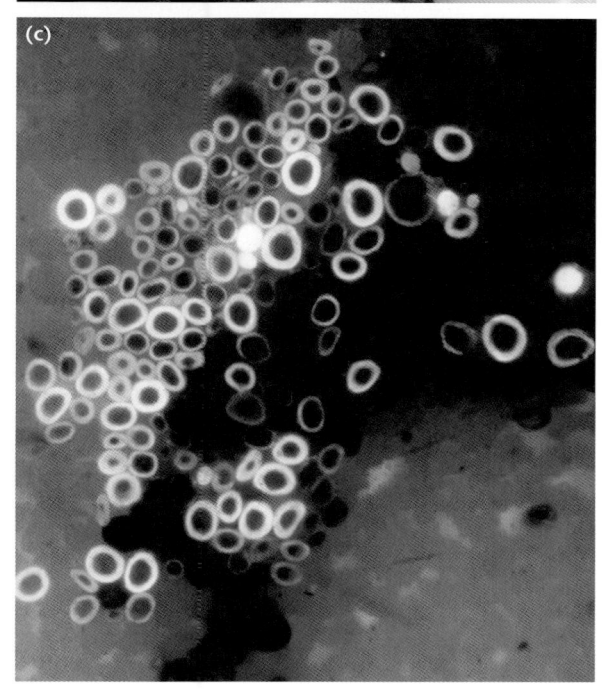

Figure 3.6: Cross-sections of hairs used in hair analyses of mammal fur. (a) Sugar Glider. (b) Greater Glider. (c) Leadbeater's Possum. (Photos by Barbara Triggs).

Figure 3.7: Hairtubes used in Victorian forests for sampling mammals. A bait chamber containing a watered-down honey and golden syrup mix hangs vertically from the bottom of the tube, where layers of double-sided stickytape have been inserted into the plastic pipe. Leadbeater's Possum, Common Ringtail Possum, Sugar Glider and Feathertail Glider were detected by these devices. (Photo by David Lindenmayer).

used to help confirm occupancy, including the physical presence of animals, hair, nesting material and scats;
- if there are any attributes of boxes or the surrounding forest stands which influence occupancy;
- the rate at which nest boxes are infested by pest animals such as bees;
- the rate of attrition of nest boxes, i.e. the rate at which nest boxes decay, fall to the forest floor and can no longer be reattached to trees.

The nest box experiments have enabled some species such as Leadbeater's Possum (which can be very difficult to capture with conventional approaches like tree trapping) to be captured and studied. For example, tissue samples taken from animals found in nest boxes have formed the basis of genetic studies of Leadbeater's Possum (Hansen *et al.* 2008; Hansen 2008).

Although nest boxes have provided valuable information on arboreal marsupials in the montane ash forests, no attempt has been made to interpret animal

- size – small boxes were 400 mm (height), 237 mm (width), 271 mm (depth), 51 mm (entrance hole diameter), internal volume (0.019 m^3). Large boxes had the following dimensions – 490 mm (height), 292 mm (width), 330 mm (depth), 103 mm (entrance hole diameter), internal volume (0.038 m^3);
- height above the ground – 3 m and 8 m above the forest floor;
- age of the surrounding forest – 1939 regrowth and 20-year regrowth.

The studies using nest boxes were specifically designed to determine:
- frequency of occupancy of nest boxes by arboreal marsupials (and other species of cavity-dependent animals). Various measures of nest box use were

Figure 3.8: One of 196 nest boxes being checked as part of a major set of experiments in the montane ash forests. (Photo by David Lindenmayer).

abundance data from their use. This is because adding artificial cavities to a forest increases the abundance of available hollows, which may provide extra breeding sites for animals. Population sizes may then be elevated well above typical abundance levels for an area. For this reason, establishing nest boxes in an area cannot provide an independent assessment of the habitat suitability or carrying capacity of a site (Gibbons and Lindenmayer 2002).

Radio-tracking

Sometimes the aim of a study is to determine the area over which an animal moves or which parts of a patch of forest or woodland it uses most frequently. One of the most effective ways is to use radio-tracking. Small radio transmitters are attached to an animal, usually via a collar fixed around its neck. A receiving system can be used to determine the precise location of the collared animal. Repeated 'fixes' of an animal's location over a set period can provide information on the location of its den sites, the size and shape of its home range and how much its territory overlaps with those of other collared individuals.

Radio-tracking has been used in two major studies of arboreal marsupials in montane ash forests. The first was on Leadbeater's Possum and involved tracking animals to their daytime nest trees in large trees with hollows (Lindenmayer and Meggs 1996). A similar kind of study focusing on den tree use was completed using radio-collared individuals of the Mountain Brushtail Possum (Lindenmayer et al. 1996c).

In both radio-tracking studies, special care was taken to prevent animals being injured. Animals were sedated under veterinary supervision to ensure that collars could be appropriately fitted to avoid injuries (Viggers and Lindenmayer 1995). This was important – if a collar were too tight it could have led to skin lesions or wounds. Conversely, if a radio-collar were too loose an animal may have been able to jam a paw underneath it and subsequently die. In both studies, the weight of radio-collars was less than 3% of an animal's body weight, limiting changes in behaviour or even risk of death.

In mid 2008, a new radio-tracking study was commenced on the population of the Mountain Brushtail Possum at Cambarville. New-generation

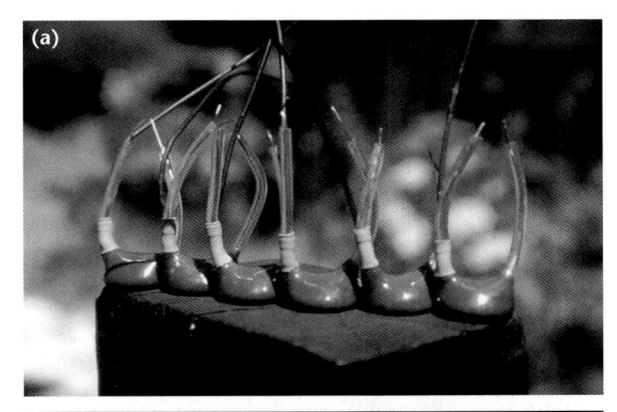

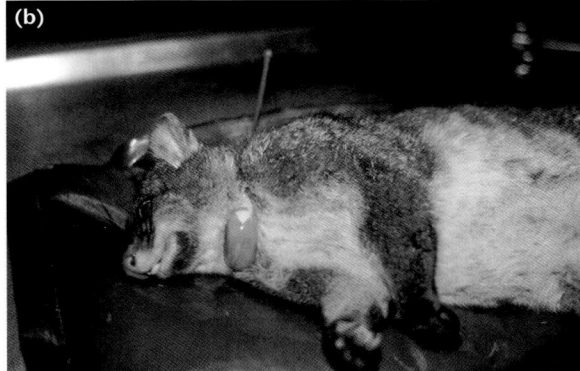

Figure 3.9: Radio-collars used in studies on (a) Leadbeater's Possum. (b) Mountain Brushtail Possum. (Photos by David Lindenmayer).

contact transmitters were fitted that record data not only on the location of a collared individual but also on the locations and identities of nearby animals in the population. The aim is to quantify interactions between animals and help resolve key questions about the social organisation and mating system of the Mountain Brushtail Possum. The study area where animals were being radio-tracked was burned by the February 2009 wildfires, and a new research trajectory will be to initially quantify how many collared individuals in the population survived the fire. It is then planned, in subsequent work, to document post-fire population dynamics and quantify the length of the process of recovery to pre-fire levels of abundance and social organisation.

Genetic tools

Genetic information gathered from wildlife populations has a range of important uses. Genetic studies are becoming an increasingly important component of the studies in the montane ash forests. For example, genetic analyses were pivotal in demonstrating that genetic differences between different populations of the Mountain Brushtail Possum mirrored differences in external morphology. The combination of genetic and morphometric information underpinned the discovery that the Mountain Brushtail Possum was not a single widespread species but comprised a northern species (the renamed Short-eared Possum, *Trichosurus caninus*) and the newly recognised southern species – the Mountain Brushtail Possum (*Trichosurus cunninghami*) (Lindenmayer *et al.* 2002d).

Genetic analyses have recently been used in quantifying aspects of the breeding and dispersal biology of the Mountain Brushtail Possum at Cambarville (Banks *et al.* 2009, 2008), as well as the impacts of forest fragmentation and patch isolation on patterns of genetic variability in Leadbeater's Possum (Hansen *et al.* 2008). Genetic analyses will be very important in quantifying some of the dynamics of animal population recovery following the 2009 wildfires, for example, by helping to elucidate the distances over which newly colonising individuals will move to join recovering populations.

Genetic analyses of arboreal marsupials have entailed gathering blood or ear tissue. This has required some form of capture and careful and humane handling to ensure animals are not injured. As in the case of fitting radio-collars or animal marking, it has been appropriate to immobilise individuals with sedative drugs or anaesthesia. Notably, it has been demonstrated that it is possible to use DNA in fur samples gathered from hairtubes to address a range of genetic questions (Banks *et al.* 2003). This exciting development has the potential to answer a range of additional kinds of questions about wildlife in the montane ash forests of the Central Highlands of Victoria.

SMALL TERRESTRIAL MAMMALS

Trapping, hairtubing and genetic methods have been used extensively over the past 25 years in studies of small terrestrial mammals in the Central Highlands of Victoria. All three field methods are briefly described.

Trapping

All trapping studies of small terrestrial mammals have been conducted using metal box Elliott traps (Elliott Scientific Equipment, Upwey, Victoria). Different trap arrays have been deployed, depending on the objectives of particular studies. For example, index lines or transects of traps were used in studies comparing animal abundance in ash-type forest stands of different ages, whereas grids of traps were established as part of a trap–recapture study (Cunningham *et al.* 2005). In all studies, traps were baited with a mixture of peanut butter, honey and rolled oats. Trapping was conducted for four consecutive days on most sites, except when it rained heavily and traps were closed after fewer days to reduce the potential for animal mortality from trapping. Elliott traps in which an animal had been captured were wiped clean, rebaited and repositioned where the initial capture had taken place. Data recorded for

Figure 3.10: An Elliott trap used for making captures of small terrestrial mammals. (Photo by David Lindenmayer).

each captured animal included species, sex and age cohort – juvenile, subadult or adult (based on size, body mass and size of genitals) (Cunningham *et al.* 2005).

Hairtubing

As outlined in the section on arboreal marsupials, hairtubing has been a well-used field survey method in studies of the terrestrial vertebrate biota of montane ash forests. The approach has been a key part of work on (1) the use of retained linear strips of forest by mammals (Lindenmayer *et al.* 1994c), (2) contrasts in the detections of mammals in different ages of ash-type forest (Lindenmayer *et al.* 1999d) and (3) the microhabitat requirements of the Mountain Brushtail Possum (Lindenmayer *et al.* 1994d). In many of these studies, different kinds of hairtubes were deployed at a given survey plot. These different types of hairtubes ranged in entrance size:

- small tubes with an entrance diameter of 32 mm (Suckling 1978);
- large tubes with an entrance diameter of 105 mm) (Scotts and Craig 1988);
- a tapered 'hair funnel' with a closed bait chamber at its narrowest end.

The different kinds of hairtubes were placed in different locations, either on the ground or 2–3 m above ground in a tree. Not surprisingly, different kinds of mammals were detected by different kinds of hair-tubes. The most effective method (yielding the highest number of detections) varied between species. For most species, detection by more than one hairtubing method at a given plot was uncommon. That is, a species was recorded by one method, not by either of the other two at the same plot (Lindenmayer *et al.* 1999e). A general conclusion was that if the aim of a field survey is to detect a wide range of species, then a suite of types of hairtubes may need to be deployed (Lindenmayer *et al.* 1999e). Another outcome was that the interpretation of field data needs to be made with careful consideration of the methods used to gather those data.

An additional part of the hairtubing studies involved the collection of scat material deposited by small mammals on the hairtubes and adhesive surfaces (Lindenmayer *et al.* 1996e). It was possible to identify which species deposited scats in or on a given hairtube and compare that with detections of animals made in the same tubes. The results of this study revealed scats from the Agile Antechinus were frequently collected from large hairtubes where the species had not been recorded by hair analysis (Table 3.3). Thus, in many cases, the species was able to enter the larger hairtubes without touching the adhesive surfaces fixed to the inside walls (Lindenmayer *et al.* 1996e). This was less likely to occur in the smaller hairtubes, which had dimensions closer to the body size of the Agile Antechinus. This result emphasised the value of using a range of types and sizes of hair-tubes if a general aim of a field survey is to detect a wide variety of species of mammals. Conversely, if the objective of a field study is to survey for a given species, the types of hairtubes used should be those that are most appropriate for the detection of the target taxon (Lindenmayer *et al.* 1996e).

Genetic tools

As in the case of arboreal marsupials, studies using genetic methods have begun to add exciting new insights to the understanding of wildlife populations in montane ash forests. The three common species of small terrestrial mammals have been the focus of genetic work – the Agile Antechinus, Dusky Antechinus and Bush Rat (Kraaijeveld-Smit *et al.* 2007, 2002b). The work has revealed new insights into the

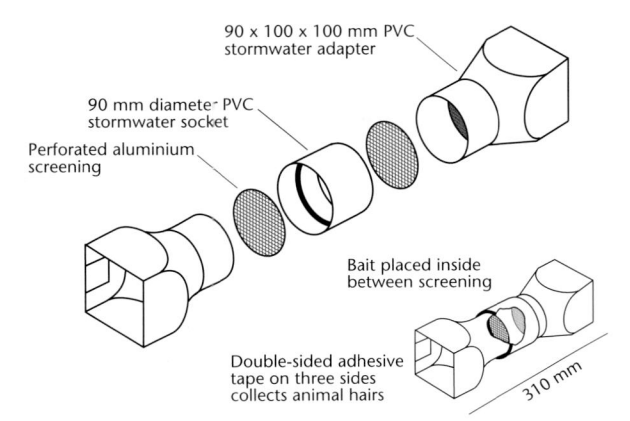

Figure 3.11: One type of hairtube used in field surveys of terrestrial mammals. (Redrawn from Scotts and Craig 1988).

Table 3.3: Detections of small mammals by hairtubes and scats, in the ash-type forests of the Central Highlands of Victoria

Hairtube type	Total no. of tubes	Hair sample positive	Hair sample negative
Agile Antechinus			
Large ground	115	14	101
Small ground	7	7	0
Small high[a]	34	28	6
Bush Rat			
Large ground	4	4	0
Small ground	4	3	1
Small high[a]	0	0	0

a These hairtubes were set approximately 3 m above the ground.
Information has not been presented on hairtubes where no scat samples were gathered.
Redrawn from Lindenmayer et al. (1996e).

dispersal biology of these species. It has indicated that the Agile Antechinus and Dusky Antechinus are characterised by strong patterns of male-biased dispersal, whereas the Bush Rat does not show such patterns (Kraaijeveld-Smit et al. 2007). The implications are important because they suggest that factors which result in landscape change, such as logging-induced habitat fragmentation, will have differing effects on population dynamics, social organisation and mating systems for the two antechinus species, compared to the native murid rodent. This is due to differences in population structure and degrees of inbreeding (Kraaijeveld-Smit et al. 2007).

BIRDS

Substantial effort has been dedicated to surveying birds in the montane ash forests, particularly in the past five years. Most emphasis has been placed on counting birds using point interval counts (Pyke and Recher 1983). Another method – dusk counts – has been used less frequently and is discussed briefly.

Point interval counts

Point interval counts involve conducting counts of all birds (seen, heard or both) around a fixed point for a specified period. The method is used widely in Australia (Pyke and Recher 1983) and overseas (Bibby et al. 1998). In the montane ash forests, a 100 m-long transect is established at any given site. Along the transects, five-minute point interval counts are completed at the permanently marked 0 m, 50 m and 100 m points. For each point count, all birds seen or heard within a site are assigned to different distance classes from a point: 0–25 m, 25–50 m, 50–100 m and more than 100 m. Surveys are typically completed in early December in any given year, which is the breeding season for the majority of species and when summer migrants have arrived.

The survey protocol is specifically designed to quantify site occupancy; for subsequent statistical analyses it is not assumed that individual counts at the three points on the same site are independent. Each site is surveyed on a different day by a different observer to reduce day effects on detection and overcome potential observer heterogeneity problems (Cunningham et al. 1999; Field et al. 2002; Lindenmayer et al. 2009a). Counts are typically pooled across the three survey points in a given site and across combinations of observers and days. Notably, data analyses have indicated that six counts on a site in a given year are sufficient to confirm occupancy of a particular species at a given site.

Bird data gathered from the point interval counts have been used in a range of studies, including:

- relationships between forest structure and bird species richness (Chapter 11);
- factors influencing the occurrence of individual species of forest birds (Chapter 9);
- the value of rainforest strips and patches for forest birds (Chapter 5);

- the impacts of variable harvest retention systems on birds (Chapter 16).

Dusk counts

During night-time stagwatch surveys for arboreal marsupials between 1997 and 2008, lists were made of forest birds that were heard calling. These surveys generated presence/absence lists from ~30 minutes of dusk counts; no data on abundance were gathered because the length of time over which surveys are conducted means that considerable multiple counts of an individual bird would be extremely likely. The data indicate that some species of birds (e.g. Scaly Thrush, *Zoothera lunalata*) are far more likely to be recorded at dusk than during dawn and early morning counts on the same sites.

Unfortunately, it is not possible to formally compare the bird data gathered using point interval counts and dusk counts because there is confounding by day and survey length. That is, the two kinds of surveys have typically been conducted on different days and for different amounts of time (15 minutes versus 30 minutes). In addition, the areas covered by the two kinds of surveys were different.

VEGETATION MEASURES

The composition and structure of the forest has been a major focus of the work in the Central Highlands of Victoria. Data gathered have been used as covariates in analyses of the habitat requirements of arboreal marsupials, birds and terrestrial mammals. Characteristics of the structure and composition of the vegetation have also been analysed in their own right, including analyses of the occurrence and characteristics of old-growth forest (Lindenmayer *et al.* 2000a) (Chapter 4), the occurrence of cool temperate rainforest (Lindenmayer *et al.* 2000b) (Chapter 7), the prevalence of large trees with hollows across forested landscapes (Lindenmayer *et al.* 1991b) (Chapter 6) and the volumes and characteristics of fallen trees or coarse woody debris in different-aged stands (Lindenmayer *et al.* 1999d) (Chapter 8).

A large number of vegetation attributes have been measured in many different kinds of field studies completed since 1983. Different attributes have been measured at different spatial scales; it is perhaps easiest to describe these attributes by categorising them at the scale at which they were measured. The measures have been made consistently in a range of studies, although on occasions additional attributes needed to be gathered for the particular needs of a given investigation. For example, following the 2009 wildfires an additional set of measures reflecting fire severity will be gathered at field sites where populations of animals and the composition, structure and cover of plant assemblages are quantified. These fire severity measures will be important in quantifying ecological recovery trajectories and their relationships with the site-level impacts of the 2009 wildfire, together with the interactions between fire severity and the age and structure of the forest at the time it was burned.

Tree-level measures

Large trees with hollows have been a particular focus of work in the Central Highlands of Victoria. Many studies have examined the features of individual trees. This required such measurements as tree condition (form), tree diameter, tree height, numbers of different kinds of cavities (fissures, hollow branches, trunk hollows) and the extent of adjacent living vegetation. A description of tree-level attributes and the way they were measured is given in Table 3.4.

Large fallen trees were measured as part of a major study of coarse woody debris (Lindenmayer *et al.* 1999d). In this case, a 100 m-long transect was established at replicate sites of different forest age. The length and diameter of each log that intersected the transect was estimated. Each log was assigned to a decay class that best represented its decomposition status. Data were gathered on the percentage cover of mosses, ferns and lichens on each log. These log-level attributes were complemented with a site-level estimate of the volume of logs per ha of forest made using the line intersect method (Van Wagner 1976; Warren and Olsen 1964) for each survey site (Lindenmayer *et al.* 1999d).

Plot- and stand-level measures

Attributes of vegetation structure and composition were typically measured on each of six 10 × 10 m plots

Table 3.4: Measured attributes of large trees with hollows in the montane ash forest of the Central Highlands of Victoria

Variable	Description
Tree species	All hollow-bearing trees were eucalypts, so buds and fruit were used to identify living tree species (Costermans 1994)
Height	Distance (m) to the highest part of a tree with cavities. Determined using a Spiegel Relaskop and verified with a clinometer
Diameter	Measured at 1.5 m height above ground using a diameter tape
Form	Hollow-bearing trees were classified into one of eight forms based on readily observable characteristics reflecting stages of advancing senescence. Classification of forms of hollow-bearing trees was as follows: Form 1, mature living tree; Form 2, mature living tree with a dead or broken top; Form 3, dead tree with most branches still intact; Form 4, dead tree with 0–25% of the top broken off, branches remaining as stubs only; Form 5, dead tree with the top 25–50% broken away; Form 6, dead tree with the top 50–75% broken away; Form 7, solid dead tree with >75% of the top broken away; Form 8, hollow stump
Holes	Determined by observation using binoculars and defined as any opening in the trunk of a tree
Fissures	Determined by observation using binoculars and defined as any narrow crack in the tree trunk >15 mm in width and >30 mm long
Hollow branches	A branch was considered to be hollow if it possessed an opening >40 mm in diameter
Total cavities	Sum of holes, fissures and hollow branches in a given tree with cavities

located randomly across each 3-ha site. These are described in Table 3.5. Table 3.6 outlines attributes measured at the 3-ha site level.

SUMMARY

This chapter has touched on some of the methods used to address key research questions in the montane ash forests of the Central Highlands of Victoria. In the case of the approaches used to study vertebrate biota, each of the methods described has particular strengths and weaknesses according to certain circumstances. Because the results of field studies can be strongly influenced by the survey methods, it is important to be aware of the potential advantages and deficiencies of each. Therefore, an understanding of the efficacy of field methods is a critical part of interpreting the data gathered.

Carefully designed research is needed to compare the effectiveness of different methods. Considerable effort has been dedicated to this issue as part of the work in the montane ash forests, not only for arboreal marsupials and small terrestrial mammals (Lindenmayer *et al.* 1999e, 1996e; Smith *et al.* 1989a) but also for birds (Cunningham *et al.* 1999; Lindenmayer *et al.*

> **Box 3.1: Postscript – after the 2009 wildfires**
>
> Many of the long-term sites established over the past 25 years in the Central Highlands of Victoria were burned in the February 2009 wildfires. This does not mean the end of the learning journey in these forests. Rather, it will be critical to quantify the ecological processes and forest patterns associated with post-fire recovery. Animal populations and vegetation cover on burned and unburned sites (which act as controls) will need to be measured repeatedly to document ecological recovery. The protocols for making these measurements will need be precisely the same as those described throughout this chapter. This is essential, to maintain the long-term integrity of the field datasets and hence make valid comparisons between animal populations and vegetation cover before and after the 2009 wildfires and to compare burned and unburned sites.

Table 3.5: Averaged measures of vegetation structure and plant species composition recorded at each of six 10 × 10 m plots on each site

Variable	Description
Bark index	Quantity of strips of decorticating bark peeling from the trunk and lateral branches of trees. Points were assigned to each clump on the basis of its size and a cumulative score was calculated for each tree on the plot (see Lindenmayer *et al.* 1990a)
Vegetation strata	Vegetation layers were assigned to classes – tree overstorey, tree understorey, tall shrub layer and low shrub layer. The number of strata was recorded
Plant matrix	A two-way height and diameter matrix was completed for all stems of all species >2 m in height on each plot. The species of stem in the plot was recorded and assigned to a height and diameter class. The height categories were 2–5 m, 6–10 m, 11–20 m, 21–30 m, 31–40 m, 41–60 m, >60 m. The diameter classes were 0–50 mm, 60–100 mm, 110–200 mm, 210–300 mm, 310–400 mm, 410–500 mm, 510–600 mm, 610–800 mm, 810–1000 mm, 1010–1400 mm, 1410–1800 mm, >1800 mm. Multi-stemmed shrubs were assigned to the class corresponding to stem with the largest diameter and height class for the plant
Number of shrubs	Data from the plant matrix were used to calculate the number of shrubs >2 m in height per plot (i.e. trees were excluded)
Understorey trees	The plant matrix was used to calculate the number and basal area of understorey trees (*Acacia* spp. and *Nothofagus cunninghamii*)
Tree ferns	The number of tree ferns (*Dicksonia antarctica* and *Cyathia australis*) was counted

2009a). In some cases, a combination of methods will be needed to address a particular research question or expand the range of useful outcomes from a project (Lindenmayer *et al.* 1999e). A good example is the use of stagwatching to identify the nest trees of arboreal marsupials, then trapping and radio-collaring gliders to study their patterns of tree use (Lindenmayer *et al.* 1996c) (Chapter 9).

Table 3.6: Attributes measured at the site level

Attribute	Description
Stand age	Measurements were taken from sites of a range of stand ages dating from 1735, 1824, 1895, 1905, 1908, 1918, 1926, 1932, 1939, 1948, after 1968. These stands typically corresponded to the date of the last major disturbance, usually wildfire or logging. In areas designated for timber production, forest age was determined at each site using assessment maps provided by the Department of Conservation, Forests and Lands. The Board of Works provided maps of the ages of forests in water catchment areas
Position	Three topographic positions were examined: gully/bottom-slope, mid-slope and top-slope/ridge
Slope	Slope was measured in degrees with a clinometer from a point in the centre of each 3-ha site
Aspect	Azimuth was measured in degrees using a compass and assigned to one of four categories: north (315–45°), east (45–135°), south (135–225°) and west (225–315°)
Radiation	A radiation index was calculated for each site from data on slope and aspect using the computer program CLOUDY. It was the ratio of the clear-day incident radiation to the clear-day horizontal radiation
Latitude, longitude, elevation	Determined for each site using 1:10 000 and 1:25 000 topographic maps

KNOWLEDGE GAPS

Although considerable effort has been dedicated to testing field methods and comparing the effectiveness of different methods, much work remains to be done on this topic. For example, an important area for future research is to further test the effectiveness of different field methods and examine issues such as the potential to underestimate the number of animals in a given area. New technologies, such as remote digital cameras, offer exciting opportunities for the study of forests and forest wildlife. New-generation contact radio-collars are another. However, much developmental work remains to be done on implementing these technologies. A major issue is that the results obtained using them will be different from those derived from previous technology. Where long-term data are important for understanding key forest ecosystem processes or wildlife population dynamics, new technologies may breach the integrity of long-term datasets. There would be serious confounding between time and field methods, making trend data un-interpretable. Careful calibration of methods is needed to overcome these kinds of problems and ensure the long-term validity of important longitudinal datasets.

LINKS TO OTHER CHAPTERS

The content of this chapter links with every other chapter in this book. This is because field-derived empirical data underpin all the studies we have conducted in the montane ash forests over the past 25 years, even those of ecological theory or where simulation models have been applied. Rigorously gathered field-derived empirical data will be critical for documenting ecological and biodiversity recovery after the February 2009 wildfires.

PART II

FOREST COVER AND COMPOSITION OF THE FOREST

Two broad kinds of forest have been the focus of work in the Central Highlands of Victoria – ash-type eucalypt forest and rainforest. They are markedly different in terms of structure, plant species composition and other attributes, such as the kinds of vertebrates they support. However, the two kinds of forest can co-occur – a rainforest understorey can often be a quintessential part of the understorey of montane ash forest, especially old-growth and multi-aged stands (Chapter 7). Nevertheless, the two kinds of forest are treated separately in the two chapters which comprise Part II of this book. In both chapters, bioclimatic and other factors influencing the broad distribution patterns of montane ash forest and rainforest, respectively, are discussed. In addition, some of the factors influencing the age and structure of these forests are outlined.

In some respects, parts of Chapters 4 and 5 have similarities with Chapter 9 in that the multi-scale factors which influence distribution patterns are discussed, although the focus in Part II is on the overstorey tree species rather than particular vertebrate taxa. However, there are strong relationships between Chapters 4 and 5 and Chapter 9, for example because animal distribution is often closely tied with forest type and tree species distribution.

4 The ash-type eucalypt forest

INTRODUCTION

Most studies over the past 25 years have targeted montane ash forests dominated by Mountain Ash, Alpine Ash or Shining Gum. This chapter addresses three broad topics associated with the composition and distribution of forests dominated by these overstorey tree species:

- bioclimatic and other environmental factors influencing the occurrence of the different species of trees which comprise montane ash forest;
- broad patterns of successional development in montane ash forest, including the development of old-growth forest;
- factors influencing the spatial distribution of different age classes of montane ash forest within the broader distributions of these tree species in the Central Highlands of Victoria.

The topics explored in this chapter are strongly interrelated with many other key ones associated with the ecology and management of montane ash forests. Therefore, to avoid too much duplication, the topics in this chapter are examined only briefly and the discussion is confined to some general perspectives. For example, initiation of succession in montane ash forests is disturbance-related – an enormous topic. Accordingly, an entire section comprising three chapters in Part V has been dedicated to natural and human disturbance processes (Chapters 12–14). Successional changes in montane ash forests are usually reflected in changes in stand structure. This is a very

Table 4.1: Studies directly or indirectly examining overstorey eucalypt trees in montane ash forests

Study description	Reference
Climate envelopes of montane ash trees	Lindenmayer et al. (1991f); Lindenmayer et al. (1996a)
Climate domains of forest types	Mackey et al. (2002)
Successional trajectories and disturbance pathways in montane ash forests	Lindenmayer (2009)
Stand structure and environmental factors	Lindenmayer et al. (1999f)
Old-growth attributes of montane ash forests	Lindenmayer et al. (2000a); Lindenmayer (2009)
Prevalence of multi-aged montane ash forests	McCarthy and Lindenmayer (1998)

large topic and the three chapters in Part III describe a wide range of aspects of the structure of montane ash forests (Chapters 6–8).

Importance of understanding distribution of the ash forest overstorey

Work over the past 25 years has shown that the dominant tree species in montane ash forests shapes the distribution and abundance of many other species and influences several key ecological processes. For example, dominant tree species has been found to significantly influence:

- arboreal marsupial species richness and the composition of arboreal marsupial assemblages (Lindenmayer *et al.* 1993a) (Chapter 9);
- the occupancy of retained linear strips by arboreal marsupials and small terrestrial mammals (Lindenmayer *et al.* 1993a) (Chapter 9);
- bird species richness and the composition of bird assemblages (Lindenmayer *et al.* unpublished data, 2009b) (Chapter 9);
- the abundance of trees with hollows (Lindenmayer *et al.* 1991b) (Chapter 6);
- the prevalence and kinds of cavities that develop in large living and dead trees (Lindenmayer *et al.* 1993b) (Chapter 6);
- the extent of development of cool temperate rainforest trees (Lindenmayer *et al.* 2000b) (Chapter 5);
- the sugar and nitrogen content of the gums produced by understorey *Acacia* spp. trees (Lindenmayer *et al.* 1994a) (Chapter 7).

These effects make it important to understand the factors influencing the broad distribution patterns of the dominant overstorey tree species in montane ash forests.

BACKGROUND DATASETS

Studies of the overstorey ash forest were underpinned by several key datasets. First, much of the environmental modelling of the distribution of particular tree species used point location data obtained from several sources, including:

- EUCALIST, a compilation of records from various Australian herbaria (Chippendale and Wolfe 1985);
- field data on forest structure and floristic composition gathered as part of field surveys for arboreal marsupials in the Central Highlands of Victoria, north-eastern Victoria and south-eastern New South Wales (Lindenmayer 1989; Lindenmayer *et al.* 1991f);
- records from botanical collections, including the National Herbarium of Victoria.

Second, extensive climate, terrain and topographic data were generated for the Central Highlands region using a range of environmental modelling tools. Surfaces of values for the environmental characteristics of the study area were derived from a suite of spatially distributed models that simulate the radiation, thermal and moisture regimes and which determine the distribution and availability of light, heat and water. The models used to derive estimates of environmental conditions were based on a digital elevation model (DEM) with a grid resolution of 20 m (Lindenmayer *et al.* 1999f).

A third major dataset comprised detailed vegetation measures at over 520 sites throughout the Central Highlands region. Measures of overstorey stand age, the presence of multiple age cohorts of trees and a wide range of other attributes of stand structure and plant species composition were recorded as part of the detailed field surveys of vegetation (Chapter 3). This information was also used to explore the existence of different natural disturbance pathways in montane ash forests (Lindenmayer 2009).

CLIMATE ENVELOPES FOR EACH SPECIES OF ASH-TYPE EUCALYPT

The results of many studies indicate that climatic factors typically set limits to the broad-scale distribution of plants (Forman 1964; Jarvis and McNaughton 1986; Woodward and Williams 1987). Because of the overarching influence of climate on key aspects of plant physiology, studies of climatic environments can make an important contribution to understanding the spatial distribution patterns of plants including overstorey trees (Austin *et al.* 1990; Fitzpatrick and Nix 1970; Mackey 1993; Specht 1981; Specht and Specht 1999). Given this, a bioclimatic analysis (using the BIOCLIM package (Nix 1986)) was completed for

Table 4.2: Bioclimatic envelope of Mountain Ash derived from BIOCLIM

Index no.	Description of bioclimatic index	Mean	Min.	5%	95%	Max.
Temperature indices						
1	Annual mean temperature	11.1	7.0	7.8	13.7	14.4
2	Mean diurnal range (mean[period max–min])	10.0	7.4	8.3	12.4	13.1
3.	Isothermality (index 2/index 7)	0.47	0.40	0.41	0.51	0.52
4	Temperature seasonality (coefficient of variation)	37.3	21.7	26.6	50.5	63.8
5	Maximum temperature of warmest month	23.5	17.2	18.4	26.8	27.9
6	Minimum temperature of coldest month	2.1	–2.4	–1.8	4.6	5.7
7	Annual range in temperature (index 5–6)	21.4	15.3	17.1	25.2	26.4
8	Mean temperature of wettest quarter	7.7	3.9	4.3	12.7	15.8
9	Mean temperature of driest quarter	15.7	4.6	11.0	19.0	19.5
10	Mean temperature of warmest quarter	16.2	11.2	12.3	19.0	19.5
11	Mean temperature of coldest quarter	6.3	1.7	3.5	9.0	9.9
Precipitation indices						
12	Annual precipitation	1280	661	787	1716	1886
13	Precipitation of wettest month	141	67	77	197	223
14	Precipitation of driest month	64	37	45	82	87
15	Precipitation seasonality	23	10	12	34	36
16	Precipitation of wettest quarter	399	187	225	557	632
17	Precipitation of driest quarter	219	130	154	279	302
18	Precipitation of warmest quarter	226	130	160	293	302
19	Precipitation of coldest quarter	385	147	203	50	632

Homocline matching was based on four sets of values in the profile: minimum to maximum and 5% to 95%. The associated predicted potential distribution of the species is presented in Figures 4.1–4.4. All values for temperature are given in °C and those for precipitation in mm.
Modified from Lindenmayer et al. (1996a).

Mountain Ash, Alpine Ash and Shining Gum (Lindenmayer et al. 1996a). These analyses derived summary information on the range of bioclimatic conditions which characterise the extant distribution of each tree species. Tables 4.2–4.4 present bioclimatic profiles containing the mean, minimum, 5%, 95% and maximum values for 19 temperature and precipitation indices derived for the three target tree species. Other estimates for the various indices that are calculated by the BIOCLIM system (i.e. the 10%, 25%, 50%, 75% and 90% levels) have been excluded.

Comparisons of cumulative frequency curves for various tree species

Figures 4.1–4.4 show the cumulative frequency plots for a subset of bioclimatic indices. They highlight clear differences in the bioclimatic regimes occupied by the three tree species. Sites supporting Mountain Ash are typically at lower elevations than the other tree species examined. In contrast, values for annual precipitation are similar for all three species. A comparison of values for annual mean temperature indicated that the lowest average temperatures were experienced by Alpine Ash and the highest by Mountain Ash. A similar trend was observed from comparisons of the mean temperature of the warmest quarter. Sites supporting Mountain Ash were characterised by the highest values for minimum temperature of the coldest month and mean temperature of the coldest quarter. The lowest mean maximum temperatures of the warmest month were for Alpine Ash (Lindenmayer et al. 1996a).

Table 4.3: Bioclimatic envelope of Alpine Ash derived from BIOCLIM

Index no.	Description of bioclimatic index	Mean	Min.	5%	95%	Max.
Temperature indices						
1	Annual mean temperature	8.6	4.1	5.8	11.5	14.4
2	Mean diurnal range (mean[period max–min])	9.6	7.1	7.8	11.8	14.0
3.	Isothermality (index 2/index 7)	0.46	0.39	0.41	0.50	0.51
4	Temperature seasonality (coefficient of variation)	50.1	22.6	31.6	70.1	90.4
5	Maximum temperature of warmest month	20.6	14.2	15.9	25.1	30.9
6	Minimum temperature of coldest month	−0.5	−3.4	−2.7	2.4	5.6
7	Annual range in temperature (index 5–6)	21.1	15.3	17.0	25.3	29.6
8	Mean temperature of wettest quarter	5.0	0.2	1.2	9.5	15.9
9	Mean temperature of driest quarter	13.3	3.5	8.9	17.0	21.7
10	Mean temperature of warmest quarter	13.9	8.7	10.2	17.4	21.7
11	Mean temperature of coldest quarter	3.5	−0.2	0.9	6.5	9.1
Precipitation indices						
12	Annual precipitation	1339	544	719	2045	3056
13	Precipitation of wettest month	153	55	72	243	326
14	Precipitation of driest month	69	29	43	96	152
15	Precipitation seasonality	23	10	12	35	38
16	Precipitation of wettest quarter	425	159	205	683	933
17	Precipitation of driest quarter	235	109	138	318	546
18	Precipitation of warmest quarter	239	116	141	320	546
19	Precipitation of coldest quarter	408	112	189	678	904

Homocline matching was based on four sets of values in the profile: minimum to maximum and 5% to 95%. The associated predicted potential distribution of the species is presented in Figures 4.1–4.4. All values for temperature are given in °C and those for precipitation in mm.
Modified from Lindenmayer et al. (1996a).

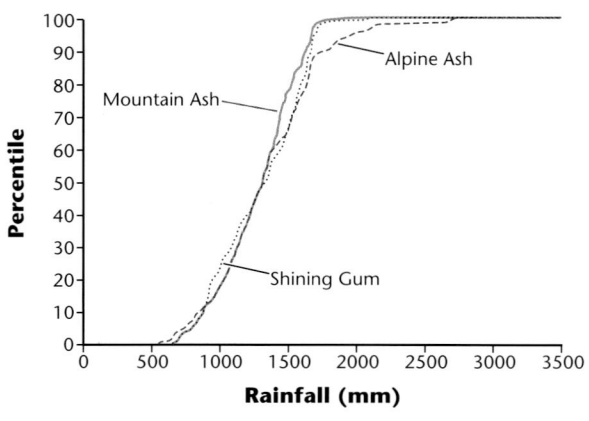

Figure 4.1: Cumulative frequency plot showing the relationship between different percentile values of the BIOCLIM-modelled annual precipitation of sites occupied by the three species of eucalypt trees. (Redrawn from Lindenmayer et al. 1996a).

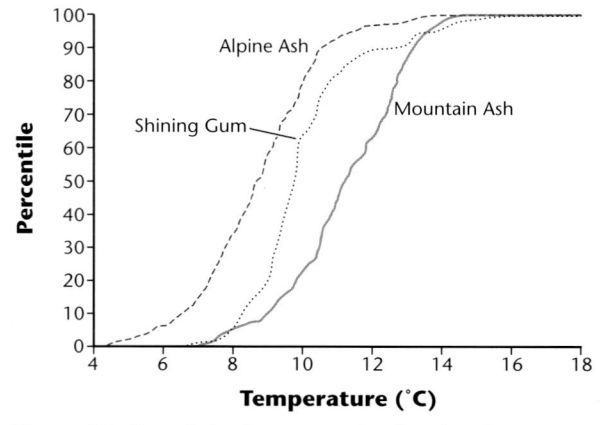

Figure 4.2: Cumulative frequency plot showing the relationship between different percentile values of the BIOCLIM-modelled annual mean temperature of sites occupied by the three species of eucalypt trees. (Redrawn from Lindenmayer et al. 1996a).

Table 4.4: Bioclimatic envelope of Shining Gum derived from BIOCLIM

Index no.	Description of bioclimatic index	Mean	Min.	5%	95%	Max.
Temperature Indices						
1	Annual mean temperature	9.9	6.7	7.8	13.6	16.1
2	Mean diurnal range (mean[period max-min])	11.3	8.4	8.8	13.9	14.1
3	Isothermality (index 2/index 7)	0.47	0.40	0.41	0.51	0.51
4	Temperature seasonality (coefficient of variation)	47.5	21.4	33.5	59.2	69.9
5	Maximum temperature of warmest month	22.9	19.3	20.2	25.9	28.9
6	Minimum temperature of coldest month	-0.9	-3.2	-3.0	2.6	7.0
7	Annual range in temperature (index 5–6)	23.9	18.2	21.3	27.4	28.0
8	Mean temperature of the wettest quarter	8.7	1.1	3.3	19.2	20.7
9	Mean temperature of driest quarter	12.0	2.4	3.4	17.8	19.7
10	Mean temperature of warmest quarter	15.6	12.5	13.5	19.6	21.0
11	Mean temperature of coldest quarter	4.3	1.1	2.2	8.3	11.9
Precipitation indices						
12	Annual precipitation	1296	489	748	1779	2162
13	Precipitation of wettest month	153	56	87	227	341
14	Precipitation of driest month	67	25	43	85	91
15	Precipitation seasonality	23	10	14	41	46
16	Precipitation of wettest quarter	405	161	227	609	909
17	Precipitation of driest quarter	234	85	147	308	317
18	Precipitation of warmest quarter	284	157	191	557	774
19	Precipitation of coldest quarter	351	85	157	525	549

Homocline matching was based on four sets of values in the profile: minimum to maximum and 5% to 95%. The associated predicted potential distribution of the species is presented in Figures 4.1–4.4. All values for temperature are given in °C and those for precipitation in mm.
Modified from Lindenmayer et al. (1996a).

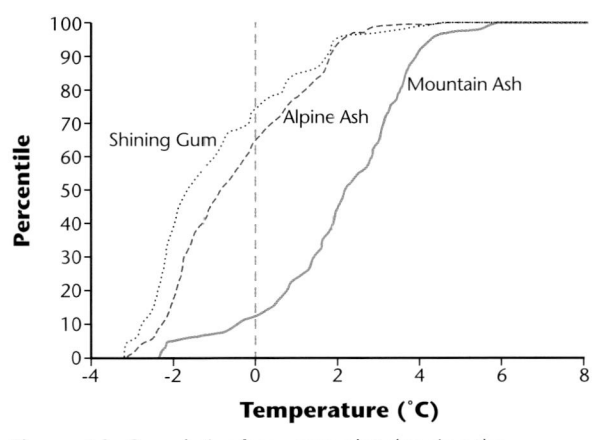

Figure 4.3: Cumulative frequency plot showing the relationship between different percentile values of the BIOCLIM-modelled minimum temperature of the coldest quarter for sites occupied by the three species of eucalypt trees. (Redrawn from Lindenmayer et al. 1996a).

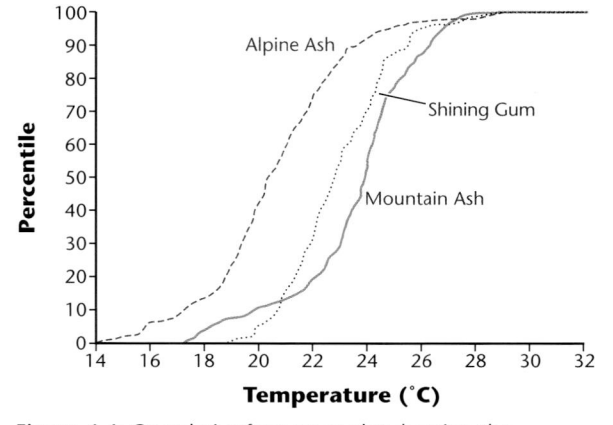

Figure 4.4: Cumulative frequency plot showing the relationship between different percentile values of the BIOCLIM-modelled maximum temperature of the warmest month for sites occupied by the three species of eucalypt trees. (Redrawn from Lindenmayer et al. 1996a).

Given the influence of extreme values and seasonal variations in climate on processes such as tree mortality and growth rates (Ashton 1975b; Cremer 1975), additional data and improved precision of estimates provided by BIOCLIM are important. For example, although stands of Alpine Ash experienced the coldest annual mean temperatures, the bioclimatic profile generated for Shining Gum had the lowest values of mean minimum temperature of the coldest month (compare Tables 4.3 and 4.4). This suggests that temperature extremes are an important factor influencing differences between the distributions of those tree species. Sites supporting Mountain Ash were characterised by the highest values for mean minimum temperature of the coldest month and the coldest quarter. They also exhibited the lowest levels of seasonal variation in temperature. These factors can have an important effect on plant growth and this more benign thermal environment may, in part, explain the very rapid rates of growth observed in Mountain Ash trees (Ashton 1975a).

Predictions of the potential bioclimatic domains of tree species

Data in the bioclimatic profiles in Tables 4.2–4.4 were used to map the predicted potential bioclimatic domains of the three species. In each case, areas were identified with marginally suitable bioclimatic conditions that extended beyond the present known distributions of the species, while the core bioclimatic domain of each species often conformed more closely to its known range. Correct interpretation of these maps requires an appreciation of the theory underpinning the BIOCLIM system. It is possible to estimate the potential physiological response of a given tree species to light, heat and water availability through controlled experiments in a growth chamber or by monitoring the outcomes of field trials. However, the long lifespan of trees presents major impediments to such investigations of the physiological response of trees to various environmental conditions.

Application of the BIOCLIM system enables the bioclimatic component of the potential physiological domain of a species to be approximated, based on samples of its extant distribution. If a species occurs naturally at a site, the corresponding environmental conditions must fall within its physiological thresholds. However, there is an additional complicating factor. The extant distribution of a species actually represents its realised niche or ecological domain. This is the potential physiological domain minus the effects of factors such as competition and disturbance. Therefore, the potential bioclimatic domain derived by the BIOCLIM system is probably an underestimate of the climatic conditions where a species can occur and successfully complete its lifecycle (Lindenmayer *et al.* 1996a).

Integrating maps of forest types with climate and environmental analyses

Mackey *et al.* (2002) substantially extended earlier work (Lindenmayer *et al.* 1996a) on the bioclimatic domains of various tree species in the Central Highlands of Victoria. They quantified the climatic regimes of the broad forest types found within several water catchments where there had been little European disturbance. They analysed the climatic and environmental correlates of Mountain Ash forest as well as rainforest and mixed-species forest. They found that mixed-species forest was typically the dominant forest type in warmer parts of the catchments and in areas on mid and upper slopes with high levels of incoming solar radiation. In contrast, rainforest was strongly associated with streamlines but not in lower reaches of the catchments characterised by warmer temperatures. Mountain Ash forest occupied an environmental domain intermediate between mixed-species forest and rainforest (Mackey *et al.* 2002). This work confirmed the environmental basis for differences in the broad distribution patterns of the major forest types across landscapes within the Central Highlands of Victoria (Colour Plate 5).

Other environmental factors influencing tree distribution patterns

Climatic factors set the broad-scale distribution limits of many organisms, including overstorey plant species. At finer spatial scales, the distribution and abundance of plant species is influenced by other interacting processes. For example, soil nutrients, changes in aspect or the incidence of rare extreme climatic events may influence the occurrence of a species in a given part of a landscape (Florence 1969; Pryor 1976). Several studies have demonstrated strong relationships between soil nutrients and changes in the dominant tree species. A transition from brown mountain soils to more weathered and nutrient-poorer krasnozems can lead to Mountain Ash forest being replaced by stands of Messmate (Ashton 1981a; Ellis 1971; Florence 1996). Similarly, an abrupt change in aspect from southerly or easterly to northerly or westerly can result in a change from Mountain Ash forest to forests dominated by Messmate or Silver-top Ash. The transition zone in climate, soil and aspect conditions may be gradual rather than abrupt; hybrids between eucalypt species can develop in these areas. For example, hybrids of Mountain Ash–Messmate and Mountain Ash–Red Stringybark are well-known in the Central Highlands region (Ashton 1981a; Ashton and Sandiford 1988).

SUCCESSIONAL TRAJECTORIES

Disturbance can be critical in shaping finer-scale distribution patterns of all species of montane ash trees. Within the broad distribution of a given ash-type overstorey eucalypt species, the abundance of stems of that species within a stand (as well as the size of individual stems) will often be a function of forest succession. Disturbance in montane ash forests triggers stand initiation and subsequent succession. As emphasised by the 2009 wildfires, the principal form of natural disturbance in montane ash forests is fire (Ashton 1981b) (see Chapter 12).

Given that the dominant tree species in montane ash forests are considered to be fire-sensitive, wildfires may be stand-replacing events because young stems belonging to a single age cohort regenerate in areas where older trees are killed by a conflagration (Ashton 1981b). Young seedlings germinate from seed released from the crowns of burned mature ash-type trees, producing a new even-aged regrowth stand. Levels of seed theft by ants is interrupted in montane ash forests following wildfire events, which facilitates mass germination of young seedlings (Ashton 1979; O'Dowd and Gill 1984).

Eucalypts in montane ash forests are self-thinning (Ashton 1981c), leading to a marked reduction in the number of stems per unit area over time (Ashton 1976; Gilbert 1959; Jackson 1968). The decline in stem density due to self-thinning is rapid, from up to several million seedlings per hectare soon after fire, to 380 stems/ha at 40 years and 40–80 stems/ha in mature forest (Ashton and Attiwill 1994). Small suppressed pole and sapling trees, which add greatly to the density of vegetation in regrowth forests, die and collapse as the forests mature (Ashton 1976). The seedlings of montane ash trees are considered shade-intolerant (Ashton 1976) and only rarely survive under the crowns of mature trees. Ashton (1962) believed that mature trees release substances in the root system which inhibit the growth of smaller trees.

Table 4.5: Heights and diameters of large individual Mountain Ash trees in the Central Highlands of Victoria

Height (m)			District	Remarks
101			Tooronga Plateau	Fallen 1948
99			Mt Baw Baw	Standing 1888; since destroyed by fire
91.8			Marysville	Standing 1928, storm-damaged 1959
Tree	Height (m)	Girth (m)	District	Remarks
Edward VII	61	34.2	Upper Yarra	Destroyed by fire
Horsfall Tree	–	27.5	Torronga Plateau	Destroyed by fire
Furmston Tree	87.5	19.5	Mt Monda	Collapsed
Unnamed	50	21.6	Marysville	Destroyed
Neerim Giant	69	17.0	Neerim	Destroyed by fire

Modified from Ashton (1975b) and updated from personal observations by DB Lindenmayer.

Harvesting by ants may remove considerable quantities of seed from the topsoil, which may prevent the development of regrowth in undisturbed forests (Ashton 1979; Cunningham 1960).

Eucalypts in montane ash forests are not only self-thinning but self-pruning (Cunningham 1960), with branches lost from the effects of wind and snow. In addition, as montane ash trees mature, the crowns of dominant eucalypts become more open and increasingly separated from those of surrounding trees (Ashton 1975a).

The processes of self-thinning, self-pruning and 'crown shyness' result in mature stands comprising large, relatively well-spaced trees with open crowns and few lateral branches on the lower trunk (Ashton 1975a).

Following disturbance and stand initiation, young montane ash trees exhibit rapid rates of growth. In the case of Mountain Ash – the world's tallest flowering plant – individual trees may reach 50 m height within 35 years of germination (Ashton 1975a). They can exceed 90 m after several hundred years, and spectacular trees over 120 m tall have been documented (Beale 2007). Table 4.5 gives the specifications of some of the largest recorded individual stems.

Many other attributes of montane ash stands exhibit marked changes as a result of succession. Some plant species decline or are almost completely lost over time, although they continue to persist in the seed bank. For example, many understorey shrubs and trees (e.g. wattles [*Acacia* spp.]) die and collapse with stand maturation (Adams and Attiwill 1984). The decline in the number of *Acacia* spp. is particularly pronounced. For example, there may be as many as 400 000 *Acacia* spp. stems in young regrowth stands but trees such as Mountain Hickory Wattle are typically absent from forests 80–100 years old (Adams and Attiwill 1984). Other plant species that are rare or absent in the early stages in stand development can take a long time to appear or become prominent in a stand. For example, the abundance of tree ferns and cool temperate rainforest trees increases significantly over time (Lindenmayer *et al*. 2000a) (Chapter 11), as do features of individual overstorey trees such as the prevalence of large hollows

Figure 4.5: The Furmston Tree on the Monda Track. This tree has now collapsed. (Photo courtesy of Pictures Collection, State Library of Victoria).

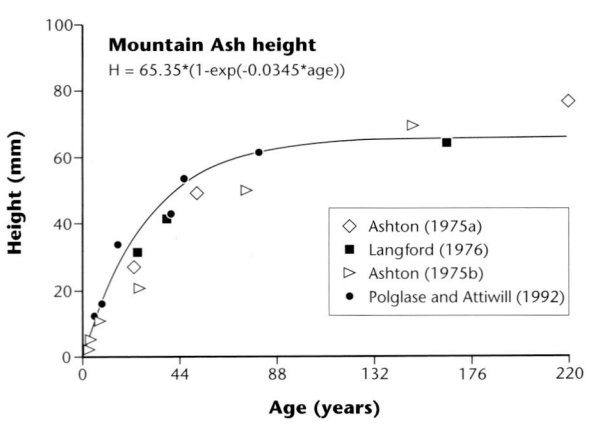

Figure 4.6: Age and tree height relationships for Mountain Ash forest. Generalised relationships have been summarised from various sources.

Mountain Ash height
$H = 65.35*(1-\exp(-0.0345*age))$

Ashton (1975a)
Langford (1976)
Ashton (1975b)
Polglase and Attiwill (1992)

(Ambrose 1982; Lindenmayer *et al.* 1993b) (Chapter 6) and clumps of mistletoe.

ENVIRONMENTAL FACTORS AND THE SPATIAL COMPOSITION OF AGE CLASSES

As outlined above, climatic conditions govern the broad-scale distribution patterns of overstorey tree species and hence forest types. Soil fertility and disturbance influence distribution patterns at finer scales. There can be significant environmental influences where stands of particular age classes of montane ash forest are most likely to occur (Colour Plate 6). Mackey *et al.* (2002) studied age class–environment relationships in water catchments where there were no confounding effects of human disturbance to obscure key patterns. They demonstrated that, prior to the 2009 wildfires in the O'Shannassy Water Catchment, old-growth Mountain Ash occupied a subset of the overall environmental domain of Mountain Ash *per se*, typically within a narrow band of mesic temperatures (~9°C) and not on ridges or steep slopes (Colour Plate 6). This environmental domain is closely linked with locations favourable for tree growth (Mackey *et al.* 2002) and with spatial differences in natural disturbance regimes (Lindenmayer *et al.* 1999f). Mesic areas are where trees will be, on average, taller than elsewhere in the landscape (Figure 4.7) and fire frequency and the intensity of past wildfires are likely to be lower (Mackey *et al.*

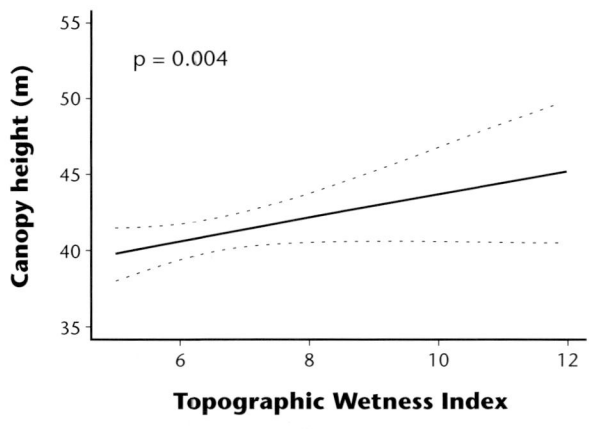

Figure 4.7: Relationships between topographic wetness and canopy height for Mountain Ash trees. (Redrawn from Mackey *et al.* 2002).

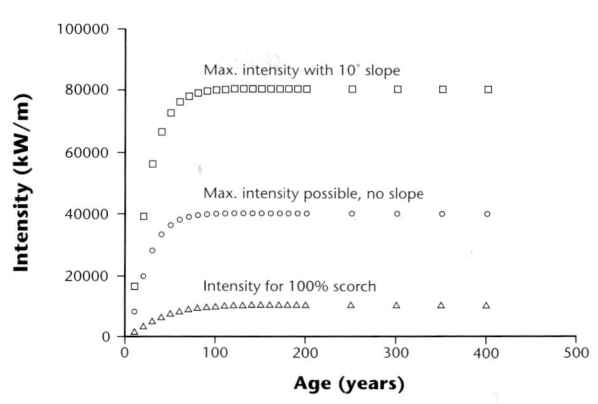

Figure 4.8: Relationships between tree age and the intensity of fires needed to kill trees, based on data on tree height and fuel load. (Redrawn from Mackey et al. 2002).

2002). Fire intensity must be very high for flames to reach the canopy of such taller trees and kill them outright (Figure 4.8). The analyses suggest that not all parts of forest landscapes are created equal in terms of growth patterns of trees, the severity of disturbance regimes and the intersection of the two. Issues associated with different natural disturbance pathways in montane ash forests and the spatial variation in them are discussed further in Chapter 12.

OLD GROWTH AND THE DIFFICULTY DEFINING IT IN MONTANE ASH FORESTS

Old-growth forests are perceived by many people as a valuable structural vegetation type (Burgman 1996; van Pelt 2007). There is, therefore, a need to fully understand what is meant by the term 'old growth'. This is also critical for attempts to map and conserve old growth (Woodgate *et al.* 1994). However, there are not-inconsiderable variations in the definitions of old-growth forest in Australia suggested by different organisations, reflecting differing ethical, social and ecological perspectives. Some examples are given below.

The Australian Conservation Foundation (in Resource Assessment Commission 1991) defined old growth as:

> *forest that has not been, or has been minimally, affected by timber harvesting and other exploitative activities by Australia's European colonisers.*

The Resource Assessment Commission (RAC 1992) considered that old-growth forests have high conservation and intangible values and defined them as forest:

both little disturbed and ecologically mature (p. 29).

The National Forest Policy Statement (Commonwealth of Australia 1992) defined old growth as:

forest that is ecologically mature and has been subjected to negligible unnatural disturbance … in which the upper stratum or overstorey is in the late mature to overmature growth phases.

The Regional Forest Agreement process in Australia (Commonwealth of Australia and Dept of Natural Resources and Environment 1997) defined old-growth forest as:

ecologically mature forest where the effects of disturbances are now negligible.

This definition has qualifications, such as old growth having 'characteristics of older growth stages', 'functional qualities expected to characterise an ecologically mature forest ecosystem' and 'negligible disturbance effects'.

These definitions suggest that old-growth forests are generally considered to be the oldest and least disturbed forests for a given ecological vegetation type. They have generic application across many forest types in Australia (Forestry Tasmania 2004; Scotts 1991; Woodgate *et al*. 1994, 1996). However, some aspects break down in particular cases. Mountain Ash forests provide an example.

First, old-growth Mountain Ash forest is rarely characterised by an absence of any kind of disturbance. This is because almost all old-growth stands include overstorey ash-type eucalypt trees of multiple ages (Lindenmayer *et al*. 2000a), indicating past disturbance and the survival of some pre-disturbance trees (Banks 1993; Beale 2007).

Second, old-growth Mountain Ash forest is not always characterised by an absence of human disturbance. This is because stand-replacing fires can radically alter stand conditions in the complete absence of human influence. Thus, a naturally occurring wildfire may convert an old-growth stand to a young regrowth one – an ecological process and disturbance pathway (Chapter 12) that presumably occurred for millions of years before humans (Aboriginal or European) colonised Australia (Attiwill 1994a).

A third issue in defining old-growth Mountain Ash forest is scale – it is possible to recognise 'old growth' at the individual tree level, the patch level, the stand level or the landscape level. This is essentially a landscape context effect (*sensu* Enoksson *et al*. 1995). An old-growth tree surrounded by a stand of other old-growth trees or located in a landscape dominated by old trees will be important for some species (e.g. the Yellow-bellied Glider); for others, a small number of old trees within a primarily regrowth-dominated patch, stand or landscape can provide suitable habitat (e.g. Mountain Brushtail Possum (Chapter 9)).

Finally, most definitions of old growth focus solely on the overstorey trees. Yet there can be old-growth understorey elements within a forest where the overstorey is predominantly regrowth. Carbon-dating work has shown that understorey plants such as tree ferns (*Dicksonia antarctica*, *Cyathea australis*) can be 350 years old in a forest dominated by 60-year-old overstorey eucalypt trees (Mueck *et al*. 1996).

Despite these issues, it was important to develop a practical and tractable approach for defining old-growth montane ash forest. In the case of Mountain Ash, old-growth stands were defined on the basis of the size and prevalence of living overstorey trees. They were considered to be stands where the majority (>70%) of living overstorey trees exceeded 180 years of age (Lindenmayer 2009). The majority of living overstorey trees in these stands were typically larger than 1 m in diameter at breast height and most contained hollows that could be readily observed from ground-level surveys.

Old-growth forest or multi-aged forest?

Old-growth montane ash forests can be broadly recognised on the basis of the dominant cohort of overstorey trees in a 3-ha stand. Many areas of forest are even-aged; stem diameter and tree age within a site are closely related in these forests (Ashton 1976; Nelson and Morris 1994). Thus, the size of trees makes it relatively straightforward to assign them to age classes, which typically correspond to major fire events. However, a detailed study of old growth in the

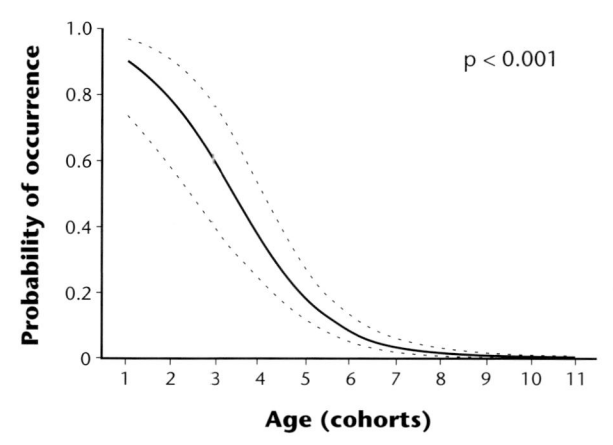

Figure 4.9: The probability of occurrence of multiple age cohorts in a stand, and stand age. Oldest age cohorts are on the far left of the x-axis (cohorts 1 and 4, ~250 and 180 years old respectively). The solid line is the mean response and the dashed lines are the 95% confidence intervals around the mean vale. (Redrawn from Lindenmayer et al. 2000a).

Central Highlands of Victoria (Lindenmayer et al. 2000a) prior to the 2009 wildfires revealed the occurrence of overstorey trees of distinctly different age cohorts in the same stand. For example, this phenomenon characterised almost every one of the 48 old-growth Mountain Ash stands surveyed in a major investigation of structural attributes of montane ash forests. There were virtually no 'pure' stands of old growth in which all overstorey trees were of the same age (Lindenmayer et al. 2000a). Instead, there was consistent evidence of past disturbance by fire, such as charcoal scars on the butts of large trees (Lindenmayer et al. 1991b) – an observation corroborated by dendrochronology of old trees by Banks (1993). This has major implications for the kinds of natural distur-

Figure 4.10: Striated Pardalote. (Photo by Julian Robinson).

> **Box 4.1: The fauna of old-growth montane ash forest**
>
> Of the array of vertebrates that occur in montane ash forests, remarkably few are solely dependent on stands or landscapes of old growth. Two species that do appear to be strongly associated with old-growth Mountain Ash forests are the Yellow-bellied Glider (Lindenmayer et al. 1999a) and the Sooty Owl (Milledge et al. 1991). Both are associated with large contiguous areas of old-growth Mountain Ash forest and, for example, are typically absent from wood production ash forests where only limited small patches of old-growth forest remain.
>
> Other species of vertebrates are most abundant in old-growth Mountain Ash stands, although they are not confined exclusively to them. An example is the Greater Glider, which is three times more dense in old-growth Mountain Ash than in younger ash forest (Lindenmayer et al. 1995b). Similarly, the Striated Pardalote (*Pardalotus striatus*) is most often detected in old-growth forests, although it is commonly detected in stands of other age (Lindenmayer et al. 2009b). The Greater Glider, Yellow-bellied Glider, Striated Pardalote and Sooty Owl are distributed throughout eastern Australia and hence inhabit many forest types in addition to those dominated by montane ash trees.
>
> Many species of vertebrates are strongly associated with old-growth Mountain Ash forest elements or attributes, although they may not be confined to old-growth stands or be most abundant in these areas. That is, old-growth elements are important for a particular species, but the spatial context for them may not be (Enoksson et al. 1995). Detailed work (Chapter 6) has repeatedly demonstrated close relationships between many species of cavity-dependent vertebrates and the abundance of large trees with hollows in a stand (including stands of regrowth) – trees that are typically 150+ years old (Lindenmayer 2000). If old-growth hollow trees are located in stands of predominantly regrowth overstorey trees, the resulting stand structure can still be suitable nesting or denning habitat for the majority of taxa.

bance pathways that may exist in montane ash forests – a topic revisited in detail in Chapter 12.

SUMMARY

Factors at multiple spatial scales shape the distribution of forest types in the Central Highlands of Victoria. Climatic conditions are the key determinant of the broad distribution patterns of Mountain Ash, Alpine Ash and Shining Gum (Lindenmayer *et al.* 1996a). At finer spatial scales, other environmental factors such as soil fertility, topography and natural disturbance have an important influence on tree species distribution (Florence 1996; Mackey *et al.* 2002).

Within the broad distribution of a forest type, climatic conditions, topography and disturbance influence the spatial distribution of forest age classes within montane ash-dominated landscapes. For example, prior to the 2009 wildfires, old-growth Mountain Ash forests typically occurred in a subset of the area where that tree species occurred. Temperature and topographic factors appear to be important in governing that subset of conditions and hence the location of old-growth forest (Mackey *et al.* 2002).

LESSONS LEARNED

Landscape position, environmental conditions, the frequency and intensity of natural disturbances, the kinds and numbers of living and dead biological legacies left by those natural disturbances, and post-disturbance successional trajectory and stand development are far more intimately related than previously recognised. This complicates otherwise simplistic perspectives of readily defined spatial arrangements of stands of uniform age classes in montane ash landscape. This additional complexity matters. It influences the suitability of forests as habitat for particular species, including ones of conservation and management concern such as Leadbeater's Possum (Lindenmayer and Franklin 1997) (see Chapter 12).

KNOWLEDGE GAPS

Current understanding of the relationships between the environment and the spatial distribution of age classes in montane ash forests is largely based on work within water catchments such as the O'Shannassy Water Catchment. These have been closed to public access for a prolonged period and human disturbances, especially logging, have been excluded. However, information is lacking about the former extent of old-growth montane ash forest throughout the Central Highlands of Victoria. Analyses of data on the size and diameter variation of large trees with hollows within 1939-aged regrowth forest can provide some insights into the age class distribution and former prevalence of old-growth stands in these forests. This important task remains to be completed. This kind of work should be supplemented with extensive landscape-wide dendrochronological studies in closed water catchments to provide more reliable information on the actual ages of stands in these landscapes.

FOREST PATTERN, ECOLOGICAL PROCESS AND LINKS TO OTHER CHAPTERS

Many ecological processes influence why forests occur where they do. Climate, terrain, geology, soils

---

Box 4.2: Postscript – old-growth forest, multi-aged forest and the 2009 wildfires

Many of the insights about the structure and location of old-growth forest and multi-aged forest have been derived from detailed studies in the O'Shannassy Water Catchment. Large areas of the catchment, including extensive stands of old-growth forest, were burned in the February 2009 wildfires. Careful studies of the recovery of burned stands in this catchment will be extremely instructive, and verify the efficacy (or otherwise) of hypothesised relationships, for example between tree survival and tree size, and between landscape position, fire severity, tree survival and the development of multi-aged stands. This emphasises the importance of long-term studies and monitoring programs and their role in elucidating key ecological processes and forest patterns – a topic that is discussed further in Chapter 17.

and disturbance are among the primary ones and are as important in montane ash forests as in forests around the world. For example, changes in soil type and associated soil fertility and water-holding capacity can result in marked changes in forest type in the Central Highlands of Victoria (Florence 1996). Several processes, particularly natural and human disturbance, also influence the structure of stands of montane ash forest. However, relationships between such processes and forest patterns are complex and not unidirectional. This is because the frequency and intensity of disturbances is influenced by the terrain and hence regional or landscape-level climatic conditions (Mackey *et al.* 2002). For example, fires are usually less intense in deep south-facing valley bottoms and on flat plateaux, where multi-aged stands are most likely to develop (Lindenmayer *et al.* 1999f). However, in some ways, this is also a dialectic interrelationship because the spatial patterns of forest type and structure within those vegetation types can influence disturbance regimes. The intensity of fire and possibly the rate and extent of spatial contagion in fire seems to be lower in old-growth stands and areas where rainforest develops. Thus, the ecological process of disturbance is modified by a pattern – the occurrence of old-growth forest and rainforest. The pattern of forest cover and its composition influences other key ecological processes such as carbon storage and hence nutrient cycling (Keith *et al.* 2009), and hydrological regimes, through stream flow (Vertessey and Watson 2001). Finally, the pattern of forest cover, together with the spatial pattern of structural elements within a given forest type, influences other forest patterns such as the distribution and abundance of wildlife.

Links to other chapters

This chapter has focused on broad patterns of overstorey cover in montane ash forests and the spatial patterns of age classes within that overall pattern of forest cover. The three chapters in Part III examine in more depth the key structural components of montane ash forest – overstorey trees with hollows (Chapter 6), the understorey and shrub layer (Chapter 7) and fallen trees on the forest floor (Chapter 8).

Parts of this chapter have touched on the influence of natural disturbance on the occurrence of forest types, the age of stands of montane ash forest and the potential for stands to develop trees from multiple age cohorts. The links between the composition of overstorey and disturbance dynamics are strong (see Chapter 12). Disturbance influences structure, a key theme of Part III.

The species of dominant overstorey tree influences the richness, abundance and/or composition of birds, arboreal marsupials and small terrestrial mammals in montane ash forests (Chapters 9 and 11). The abundance of trees with hollows and the prevalence of kinds of cavities also varies among the overstorey tree species in montane ash forests (Lindenmayer *et al.* 1993b, 1991b) (Chapter 6).

The size of areas of montane ash forest in different age classes can have profound impacts on particular elements of the biota. It can shape the distribution patterns of wide-ranging species such as the Sooty Owl (Milledge *et al.* 1991) and the Yellow-bellied Glider (Lindenmayer *et al.* 1999a) (Chapter 9). It can influence the medium- and long-term viability of populations of other species, such as the Greater Glider (McCarthy and Lindenmayer 1999) and Leadbeater's Possum (Lindenmayer and Possingham 1995) (Chapter 10). Knowledge about these kinds of relationships is important for informed management, including the size and location of meso-scale reserves within areas broadly designated for wood production (Chapter 15), and for designing strategies for mitigating the effects of logging (Chapter 16).

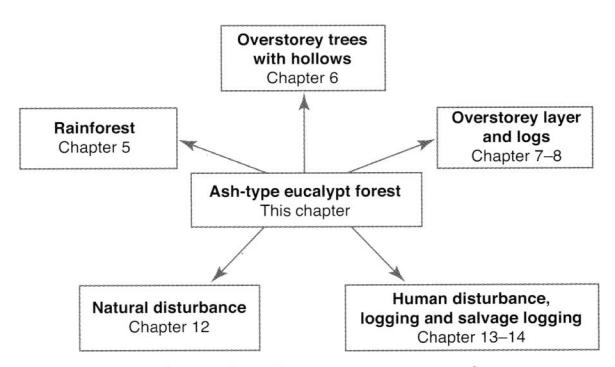

Figure 4.11: Relationships between overstorey forest cover and key topics in other chapters.

5 The rainforest

INTRODUCTION

Australia supports several types of rainforest including tropical, subtropical, dry, warm temperate and cool temperate rainforests (Bowman 1999; Mackey 1993; Price *et al.* 1995). The cool temperate rainforests of south-eastern Australia have ancient biogeographic origins that can be traced to Gondwana. Many areas of cool temperate rainforest in south-eastern Australia are dominated by the genus *Nothofagus* (Busby and Brown 1994) which has three representatives on the continent – Antarctic Beech (*Nothofagus moorei*), Myrtle Beech (*Nothofagus cunninghamii*) and Deciduous Beech (*Nothofagus gunnii*) (Busby and Brown 1994; Read and Brown 1996). The distribution of Myrtle Beech is limited to Tasmania and Victoria, although Busby (1986) showed that the potentially suitable climatic conditions for the species exist in southern New South Wales. In Victoria, Myrtle Beech occurs in three key areas – the Central Highlands, the Strzelecki Ranges in southern Gippsland and the Otway Ranges south-west of Melbourne (Busby 1986; Howard and Ashton 1973).

In the Central Highlands of Victoria, Myrtle Beech is often an understorey component of montane ash forests (Lindenmayer *et al.* 2000b). The conservation of Myrtle Beech in these forests has been of some

Figure 5.1: Cool temperate rainforest understorey with an overstorey of montane ash forest. (Photo by Esther Beaton).

Figure 5.2: Extensive patch of cool temperate rainforest dominated by Myrtle Beech and Sassafras. (Photo by David Lindenmayer).

concern because of the potentially negative (and interacting) effects of the disease Myrtle Wilt (*Chlara australis*), changed fire regimes linked with European settlement, and the impacts of logging operations that harvest timber and pulpwood from ash-type eucalypt stands overtopping or adjacent to Myrtle Beech-dominated rainforest (Burgman and Ferguson 1995).

Work over the past 25 years has focused on the Myrtle Beech-dominated rainforests of the Central Highlands of Victoria. Studies have been made of:

- the range of climatic and other factors influencing the occurrence of Myrtle Beech (Lindenmayer *et al.* 2000b). These include the temporal dynamics of the species, particularly as they relate to natural and human disturbance regimes (Lindenmayer *et al.* 2000a);
- the use of Myrtle Beech-dominated rainforest by arboreal marsupials and birds.

A summary and synthesis of these broad and related areas of research are the primary topics of this chapter.

BACKGROUND DATASETS

Several major datasets have facilitated studies of cool temperate rainforest in the Central Highlands of Victoria. The occurrence and prevalence of Myrtle Beech was recorded as part of detailed vegetation surveys at over 520 sites throughout the study region (see Chapter 3).

An extensive array of climatic, topographic and radiation variables were derived for these sites using a range of environmental modelling programs. This facilitated exploration of the environmental factors influencing the occurrence of rainforest (Lindenmayer *et al.* 2000b). The age of the overstorey ash forest varied substantially among the 520 field sites. This enabled the influence of stand age on

Table 5.1: Studies directly or indirectly examining cool temperate rainforest in the Central Highlands of Victoria

Study description	Reference
Relationships between environmental factors and occurrence of rainforest	Lindenmayer *et al.* (2000b)
Environmental domain of rainforest	Mackey *et al.* (2002)
Rainforest as part of the understorey of old-growth montane ash forest	Lindenmayer *et al.* (2000a)
Presence and abundance of arboreal marsupials and occurrence of rainforest	Lindenmayer *et al.* (1993a)
Influence of rainforest on the life history attributes of the Mountain Brushtail Possum	Banks *et al.* (2008)
Influence of rainforest on the probability of capture of small terrestrial mammals	Cunningham *et al.* (2005)
Rainforest and the occurrence of individual species of birds	Lindenmayer *et al.* (2009b)

rainforest occurrence to be quantified (Lindenmayer et al. 2000a).

A subset of the 520 vegetation sites was used in detailed surveys of forest birds. Relationships between rainforest and the occurrence of individual bird species were explored (Lindenmayer et al. 2009b).

Data from a 17-year trap–recapture study of the Mountain Brushtail Possum were gathered from a 35-ha site at Cambarville, near Marysville. The amount of Myrtle Beech surrounding each trapping point was measured and used in analyses of the factors affecting the body size, fecundity and offspring sex ratio of the Mountain Brushtail Possum (Banks et al. 2008). A second trap–recapture study, focused on small terrestrial mammals at Mt Donna Buang, also involved the collection of data on the prevalence of Myrtle Beech as part of the forest understorey. Analyses of the factors influencing the probability of animal capture included quantifying the effects of Myrtle Beech prevalence as a significant explanatory covariate (Cunningham et al. 2005).

FACTORS INFLUENCING THE OCCURRENCE OF MYRTLE BEECH

The presence of Myrtle Beech in the Central Highlands of Victoria was measured in field surveys of over 470 sites (more than 2840 individual plots) distributed widely throughout Mountain Ash, Alpine Ash and Shining Gum forests. Each individual Myrtle Beech stem more than 2m in height and more than 50 mm in diameter was recorded as part of a height-by-diameter matrix completed in each of six 10 × 10 m plots measured on the 470 sites. The occurrence of Myrtle Beech within any one of the six plots was taken as a positive record of the species' presence on site. Statistical modelling was employed to examine relationships between the occurrence of Myrtle Beech and measured site attributes such as slope, aspect and disturbance history as well as other environmental variables.

Statistical modelling showed that the presence of Myrtle Beech was significantly related to a combination of five variables. These were (after Lindenmayer et al. 2000b):

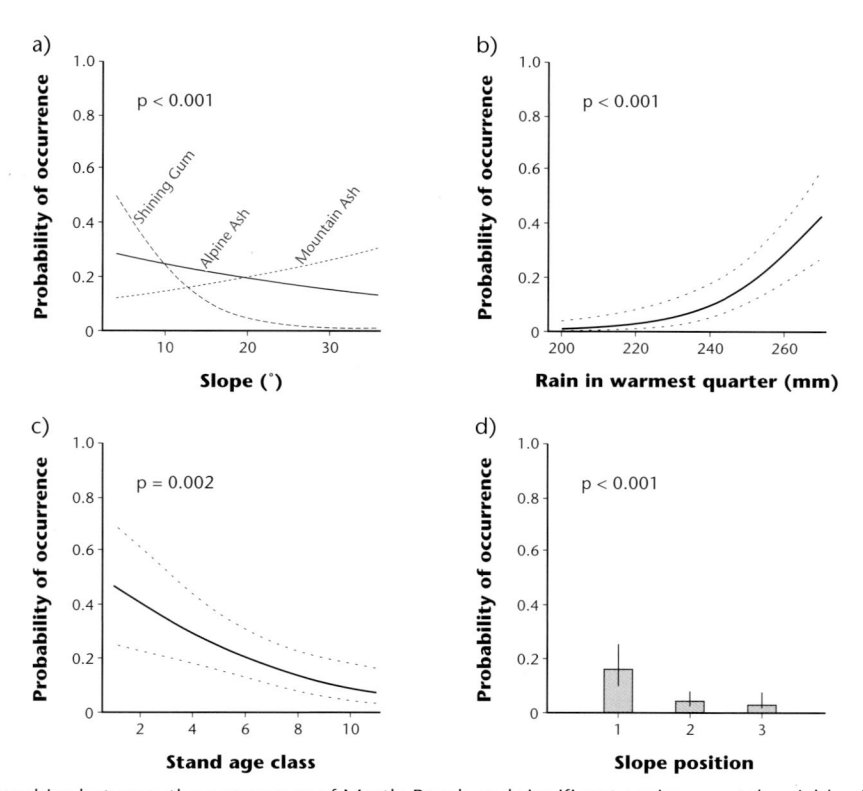

Figure 5.3: Relationships between the occurrence of Myrtle Beech and significant environmental variables (Redrawn from Lindenmayer et al. 2000b). Old-growth stands are age classes 1 + 2; young (≈ 20-year stands) are age classes 9 + 10. In (d), 1 = gully; 2 = midslope; 3 = ridge.

- the age of the overstorey eucalypt stand;
- dominant eucalypt tree species;
- topographic position in the landscape;
- slope;
- the estimated quantity of precipitation in the warmest quarter of the year.

Myrtle Beech was significantly more likely to occur in old-growth eucalypt stands, in gullies and in locations with high estimated values for rainfall in the warmest quarter. In addition, there was a slope by dominant tree species interaction. Within forests dominated by Mountain Ash, Myrtle Beech was more likely to occur on steep slopes. In contrast, within Shining Gum and Alpine Ash-dominated stands Myrtle Beech was significantly more often detected on flatter terrain.

Temporal dynamics

In many parts of the Central Highlands of Victoria, there is often not a strong boundary separating the start of Myrtle Beech-dominated rainforest from the edge of montane ash forest. Rather, its occurrence appears to be strongly related to natural and human disturbance. This was highlighted above: Myrtle Beech was significantly more likely to occur in older forest – the species was virtually absent from young stands (Figures 5.3, 5.4). These results for the Central Highlands are broadly consistent with those from Tasmania, in which there is an increasingly luxuriant development of cool temperate understorey elements with increasing age of overstorey eucalypt forests (Gilbert 1959). The paucity of Myrtle Beech in young stands may be related to the effects of clearfelling operations. Myrtle Beech is a vegetatively resprouting species and appears to recover poorly after timber harvesting (Lindenmayer and Ough 2006; Ough 2002). The species has limited dispersal capability (Howard 1973) and may be slow to re-establish on intensively clearfelled sites (Chapters 13 and 14).

COOL TEMPERATE RAINFOREST AND VERTEBRATE BIOTA

As part of studies of vertebrates in the Central Highlands of Victoria, several investigations have examined the relationships between Myrtle Beech and the occurrence of small terrestrial mammals, arboreal marsupials and birds.

Rainforests and small terrestrial mammals

In the case of small mammals, a detailed study of the capture probabilities of different species was completed over a three-year period at Mt Donna Buang near Warburton (Cunningham *et al.* 2005). Animal capture data were analysed in relation to forest structure and composition. Areas with high cover of Myrtle Beech produced significantly lower numbers of cap-

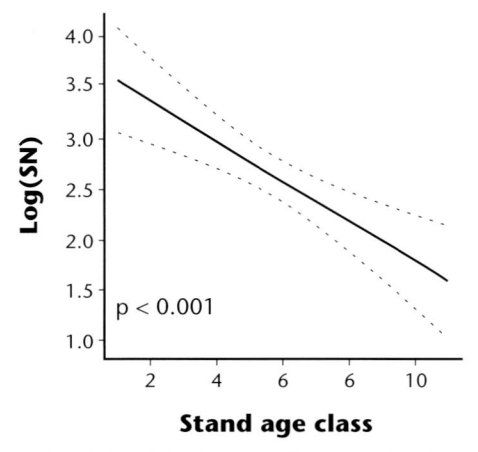

Figure 5.4: Relationships between the amount of Myrtle Beech in a stand (reflected by the stocking rate (SN) of trees) and the age of the eucalypt overstorey. (Redrawn from Lindenmayer *et al.* 2000b).

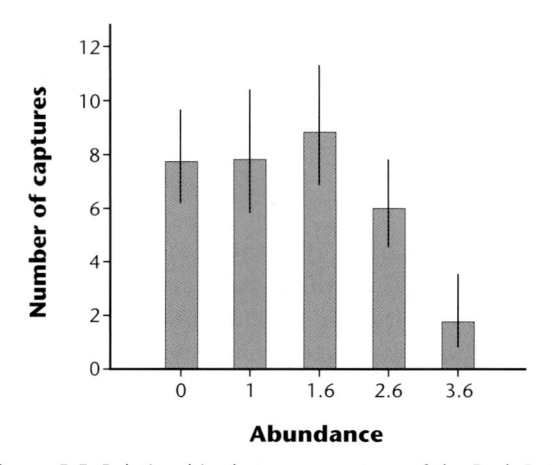

Figure 5.5: Relationships between captures of the Bush Rat and the abundance of Myrtle Beech stems in trapping plots. (Redrawn from Cunningham *et al.* 2005).

tures of the Bush Rat (Cunningham *et al.* 2005) (Figure 5.5). This result may be due to the fact that the dense canopies of these trees are often associated with an open ground layer supporting few understorey, shrub layer or ground layer plants that are important for small terrestrial mammal species.

Rainforests and arboreal marsupials

A long-term trap–recapture study of the Mountain Brushtail Possum explored relationships between the fecundity and body condition and microhabitat features including the prevalence of Myrtle Beech trees (Banks *et al.* 2008) (Figure 5.6). The work clearly demonstrated that animals living in areas dominated by stands of Myrtle Beech are in poorer condition (reflected in their body weight). Breeding females in these areas typically produce sons, which subsequently disperse. This is in contrast with (larger) females in eucalypt-dominated areas that are more likely to give birth to daughters that often remain in, or close to, the natal territory. Kinship structures comprising sets of closely related animals are less complex in areas where microhabitat structure and composition is dominated by stands of Myrtle Beech (Banks *et al.* 2008).

The occurrence of rainforest is known to have a negative influence on the occurrence of other species of arboreal marsupials. A detailed survey of retained linear strips in montane ash forests showed that sites dominated by streams with a significant component of cool temperate rainforest were significantly less likely to support the Greater Glider than strips located

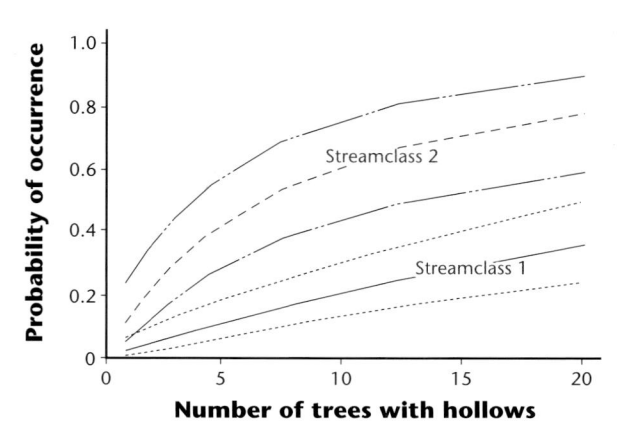

Figure 5.7: The probability of occurrence of the Greater Glider in retained linear strips with and without a stream. (Modified from Lindenmayer *et al.* 1993a.) Streamclass 1 corresponds to linear strips with a stream or drainage line; Streamclass 2 lacks these features. Each prediction shows the mean response and 95% confidence intervals around the prediction.

elsewhere in forest landscapes (Lindenmayer *et al.* 1993a). Myrtle Beech and other rainforest tree species do not provide suitable food resources for this highly specialised eucalypt-dependent arboreal folivore.

Rainforests and birds

Relationships between birds and cool temperate rainforest can be complex. Overall bird species richness is lower in cool temperate rainforest than in surrounding areas of montane ash forest (12.5 vs 14.4 species per site respectively). There is evidence that some species of birds are significantly more likely to occur in either cool temperate rainforest or montane ash forest. For example, species significantly less likely to occur in cool temperate rainforest include the Grey Shrike-thrush (*Colluricincla harmonica*) and the Brown-headed Honeyeater (*Melithreptus brevirostris*). In contrast, the Pink Robin (*Petroica rodinogaster*) was significantly more often recorded from cool temperate rainforest than Mountain Ash forest. However, no species of birds are confined solely to stands of cool temperate rainforest. Thus, birds most often found in cool temperate rainforest are not confined to these areas, nor are they rare in montane forest. Rather, there is a gradient in the composition of the bird species assemblage from such taxa as the Pink Robin and the exotic European Blackbird (*Terdus merula*),

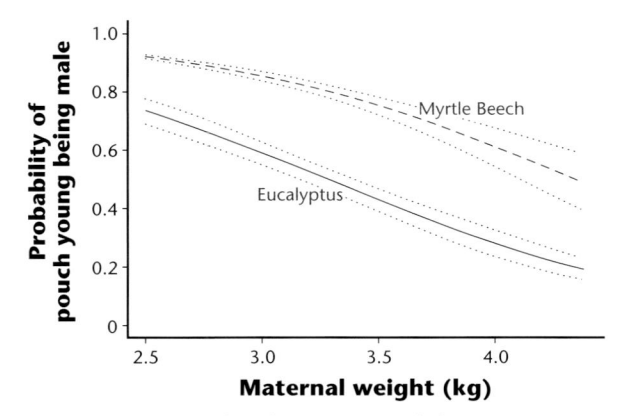

Figure 5.6: Relationships between microhabitat structure (eucalypt vs Myrtle Beech) and offspring sex and microhabitat structure. (Modified from Banks *et al.* 2008).

which tended to be recorded most often in cool temperate rainforest, to bird species such as Horsefield's Bronze Cuckoo (*Chrysococcyx basilis*), Spotted Pardalote *(Pardalotus punctatus)* and Laughing Kookaburra (*Dacelo novaeguineae*) that were most often detected in Mountain Ash forest.

Studies of birds were expanded to quantify relationships between vegetation structure and the occurrence of individual taxa. Of the 70 species recorded relatively routinely in montane ash forests, the occurrence of only one, the Crimson Rosella (*Platycercus elegans*) was positively associated with the amount of Myrtle Beech in the understorey (Lindenmayer et al. 2009b). Only one species, the Eastern Yellow Robin (*Eopsaltria australis*) was negatively associated with the amount of Myrtle Beech in the understorey (Lindenmayer et al. 2009b). No other species of birds show strong positive or negative relationships with the presence of Myrtle Beech stands. An earlier qualitative investigation (Loyn 1985) found that, within Mountain Ash forest, stands of Myrtle Beech are critical areas for Pink Robin nesting.

SUMMARY

Although some kinds of rainforests in many parts of the world are characterised by high levels of biodiversity, not all types of Australian rainforest support diverse vertebrate faunas. Myrtle Beech-dominated rainforests in the Central Highlands of Victoria exemplify this. Although they support a rich array of mosses and ferns (Ashton 1986 ; Ough and Ross 1992), no vertebrate animals appear to be confined solely to them. Moreover, the occurrence of small mammals and arboreal marsupials appears to be negatively related to the presence of Myrtle Beech forests. Breeding biology (reflected in offspring sex ratios and patterns of social organisation) also appears to be influenced by the prevalence of Myrtle Beech. Four factors may explain these influences on vertebrate biota. First, cool temperate rainforest trees often do not develop cavities (except in the very largest and oldest stems), making them unsuitable for many hollow-dependent animals (Gibbons and Lindenmayer 2002). Second, Myrtle Beech is a wind-pollinated species, so flower-derived food resources such as nectar, pollen, fruit and seeds, which are important for many vertebrates and invertebrates, are extremely limited. Myrtle Beech rainforests are also unsuitable habitat for many species of arboreal marsupials that depend on insects and plant gums for food. Third, they are dominated by only a few species of trees, and their relative floristic and structural simplicity means that they may not contain a wide array of foraging substrates for animals. Finally, ground cover is often limited in areas dominated by overstorey Myrtle Beech trees, leading to unsuitable microhabitat conditions for cover-dependent small mammal taxa.

LESSONS LEARNED

Studies of cool temperate rainforest in the Central Highlands of Victoria have demonstrated that their distribution patterns are more complex than previously recognised. Multiple factors ranging from climatic conditions, disturbance processes (reflected in the age of the dominant overstorey eucalypt stands) and topography have a significant impact on species occurrence. There can be complex interactions between age cohorts of different tree species in the same stand – in this case, the dominant overstorey eucalypts and the understorey which includes Myrtle Beech stems (Lindenmayer et al. 2000b).

Informed conservation management of Myrtle Beech will be partly predicated on an understanding of the factors influencing the distribution of the species (Burgman and Ferguson 1995). Appropriate management could have positive benefits for other taxa which often co-occur with it, such as the Pink Robin (Loyn 1985) and several non-vascular plants (e.g. Slender Tree Fern [*Cyathea cunninghamii*]) (Mueck 1990).

While stands of cool temperate Myrtle Beech-dominated rainforest are important places for conservation, they do not support high levels of vertebrate biodiversity. Thus, the conservation of many elements of the biota in the Central Highlands of Victoria will depend on the maintenance of the spatially adjacent eucalypt-dominated montane ash forests. This has implications for the width of streamside reserves within wood production forests, which may need to be widened to encompass rainforest areas and adjacent eucalypt stands.

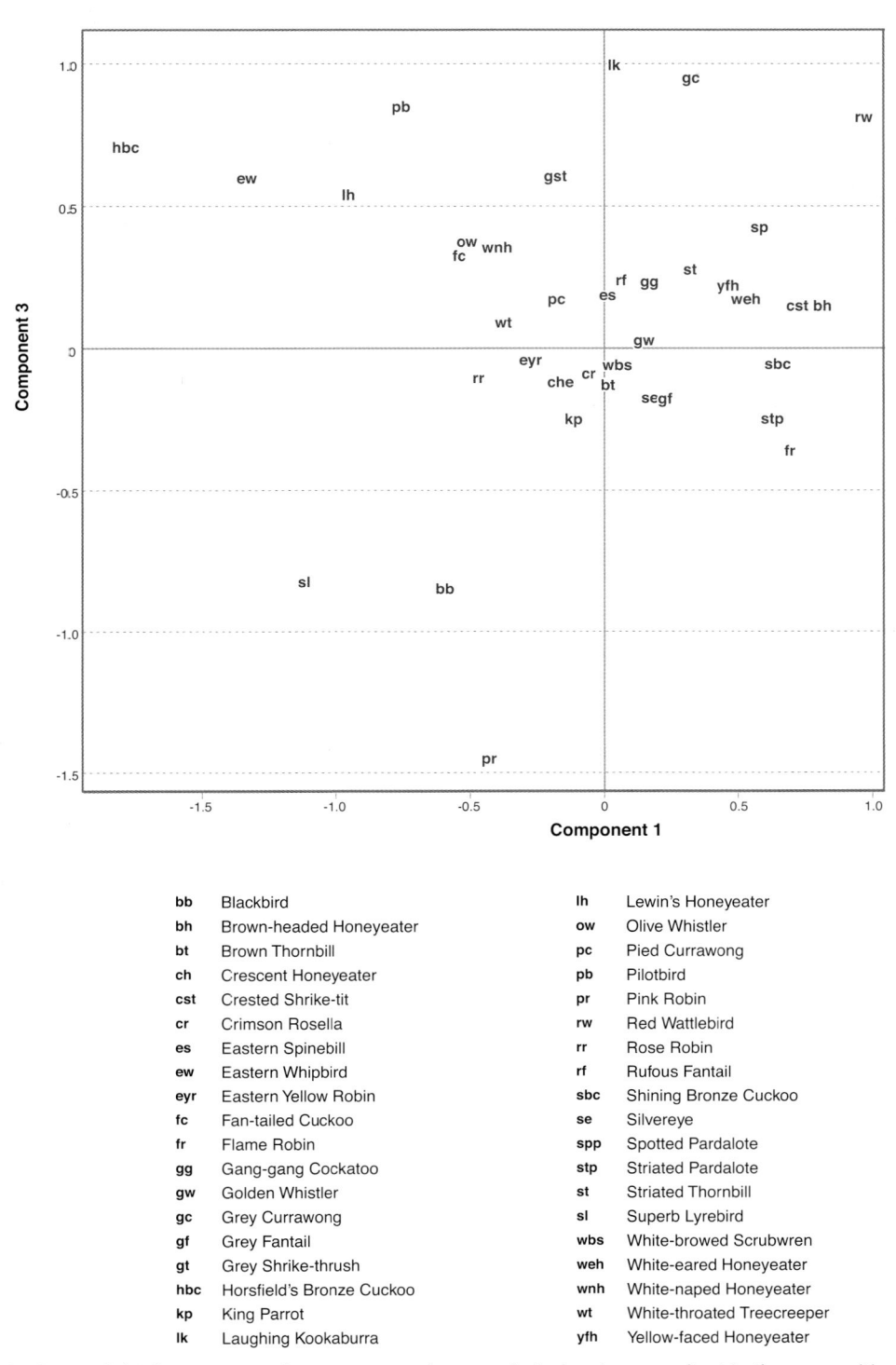

bb	Blackbird	lh	Lewin's Honeyeater
bh	Brown-headed Honeyeater	ow	Olive Whistler
bt	Brown Thornbill	pc	Pied Currawong
ch	Crescent Honeyeater	pb	Pilotbird
cst	Crested Shrike-tit	pr	Pink Robin
cr	Crimson Rosella	rw	Red Wattlebird
es	Eastern Spinebill	rr	Rose Robin
ew	Eastern Whipbird	rf	Rufous Fantail
eyr	Eastern Yellow Robin	sbc	Shining Bronze Cuckoo
fc	Fan-tailed Cuckoo	se	Silvereye
fr	Flame Robin	spp	Spotted Pardalote
gg	Gang-gang Cockatoo	stp	Striated Pardalote
gw	Golden Whistler	st	Striated Thornbill
gc	Grey Currawong	sl	Superb Lyrebird
gf	Grey Fantail	wbs	White-browed Scrubwren
gt	Grey Shrike-thrush	weh	White-eared Honeyeater
hbc	Horsfield's Bronze Cuckoo	wnh	White-naped Honeyeater
kp	King Parrot	wt	White-throated Treecreeper
lk	Laughing Kookaburra	yfh	Yellow-faced Honeyeater

Figure 5.8: The first and third components from correspondence analysis showing a gradient in the composition of the bird assemblage between cool temperate rainforest and Mountain Ash forest. Birds more likely to be found in areas of cool temperate rainforest had a third component score of -0.5 to -1.5; those more likely to be found in Mountain Ash forest had a third component score of 0.5 to 1.0.

Figure 5.9: Pink Robin. (Photo by Graeme Chapman).

KNOWLEDGE GAPS

The potential impoverishment of vertebrate assemblages in rainforest patches needs further exploration. Bird and mammal abundance should be lower in larger patches of this kind of forest – the converse pattern for patch-size relationships typically observed in ecology and conservation biology (Gaston and Spicer 2004; Rosenweig 1995).

A second knowledge gap relates to relationships between the occurrence of cool temperate rainforest and natural disturbance regimes. Many authors have suggested that cool temperate rainforest should replace montane ash forest in the absence of wildfire (Gilbert 1959). However, wildfires do not burn all parts of landscapes with equal frequency or intensity (Lindenmayer *et al.* 1999f; McCarthy *et al.* 1999) (see Chapter 12), suggesting the potential for interesting relationships between the occurrence of cool temperate rainforest and a wider range of factors than disturbance alone – a possibility that has received some study (Lindenmayer *et al.* 2000b) but warrants further exploration.

FOREST PATTERN, ECOLOGICAL PROCESS AND LINKS TO OTHER CHAPTERS

The occurrence of stands of rainforest is a spatial pattern (of tree and associated plant community distribution), often located within or immediately adjacent to another vegetation type (montane ash forest). Rainforest distribution, like that of ash-type forest (see Chapter 4), is influenced by many ecological processes ranging from climatic conditions to disturbance regimes. However, rainforest also influences ecological processes such as disturbance regimes. For example, the low levels of light that penetrate to the floor of cool temperate rainforests creates moist microclimatic conditions conducive to the development of luxuriant bryophyte communities, often on large pieces of coarse woody debris (Howard 1973). These conditions may influence the intensity, spread and hence spatial pattern of wildfires – the principal form of natural disturbance in landscapes within the Central Highlands of Victoria. The occurrence of rainforest also influences logging – the other key kind of disturbance process in montane ash forests. This is because logging is excluded from rainforests as well as from riparian areas where rainforest is most likely to develop (Burgman and Ferguson 1995). Of course, the spatial pattern of rainforest occurrence influences many other patterns in the Central Highlands, such as the distribution of other plants (Howard and Ashton 1973) and some animals such as communities of birds (Loyn 1985). There are relationships between the occurrence of rainforest and fine-scale ecological processes within some animal populations. Banks *et al.* (2008) has shown links, for example, between the breeding biology and social organisation of the Mountain Brushtail Possum and the occurrence of cool temperate rainforest.

This discussion emphasises the point made at the close of Chapter 1, that the body of work in montane ash forests encompasses the influences of ecological processes on forest patterns, the influences of forest patterns on ecological processes, and the relationships between particular patterns and other patterns. It

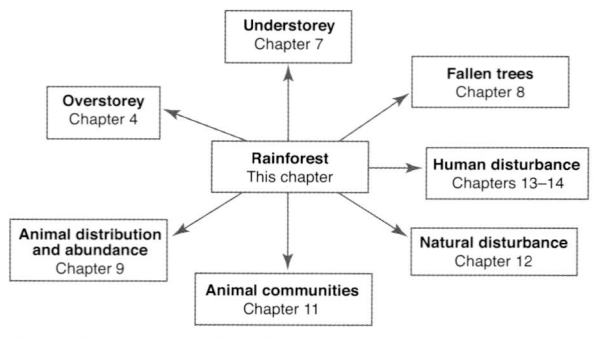

Figure 5.10: Relationships between rainforest cover and key topics in other chapters.

underscores the close connections between topics discussed in various chapters. For example, the occurrence of Myrtle Beech is partly linked to natural and human disturbance regimes. These are core topics in the three chapters that comprise Part V (Chapters 12–14). Myrtle Beech can be a prominent part of the understorey of montane ash forests, which is examined in Chapter 7.

PART III

THE STRUCTURE OF THE FOREST

Part II focused on the broad distribution of montane ash forest and the rainforest that occurs within it. Part III examines the variation in structure within stands of montane ash forest. One of the most straightforward ways to think about forest structure is to work through layers – canopy to forest floor. This is done in Part III: its three component chapters are the canopy, the understorey and the ground layer.

The structure of a forest has major effects not only on its suitability as habitat for biota, but on its ecological functions such as carbon storage. The aim of Part III is to highlight the importance of different layers' attributes for particular elements of biodiversity and ecosystem processes. This material is a prelude to understanding the factors underpinning animal distribution and occurrence that are discussed in subsequent chapters. Much of the way the forest is structured is an outcome of forest disturbances, which is explored in Part V.

6 Key structural features: overstorey trees with hollows

INTRODUCTION

Large trees, particularly those with cavities or hollows, are a key component of stand structure in forests around the world (Fischer and McClelland 1983; Franklin *et al.* 1981; Gibbons *et al.* 2008; Linder and Östlund 1998; Rose *et al.* 2001). Living and dead trees with hollows are a key structural feature of the overstorey of montane ash forests. They play many key ecological roles in forests generally, and montane ash forests in particular. These roles include:

- providing perching, roosting, denning and nesting sites for a wide range of cavity-dependent species (Gibbons and Lindenmayer 2002). Other important behaviour takes place inside hollows, such as group huddling to promote thermoregulation, protection from predators and removal of parasites;
- supporting important foraging substrates for animals by providing large numbers of flowers and clumps of mistletoe (Ashton 1975a; Watson 2001);
- storing large amounts of carbon (Mackey *et al.* 2008).

Large trees with hollows can create problems for forest management, including acting as pipes to disperse sparks during wildfires (Crowe *et al.* 1984), harbouring insect pests (Neumann and Marks 1976), creating a safety hazard for forest workers (Bunnell *et al.* 2003) and suppressing the regeneration of young regrowth trees (Incoll 1979; Rotherham 1983).

Large trees with hollows have been a major and ongoing research focus in the Central Highlands ash-type forests since work commenced there in 1983. There are four good reasons for this. First, large trees with hollows can take well over 120 years to develop in montane ash forests (Ambrose 1982). Second, forestry practices such as clearfelling result in a significant long-term reduction in the abundance of trees with hollows (Lindenmayer *et al.* 1991b). Third, populations of cavity-dependent vertebrates such as arboreal marsupials appear to be limited by the abundance of trees with hollows and have long been included among the species most sensitive to logging operations. Finally, the abundance of trees with hollows forms a key part of the zoning system for Leadbeater's Possum in wood production montane ash forests (Lindenmayer and Cunningham 1996b; Macfarlane *et al.* 1998).

This chapter summarises the many kinds of research on trees with hollows conducted in the montane ash forests over the past 25 years. The key themes of work discussed include:

- the different kinds of trees with hollows that occur in montane ash forests;
- the recruitment, decay and collapse of trees with hollows in montane ash forests;
- where trees with hollows are most likely to occur in landscapes;
- the relationships between the occurrence of trees with hollows and the presence and abundance of arboreal marsupials;
- partitioning of the tree hollow resource among arboreal marsupials;
- multiple den tree use by arboreal marsupials;
- the use of artificial tree hollows (nest boxes) by cavity-dependent vertebrates.

The focus of work on vertebrates has been on arboreal marsupials. Very limited research has occurred on other taxa with cavity-using representatives, such as birds, bats and reptiles.

This chapter is heavily biased toward discussions about trees with hollows as a key part of the forest overstorey. Nothing has been written about other parts of the overstorey, such as the canopy of upper lateral branches. This is for good reason – almost nothing is known about them. Tree canopies are likely to be very species-rich environments for groups such as invertebrates (as documented in other types of eucalypt-dominated vegetation, Majer *et al.* 1994). These areas are important foraging substrates for arboreal marsupials, such as the Greater Glider, and some species of birds (Loyn 1985). However, beyond these broad but simple statements, quantitative research is lacking on the canopies and other attributes of overstorey trees in montane ash forests.

BACKGROUND DATASETS

Several major datasets have been gathered as part of studies of trees with hollows in montane ash forests. First, all trees with hollows were measured on over 520 sites, each 3 ha in size, located throughout the Central Highlands of Victoria. The sites covered a wide range of environmental conditions such as dominant tree species, stand age, topographic position, slope angle, aspect and elevation (Lindenmayer *et al.* 1991b). A tree with hollows was defined as any tree greater than 0.5 m in diameter containing one or more obvious cavities (determined from careful ground-based observation using binoculars). Thus, the measured population of trees with hollows encompassed living and dead trees with hollows – the reason why the commonly employed North American term 'snag' was not used. All trees with hollows on each site were marked with a large fluorescent number. The

Table 6.1: Studies directly or indirectly examining trees with hollows in the montane ash forests of the Central Highlands of Victoria

Study description	Reference
Attributes and condition of trees with hollows	Lindenmayer *et al.* (1993b)
Factors affecting the distribution and abundance of trees with hollows	Lindenmayer *et al.* (1991b)
Decay, collapse and predicted future abundance of trees with hollows	Lindenmayer *et al.* (1990b); Lindenmayer *et al.* (1997b); Lindenmayer and Wood (2009)
Animal presence and abundance and the occurrence of trees with hollows	Lindenmayer *et al.* (1990a); Lindenmayer *et al.* (1990e); Lindenmayer *et al.* (1991e); Lindenmayer *et al.* (1994b)
Partitioning of the tree hollow resource by arboreal marsupials	Lindenmayer *et al.* (1991c); Lindenmayer *et al.* (1991d); Lindenmayer (1997)
Inter-specific co-occupancy of trees with hollows by arboreal marsupials	Lindenmayer *et al.* (1990c)
Den-swapping by arboreal marsupials	Lindenmayer and Meggs (1996); Lindenmayer *et al.* (1996c); Lindenmayer *et al.* (1996d); Lindenmayer *et al.* (1997a)
Use of artificial trees hollows – nest boxes	Lindenmayer *et al.* (2003c); Lindenmayer *et al.* 2009c)

Figure 6.1: Marked trees with hollows used in studies of tree fall and tree occupancy by arboreal marsupials. (Photo by David Lindenmayer).

- the relationships between the presence and abundance of different species of arboreal marsupials on a site and the abundance of trees with hollows on those sites (Lindenmayer *et al.* 1994b, 1991d).

A detailed set of radio-tracking studies over a period of ~18 months was completed on a single large (35 ha) site at Cambarville. That study quantified the use of trees with hollows by a single population of Mountain Brushtail Possum (Lindenmayer *et al.* 1996c) and a single population of Leadbeater's Possum (Lindenmayer and Meggs 1996).

Finally, two major nest box studies examined patterns of occupancy by vertebrates, infestation by pest invertebrates (primarily the Feral Honeybee [*Apis mellifera*]) and rates of nest box attrition (Lindenmayer *et al.* 2003c, 2009c).

DIFFERENT KINDS OF HOLLOW TREES IN MONTANE ASH FOREST LANDSCAPES

As outlined above, the definition of a tree with hollows included living and dead trees. The trees were classified into different decay states which reflect tree condition based on readily observable external characteristics. This broad classification system was

precise location of each numbered tree on each site was carefully recorded, to help subsequently relocate all trees. Various subsets of the 520 sites have been used over the past 25 years to quantify patterns of cavity development (Lindenmayer *et al.* 1993b), as well as rates of collapse of the trees with hollows (Lindenmayer *et al.* 1997b, 1990b; Lindenmayer and Wood 2009).

Detailed stagwatching surveys (see Chapter 3) have been completed at approximately 200 of the 520 sites. Data on animal occurrence on these sites and within individual trees with hollows on the sites have been used in quantifying a range of relationships. These include:

- the factors influencing the occupancy of particular trees with hollows (Lindenmayer *et al.* 1990a);
- the characteristics of trees with hollows most likely to be used by different species of arboreal marsupials (Lindenmayer *et al.* 1991e);

Box 6.1: Bark streamers – a critical resource provided by overstorey ash-type eucalypt trees

Strips of decorticating bark are a vital attribute of overstorey montane ash trees for a number of animal species. Bark is shed annually and forms spectacular streamers on Mountain Ash and, to a lesser extent, Alpine Ash trees. Several tonnes of bark per hectare are shed annually, adding considerably to the amount of litter in montane ash forests. Large old trees produce significantly more bark than young stems (Lindenmayer *et al.* 2000a). Bark streamers provide habitat for a wide range of invertebrates such as spiders and predatory wingless tree crickets. These invertebrates are, in turn, prey for several species of marsupials (Smith 1984b) and birds. The presence and abundance of Leadbeater's Possum has been found to be significantly related to the quantity of bark in montane ash forests (Lindenmayer *et al.* 1991d).

Figure 6.2: Bark streamers in montane ash forests. (Photo by Esther Beaton).

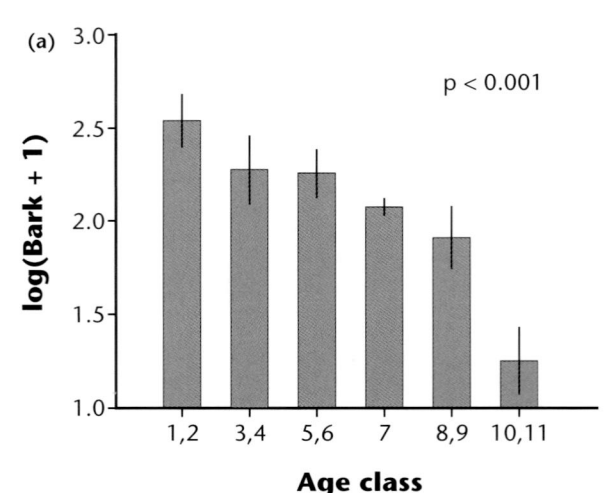

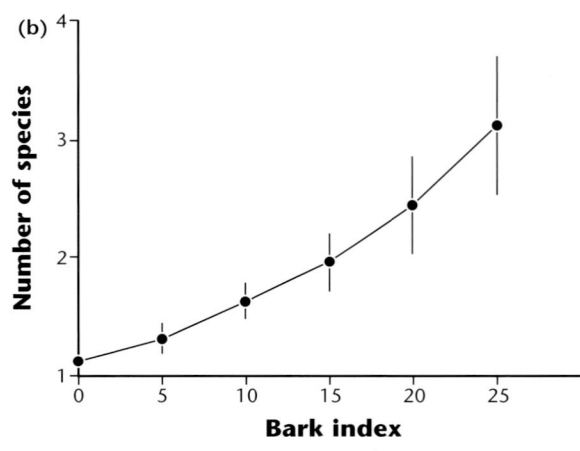

Figure 6.3: Relationships between (a) stand age and the amount of bark old-growth forest corresponds to age classes 1 + 2 and (b) number of species of arboreal marsupials and amount of bark.

borrowed from one developed in the wet coniferous Douglas Fir (*Pseudostuga menziesii*) forests in the Pacific north-west of the US (Cline *et al.* 1980).

Figure 6.4 shows a sequence of changes in the external characteristics of hollow trees. The trees progress from being tall living stems to tall dead trees, then shorter dead stems and finally they decay and collapse onto the forest floor. Very tall living montane ash trees may experience problems in pumping water from the root system to the tree crown. The tree crown eventually dies, leaving dead exposed branches at the top. Pools of rainwater may accumulate in the tops of the dead branches and accelerate the rate of decay inside the main trunk or centre of the tree. Eventually the tree dies and begins to break up, shedding dead branches and the upper sections of the trunk. Finally, the tree becomes so decayed it collapses. Fallen trees are still important for many forest animals. They provide shelter for small mammals such as Bush Rats, or a place to lay eggs, as in some species of reptiles (Lindenmayer *et al.* 2002a) (see Chapter 8).

The classification of tree forms in montane ash forests has proved valuable in a number of ways, including being a significant predictor of the probability of tree fall (Lindenmayer *et al.* 1997b, 1990b; Lindenmayer and Wood 2009) and the development of different kinds of cavities in trees (Lindenmayer *et al.* 1993b).

Figure 6.4: Forms of trees with hollows in montane ash forests. Form 1: Mature living tree. Form 2: Mature living tree with a dead or broken top. Form 3: Dead tree with most branches still intact. Form 4: Dead tree with 0–25% of the top broken off; branches remaining as stubs only. Form 5: Dead tree with top 25–50% broken away. Form 6: Dead tree with top 50–75% broken away. Form 7: Solid dead tree with ≥75% of the top broken away. Form 8: Hollow stump. Form 9: Collapsed tree. (Redrawn from Lindenmayer et al. 1997b).

WHERE DO HOLLOW TREES OCCUR IN THE LANDSCAPE?

Considerable effort has been dedicated to studying patterns of distribution and abundance of trees with hollows in montane ash forests. Extensive surveys of over 520 field sites involved measuring and counting literally thousands of hollow trees throughout large areas of the Central Highlands of Victoria (Lindenmayer et al. 1991b). The work has shown that hollow trees are now absent from very large areas of the region. However, there are some parts of montane ash forests where hollow trees are significantly more likely to be found. These are:

- forests in the O'Shannassy Water Catchment which are closed to human access and disturbances such as logging;
- stands of old-growth forest more than 90 years old;
- areas on flat terrain;
- forests where clearfelling has not occurred;
- sites in gullies;
- forests where the dominant tree species was Mountain Ash (compared with Alpine Ash and Shining Gum).

There are good reasons for these findings. As trees age they are more likely to develop hollows (Lindenmayer et al. 1993b). Therefore, trees in stands of older

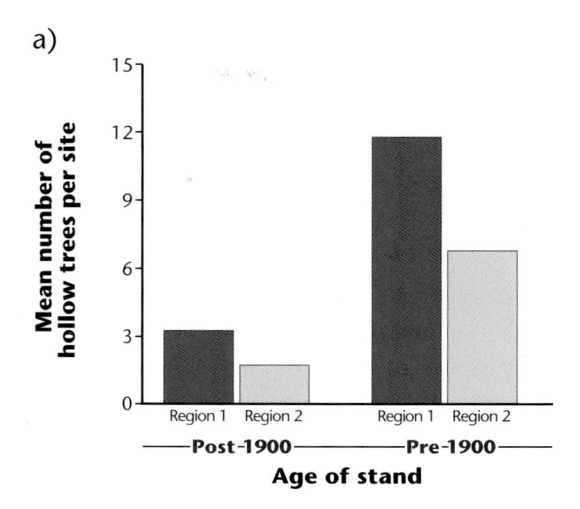

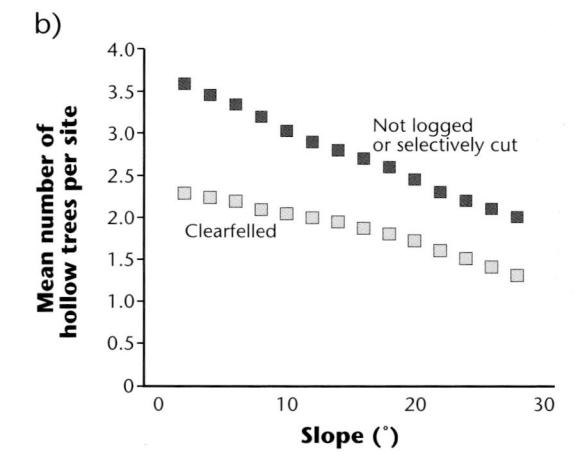

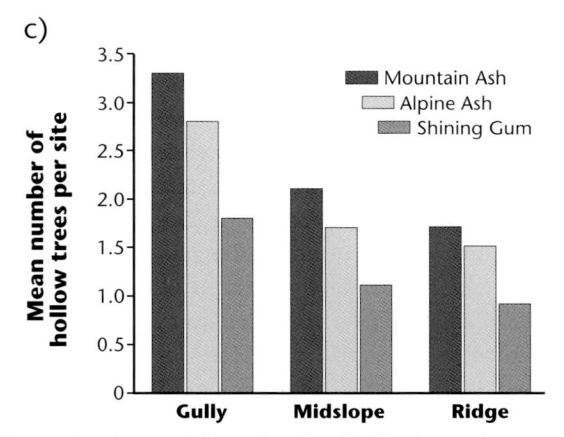

Figure 6.5: Factors influencing the distribution and abundance of trees with hollows on sites within montane ash forests. (Redrawn from Lindenmayer 1996; Lindenmayer et al. 1991b).

forest would be expected to have more hollows than trees in young forest. Most of the remaining old-growth montane ash forest in Victoria occurred in the O'Shannassy Water Catchment, which explains why this area had more hollow trees than elsewhere in the region.

There are differences between tree species in the rate of development of hollows. Mountain Ash seems more likely to contain cavities than other types of trees in montane ash forests (Lindenmayer *et al.* 1993b). Sites on flatter terrain and in gullies are more likely to escape the effects of intense wildfires than are places on steep exposed ridges (Mackey *et al.* 2002). High-intensity wildfires can kill and destroy hollow trees, which explains why they are less common in such parts of the landscape. Finally, activities such as clearfelling remove all, or nearly all, the trees on a logged site. Therefore, it is not surprising that forests cut in this way support fewer trees with hollows (Lindenmayer *et al.* 1991b).

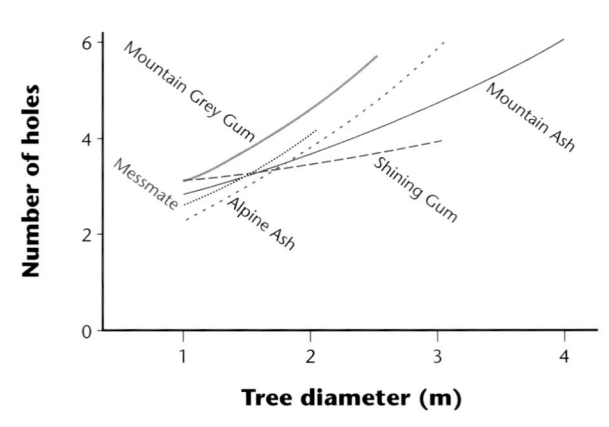

Figure 6.6: Relationships between tree diameter and number of holes in different species of montane ash trees in the Central Highlands of Victoria. (Redrawn from Lindenmayer *et al.* 1993b).

DYNAMICS OF POPULATIONS OF TREES WITH HOLLOWS: RECRUITMENT

The development of cavities in eucalypt trees is complex. It is a slow process, often taking several hundred years. Many factors contribute to the formation of cavities in eucalypt trees. They include fungal and termite activity, incomplete breakage of branches leaving weakened areas on the tree trunk, tree age and the associated propensity to develop defects in the wood and damage by lightning and wildfires (Lindenmayer *et al.* 1993b). These processes may interact. For example, the process of shedding very large branches may be impaired in old trees, giving rise to wounds on the trunk where the branch was formerly attached. This provides a point of entry for fungi and termites, which can promote the excavation of hollows.

The time at which hollows begin to develop varies substantially between different species of eucalypts (Gibbons and Lindenmayer 2002). In montane ash trees, cavities appear to begin clear development at ~120 years of age (Ambrose 1982). Larger hollows suitable for use by species such as Leadbeater's Possum generally do not begin to develop until the trees exceed 190 years. Older and bigger trees contain sig-nificantly more hollows than do younger ones. There tends to be a strong relationship between the number of cavities in a tree and its diameter, which is a function of tree age (Lindenmayer *et al.* 1993b) (Figure 6.6). The processes of cavity formation and changes in tree morphology vary over time and continue well after a tree has died. This, in turn, leads to differences in the abundance of kinds of cavities in trees in different condition classes (Figure 6.7).

DYNAMICS OF POPULATIONS OF TREES WITH HOLLOWS: TREE DECAY AND COLLAPSE

Many of the large trees with hollows in montane ash forests are trees that were killed or fire-scarred by the 1939 Black Friday wildfires (Gill 1981; Noble 1977). Extensive field studies have been monitoring the deterioration and fall of these trees since 1983, based on repeated condition assessments of marked and mapped trees across a wide range of sites throughout the Central Highlands region (Lindenmayer *et al.* 1997b, 1990b; Lindenmayer and Wood 2009). Assessments were made in 1983, 1988, 1993 and 2007. Each time a tree was visited, information was recorded on its form (see Figure 6.4) and whether it was still standing. These data were compared with similar information from earlier surveys. This made it feasible to determine the rate of decay among trees with hollows.

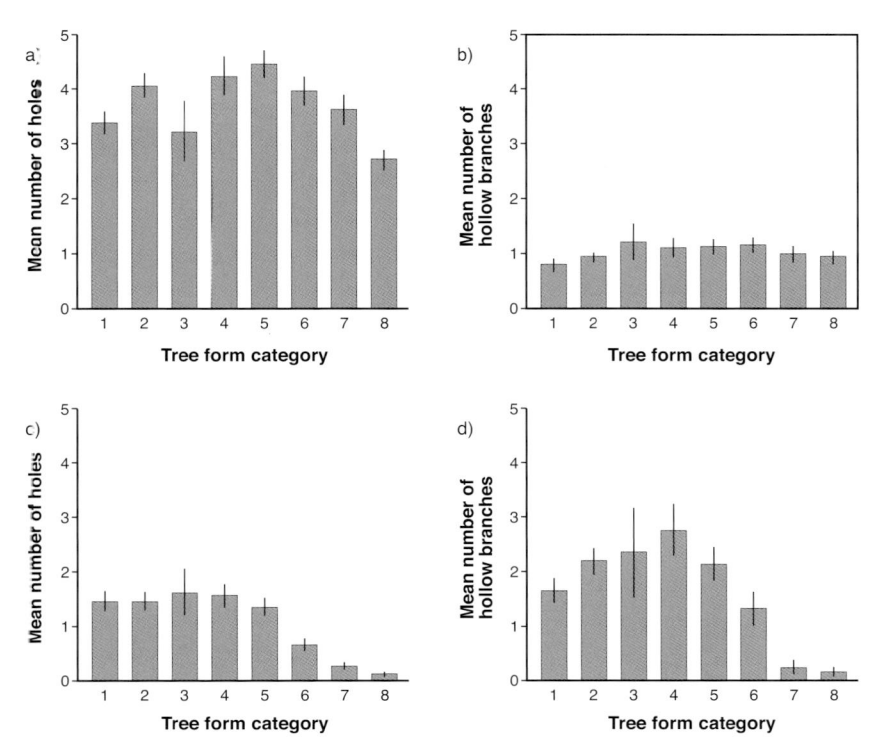

Figure 6.7: Relationships between tree form and abundance of kinds of cavities in montane ash trees. (Redrawn from Lindenmayer *et al.* 1993b).

It was also possible to use current fall rates to predict which trees are most likely to collapse in the near future. This work enabled predictions of the likely future abundance of trees with hollows (Lindenmayer *et al.* 1997b; Lindenmayer and Wood 2009).

Collapse rates during different measurement periods

Long-term studies of tree fall have quantified rapid rates of decay and collapse. Initial measurements between 1983 and 1993 suggested that about 4.4% of the population of trees with hollows was collapsing annually. In subsequent measurement periods, the rate of collapse of trees with hollows slowed to ~2.2% pa, approximately half that of the initial periods of measurement. Thus, the trend patterns observed over the first two five-year measurement periods were not congruent with those quantified in the following ~14-year period.

The reasons why the rate of tree collapse slowed over the 25-year duration of the study remain unclear. Anecdotal observation has showed that 'wet' trees with hollows appear to decay and collapse most rapidly. South-eastern Australia has experienced a

Figure 6.8: A collapsed tree with hollows. (Photo by David Lindenmayer).

prolonged dry period with many years of below-average rainfall during the past two decades (Cai and Cowan 2008; McAlpine *et al*. 2007). Sites, and trees within sites, are now drier than they were during the early periods of study. This may explain the slowing collapse rate of trees with hollows. An alternative or additional explanation may be that the trees susceptible to collapse have already fallen over and those least likely to collapse remain standing. For example, small-diameter trees with hollows were significantly more likely to collapse. Thus, many of the remaining trees are the larger-diameter, more persistent trees.

Factors influencing the probability of collapse

Initial studies indicated that the trees most likely to collapse were smaller-diameter stems and those already very decayed (e.g. tree forms 6, 7 and 8 in Figure 6.4). Recent work has shown that these significant predictors of tree fall remained important in the most recent survey period. The relationship between the form of a tree and its susceptibility to collapse was expected, as the different decay classes represent different stages of deterioration in condition (Cline *et al*. 1980). Tree condition and tree fall relationships have been well-documented for a long time in the forest ecology literature in a wide range of forest types worldwide (Sicamma *et al*. 2007; Aakala *et al*. 2008). The relationship between tree diameter and probability of collapse (Figure 6.9) in montane ash forests is common to many studies (Keen 1955; Garber *et al*. 2005; Russell *et al*. 2006; Vanderwel *et al*. 2006).

Predicted future abundance of trees with hollows

Repeated measurements of trees with hollows enabled the development of a transition matrix for trees, moving through the different stages of decay to collapse. A probability value could be assigned to the movement of a tree of a given form to a more advanced state of decay. Given this, it was possible to assume the continuation of the transition process and predict the future abundance of trees with hollows on sites. Because most trees in existing regrowth forests are younger than 70 years old and date from the 1939 wildfires, there will be very limited (if any) recruitment of new trees with hollows for another 50–120 years on these sites. Tree fall will result in most sites supporting very few trees with hollows in the coming decades. This is highlighted in Figure 6.10, which shows the number of trees with hollows in different forms (including collapsed trees) for each of the four measurement periods (1983, 1988, 1993 and 2007) and the number predicted by 2049 (assuming continuation of the rates of transition in 1993–2007) and 2047 (assuming continuation of the rates of transition in 1983–93). Both plots clearly show the increasing numbers of trees in the collapsed tree group.

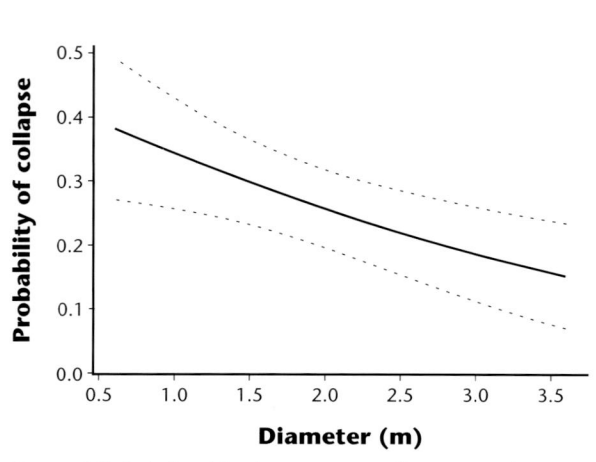

Figure 6.9: Relationships between tree diameter and probability of collapse of trees with hollows. (Redrawn from Lindenmayer *et al*. 2009c).

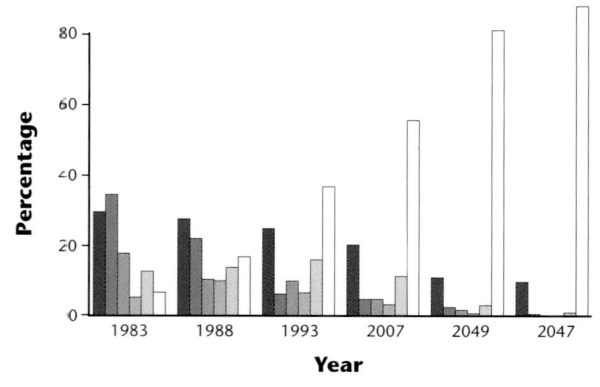

Figure 6.10: Trees with hollows in amalgamated decay classes in 1983, 1988, 1993 and 2007 and projections to 2049 and 2047. Standing living trees (forms 1 and 2 from Figure 6.4) correspond to the darkest columns on the extreme left-hand side of each period. The lightest columns on the extreme right-hand side of each period are those that had collapsed (form 9 in Figure 6.4).

Landscape context and accelerated rates of tree fall

The studies of tree fall summarised above are based on sites mostly located within contiguous areas of ash-type forest. A subsidiary investigation of tree fall involved gathering an additional dataset on the decay and collapse of trees with hollows in 49 linear strips of forest (15–200 m wide) that had been retained within areas where the surrounding forest had been clearfelled (Lindenmayer et al. 1997b).

The work showed that rates of tree collapse in the narrow strips were twice as rapid as those quantified in contiguous forest. The probability of collapse was significantly related to both the characteristics of trees (Figure 6.12) and the attributes of strips (Figure 6.13). As in the case of contiguous forest sites, trees in more decayed forms were more likely to collapse (Figure 6.8). In addition, sites where the surrounding clearfelled coupes were large experienced greater levels of tree fall than strips where adjacent logged areas were small (Lindenmayer et al. 1997b). Greater wind fetch and other altered micro-meteorological conditions are the likely reasons for accelerated tree fall in strips where neighbouring logged areas are large (Lindenmayer et al. 1997b).

Accelerated tree fall on logged sites

Accelerated rates of tree fall characterise linear strips retained within logged coupes. Fall rates are even higher for trees with hollows that are retained on logged coupes (Ball et al. 1999; Lindenmayer et al.

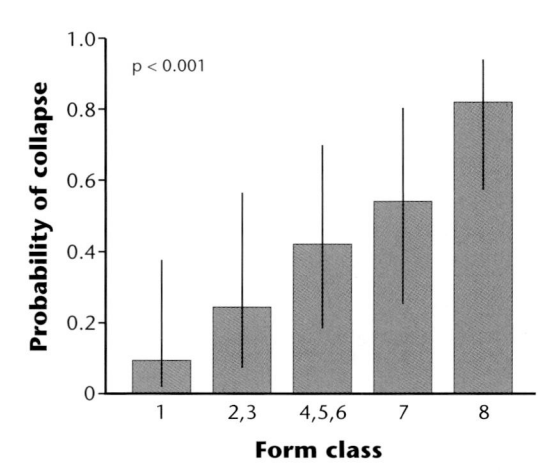

Figure 6.12: Relationship between tree collapse in narrow retained strips and tree form. (Redrawn from Lindenmayer et al. 1997b).

1990b). For example, such trees are often badly burned as part of the intensive slash burn applied after logging to promote the regeneration of cutblocks (Figure 6.14). A simulation modelling study by Ball et al. (1999) highlighted the rates of attrition of such kinds of retained trees and demonstrated the need to find new ways to better protect trees with hollows as part of logging operations, and substantially increase the number of trees retained in logged areas. These issues are discussed further in Chapter 16, that focuses on mitigating logging effects.

Figure 6.11: Tree collapse in a retained strip of forest surrounded by recently clearfelled forest. (Photo by David Lindenmayer).

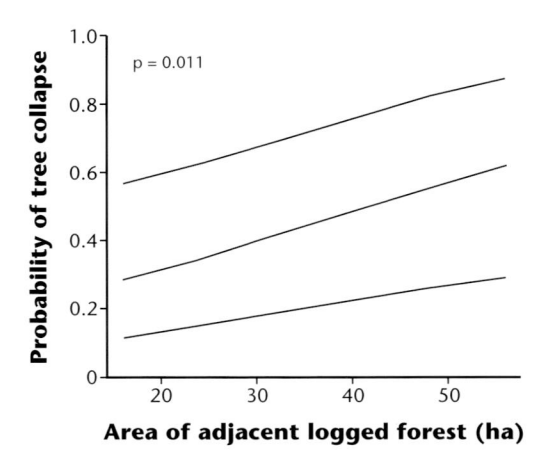

Figure 6.13: Relationship between tree collapse in narrow retained strips and the size of surrounding logged coupes. (Redrawn from Lindenmayer et al. 1997b). The middle line represents the mean response and the lines above and below it are the 95% confidence intervals.

Figure 6.14: Burning trees with hollows on logged sites in the Central Highlands of Victoria. (Photo by David Lindenmayer).

IMPORTANCE OF TREES WITH HOLLOWS FOR ANIMAL DISTRIBUTION AND ABUNDANCE

Almost all species of arboreal marsupials in montane ash forests are dependent on trees with hollows as den and nest sites. Given such dependence, it is perhaps not surprising that strong relationships have been identified between the prevalence of trees with hollows in a forest area and the presence and abundance of these species (Lindenmayer *et al.* 1994b, 1991d, 1990c) (Figure 6.15). The Sugar Glider, Greater Glider, Leadbeater's Possum and Mountain Brushtail Possum are significantly more likely to be found in forest stands that support a large number of trees with hollows.

Statistical relationships between the abundance of trees with hollows and occurrence of animals have been particularly important in setting the management zoning system for Leadbeater's Possum (Lindenmayer and Cunningham 1996a; Macfarlane *et al.* 1998) (see Chapter 16). It has been critical to test the performance of these statistical models. An additional dataset comprising 55 new stagwatched sites was gathered to test models developed from earlier work (Lindenmayer *et al.* 1994b). The tests indicated that the original model for Leadbeater's Possum, and models for other species of arboreal marsupials, performed well when tested in this way. That is, the trends in relationships between the response variable (the abundance of trees with hollows) and the presence/absence and abundance of various species of arboreal marsupials was generally consistent between the initial datasets and the new dataset collected to assess original model performance

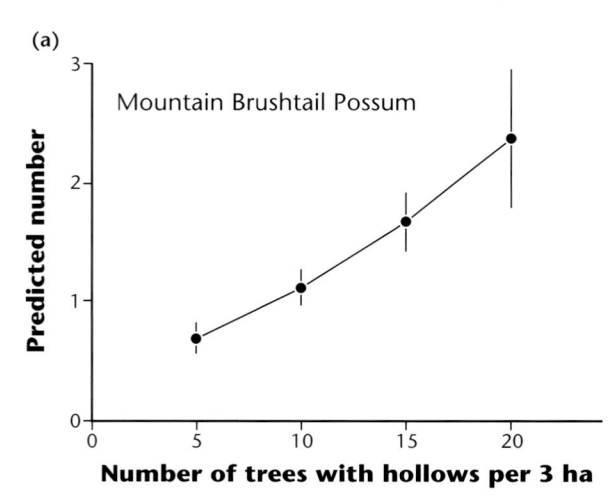

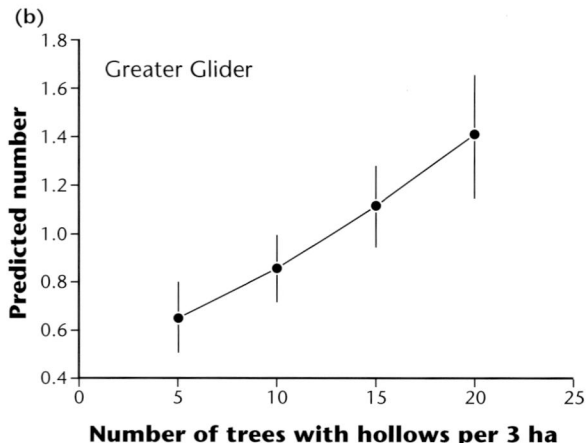

Figure 6.15: Relationships between the number of trees with hollows and the predicted numbers of (a) Mountain Brushtail Possum and (b) Greater Glider.

(Lindenmayer *et al.* 1994b). There is important congruence between these results and those from a separate study of arboreal marsupials in retained linear strips, which also showed strong positive relationships between the abundance of trees with hollows and the probability of occurrence of arboreal marsupials such as the Mountain Brushtail Possum and the Greater Glider (Lindenmayer *et al.* 1993a).

There are good ecological reasons for the relationships between animal occurrence and the abundance of hollow trees. First, as discussed below, not all trees with hollows will be suitable for occupancy. Therefore, areas that support more trees with hollows have a greater chance of having some that can be used. Second,

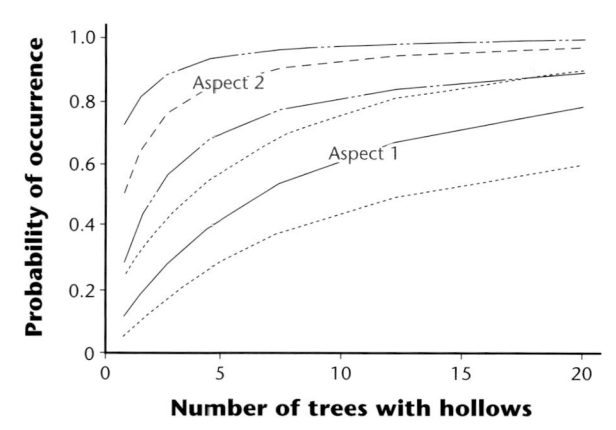

Figure 6.16: Relationships between the abundance of trees with hollows and the probability of occurrence of the Greater Glider in retained linear strips (Lindenmayer et al. 1993a). The relationship for the Greater Glider includes predictions for two different aspect classes, as aspect was a significant explanatory variable in the final model for the species. The label shows the mean response for each aspect class with associated confidence intervals.

requirements for multiple den sites (as part of den-swapping behaviour, see below) are more likely to be met in areas that support numerous trees with hollows.

Given the spatial variation in the abundance of trees with hollows in montane ash forests and the strong associations between animal occurrence and tree hollow abundance, it is not surprising that the distribution and abundance of arboreal marsupials is also patchy and discontinuous. This is exemplified in maps of the distribution of species such as the Greater Glider when key attributes of the habitat requirements of the species are well-known (the abundance of tree hollows and the age of the forest) and can be mapped spatially (Lindenmayer et al. 1995b) (see Chapter 9).

PARTITIONING OF TREE HOLLOW RESOURCES BY ANIMALS
Characteristics of occupied trees

Extensive stagwatching surveys have provided a wealth of information on the use of different kinds of nest trees by the various species of arboreal marsupials inhabiting the montane ash forests of Victoria. This work has shown that the nest tree resource appears to be partitioned between the different species (Table 6.2). There is limited apparent overlap between the various species in:

- the morphological characteristics of trees used as nest sites;
- the type of entrance to the nest site (e.g. a hole compared with a hollow branch);
- the time after dusk that animals emerge from the nest site to commence foraging;
- the height above ground of the entrance to the tree hollow.

Table 6.2: Partitioning of the nest-tree resource by possums and gliders in montane ash forests

Animal species	Features of trees used for nesting	Type of nest entrance	Height of nest entrance (m)	Time of emergence after dusk (mins)
Leadbeater's Possum	Short, large diameters with many holes. Dense vegetation near nest entrance	Hole	~15	~20
Greater Glider	Tall, large diameters	Spout or hollow branch	~40	~35
Mountain Brushtail Possum	Short, large diameter with few holes	Spout or hollow branch	~25	~25
Sugar Glider	Many narrow cracks	Hole	~30	~15
Yellow-bellied Glider	Not known	Spout or hollow branch	~20	~45
Feathertail Glider	Not known	Hole	~25	~5
Common Ringtail Possum	Not known	Not known	~30	~45

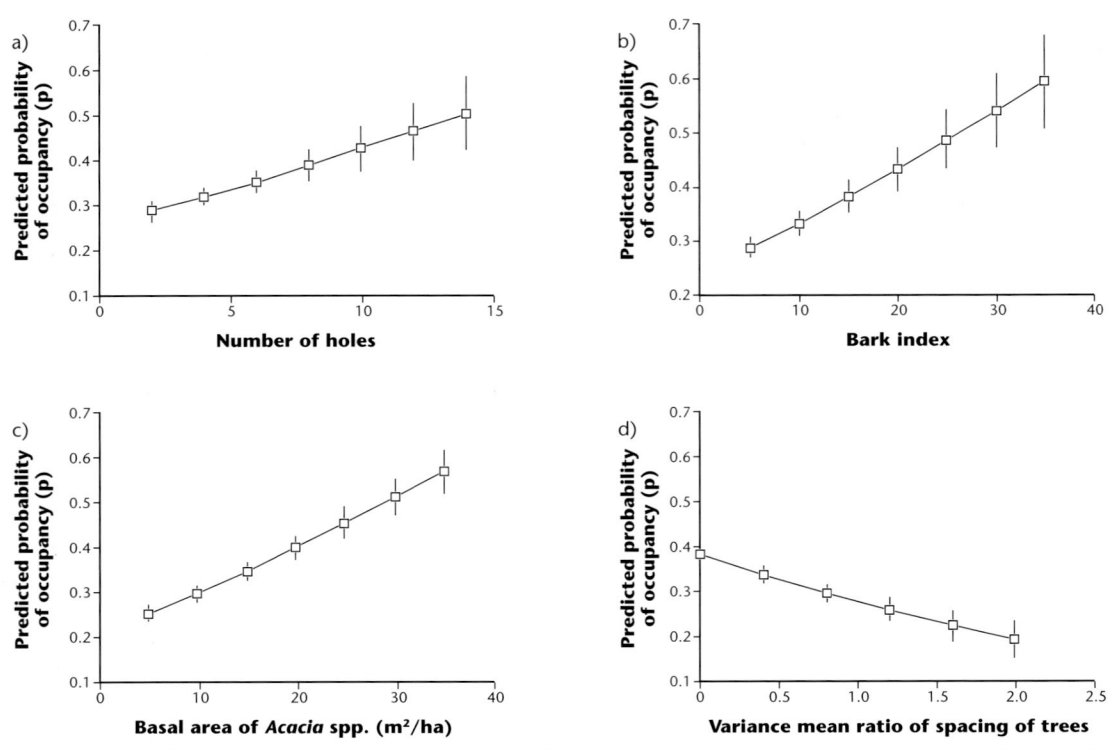

Figure 6.17: Relationships between occupancy of a tree with hollows by an arboreal marsupial and a range of key attributes of surrounding trees and vegetation. (Redrawn from Lindenmayer et al. 1990a).

For example, trees typically occupied by the Greater Glider are very tall, living or recently dead. The entrance to the nest is often a hollow branch or broken branch stub and is usually 40 m or more above the ground. This height above the forest floor may assist animals to glide away from the nest site prior to commencing foraging. The Greater Glider is one of the last animals to emerge from the nest tree. In contrast, trees most likely to be inhabited by a colony of Leadbeater's Possum are large dead and highly decayed stems, characterised by a very large diameter (typically 1–2 m or bigger), being relatively short in height (often less than 25 m), many obvious holes in the main trunk (usually more than six), and large amounts of dense vegetation close to, or touching, the main trunk (Lindenmayer et al. 1991e).

Factors influencing occupancy

Not all trees with hollows may be occupied by arboreal marsupials. There are many reasons for this. First, different species prefer different types of nest sites and some hollow trees may not meet those requirements. Second, trees with hollows are only one component of an animal's requirements. Other resources such as the availability of food and suitable habitat in the surrounding forest affect whether an animal can persist on a given site and occupy a particular hollow-bearing tree. For example, the prevalence of *Acacia* spp. trees in the understorey and the prevalence of bark strips on living eucalypts surrounding a nest tree can significantly influence its suitability for occupancy (Lindenmayer et al. 1990a). Other factors, such as the aspect of the forest patch in which a hollow tree occurs, can be important. Differences in the aspect of a site will influence microclimatic conditions in the hollows within a tree. Extremes of temperature within hollows can, in turn, determine if it is suitable for use by arboreal marsupials (Gibbons and Lindenmayer 2002). Studies have shown that hollow trees on northerly and westerly aspects have the lowest rates of occupancy (Lindenmayer et al. 1990a).

Forest Pattern and Ecological Process | 85

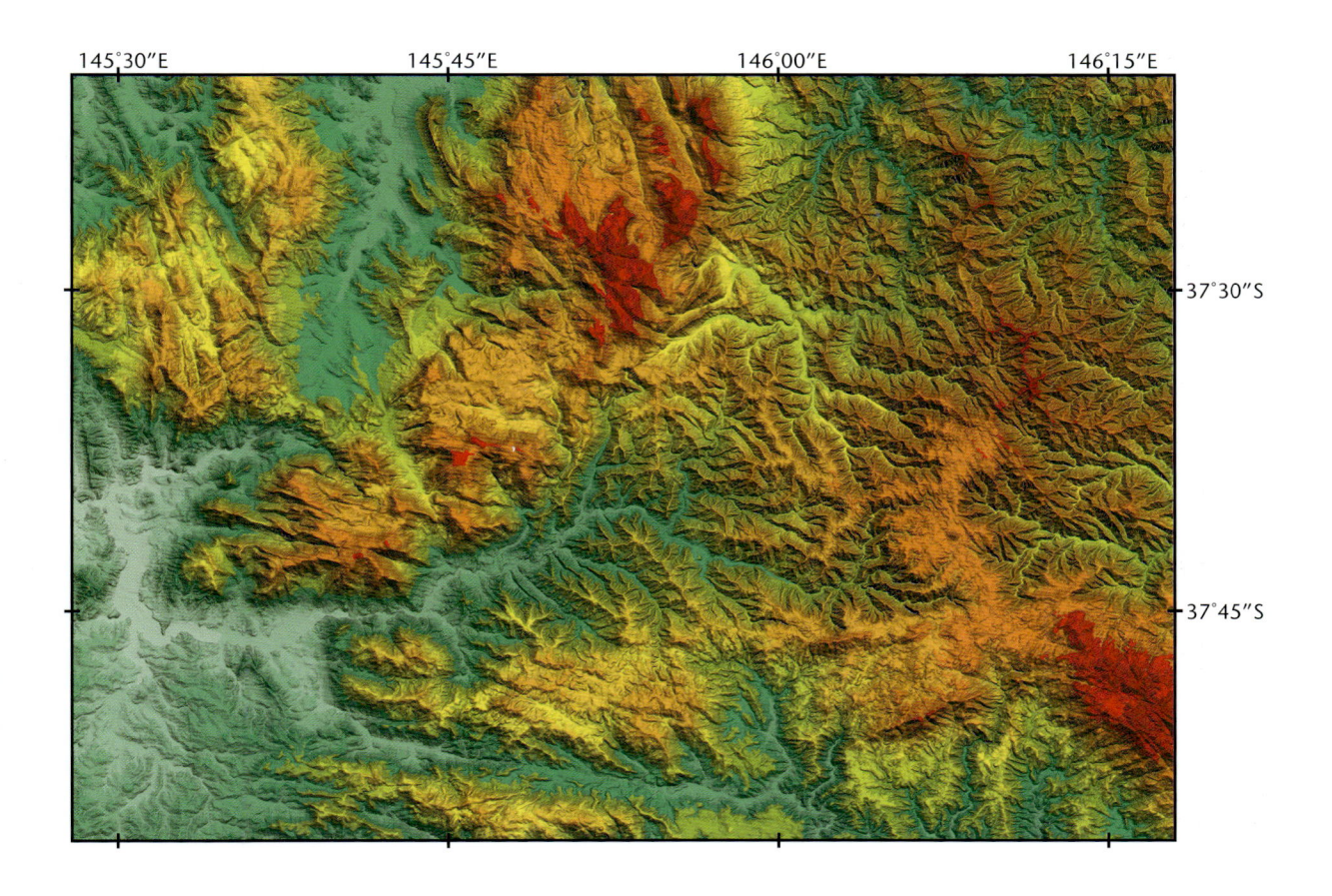

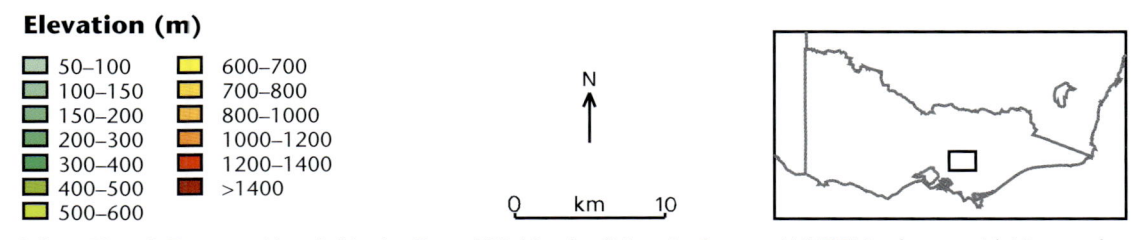

Elevation (m)

- 50–100
- 100–150
- 150–200
- 200–300
- 300–400
- 400–500
- 500–600
- 600–700
- 700–800
- 800–1000
- 1000–1200
- 1200–1400
- >1400

Colour Plate 1: Topographic relief in the Central Highlands of Victoria. Source: ANUDEM software with Vicmap data.

Annual mean temperature (°C)

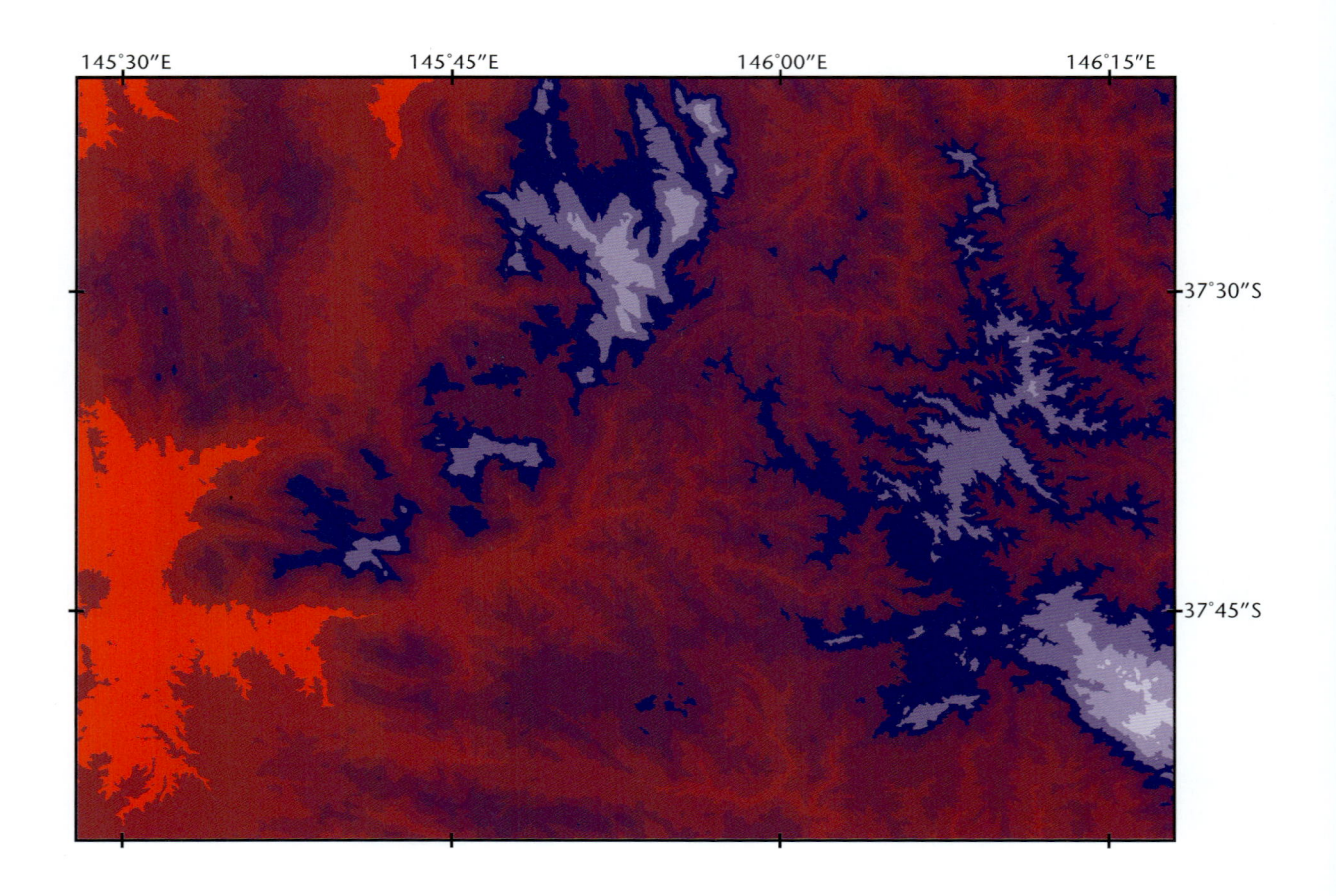

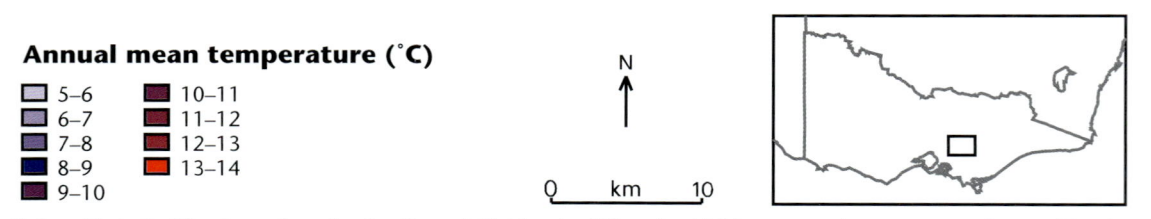

- 5–6
- 6–7
- 7–8
- 8–9
- 9–10
- 10–11
- 11–12
- 12–13
- 13–14

Colour Plate 2: Climate surfaces for the Central Highlands of Victoria. (a) Mean annual temperature. Source: The Australian National University Climate Surfaces and ANUCLIM software.

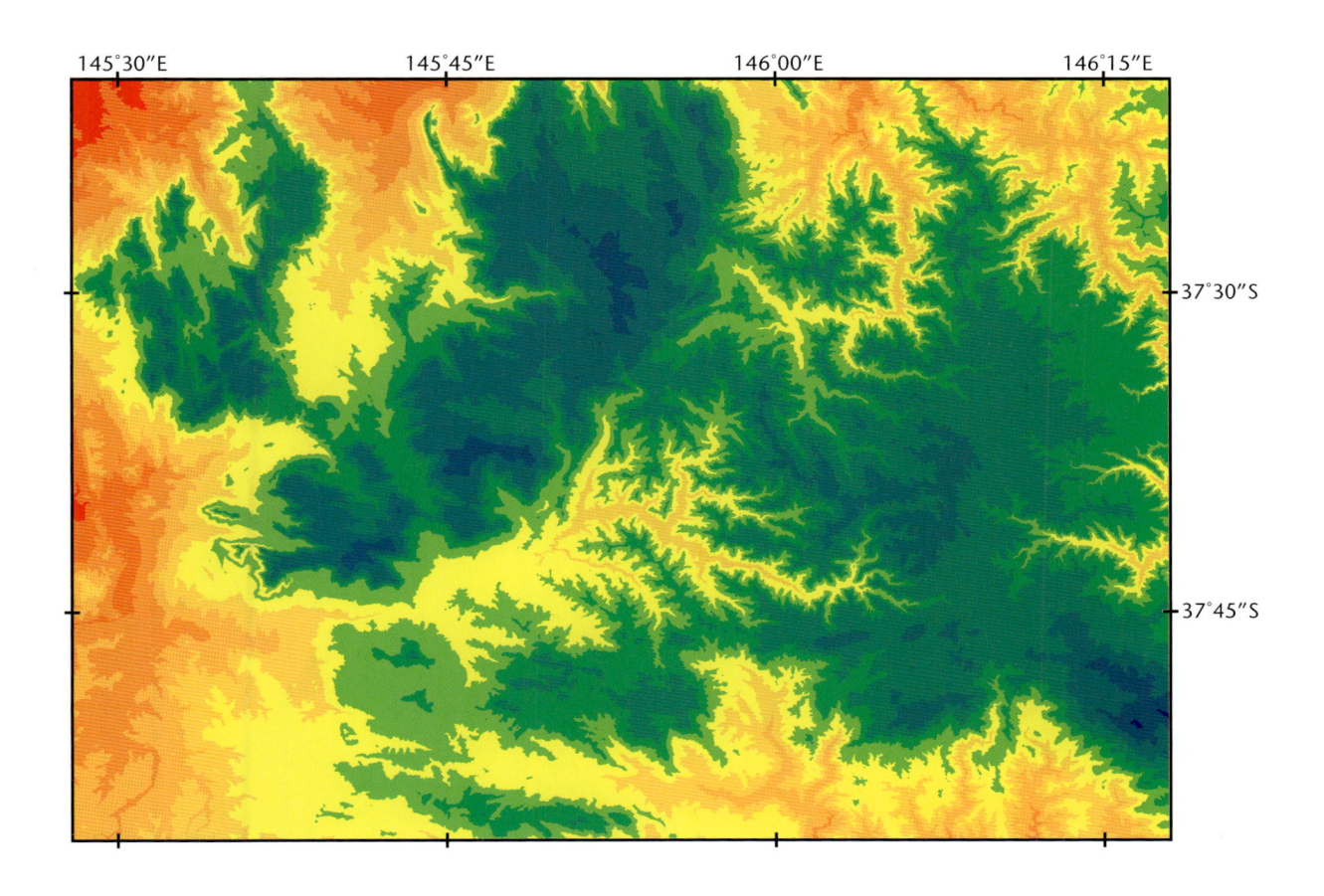

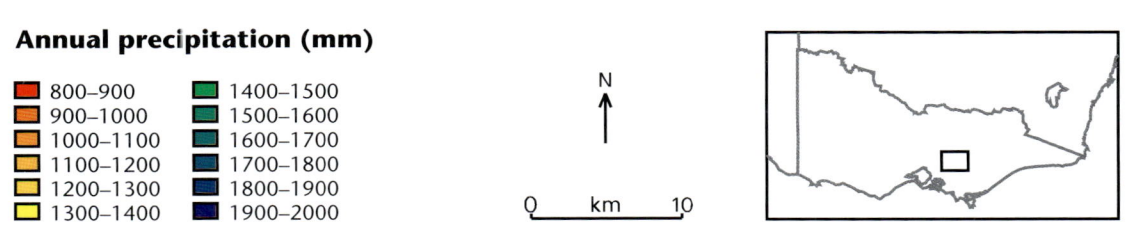

Annual precipitation (mm)

- 800–900
- 900–1000
- 1000–1100
- 1100–1200
- 1200–1300
- 1300–1400
- 1400–1500
- 1500–1600
- 1600–1700
- 1700–1800
- 1800–1900
- 1900–2000

Colour Plate 2 (continued): Climate surfaces for the Central Highlands of Victoria. (b) Annual precipitation. Source: The Australian National University Climate Surfaces and ANUCLIM software.

Geology

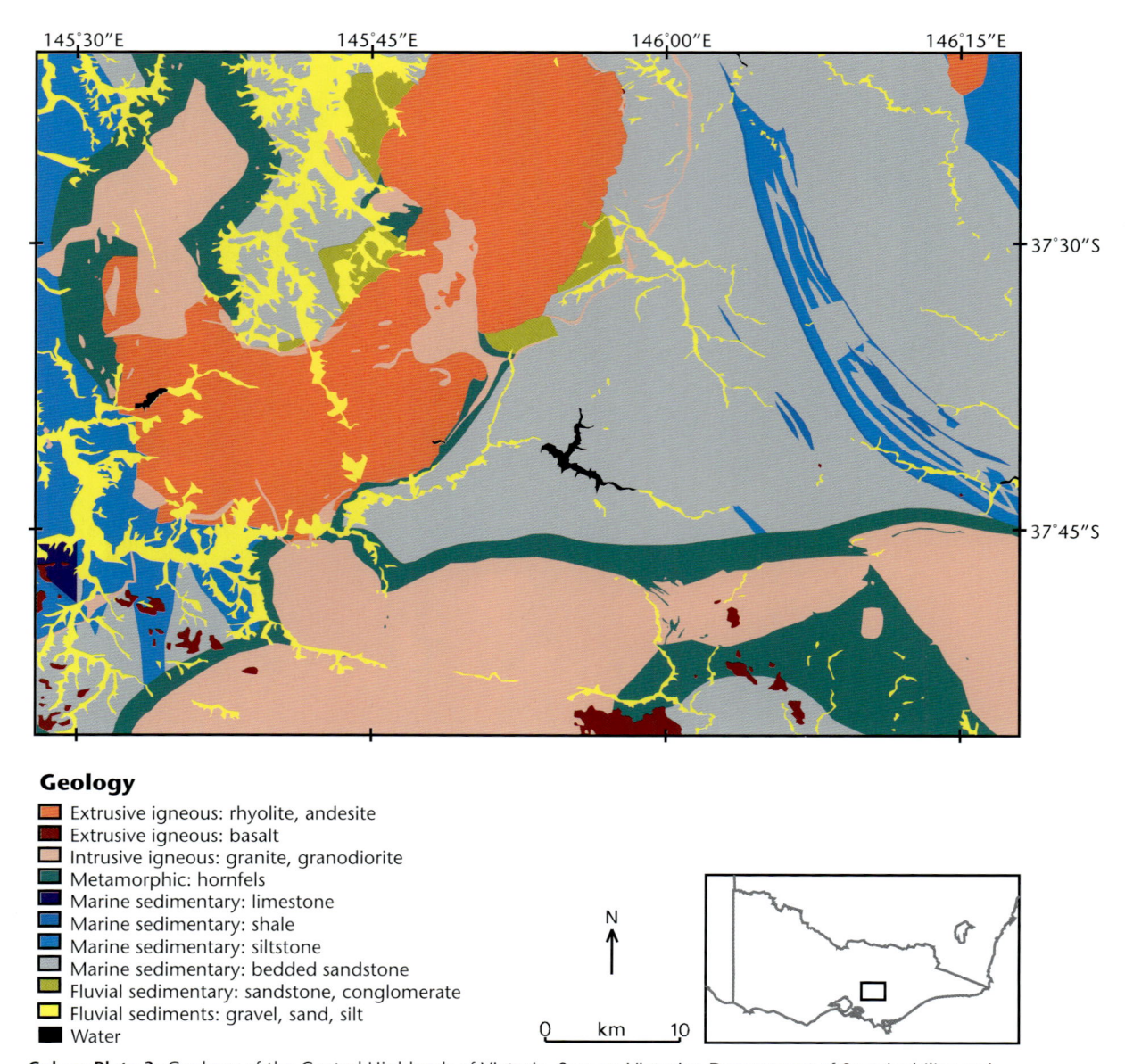

- 🟧 Extrusive igneous: rhyolite, andesite
- 🟥 Extrusive igneous: basalt
- 🟪 Intrusive igneous: granite, granodiorite
- 🟩 Metamorphic: hornfels
- 🟦 Marine sedimentary: limestone
- 🟦 Marine sedimentary: shale
- 🟦 Marine sedimentary: siltstone
- ⬜ Marine sedimentary: bedded sandstone
- 🟨 Fluvial sedimentary: sandstone, conglomerate
- 🟨 Fluvial sediments: gravel, sand, silt
- ⬛ Water

Colour Plate 3: Geology of the Central Highlands of Victoria. Source: Victorian Department of Sustainability and Environment.

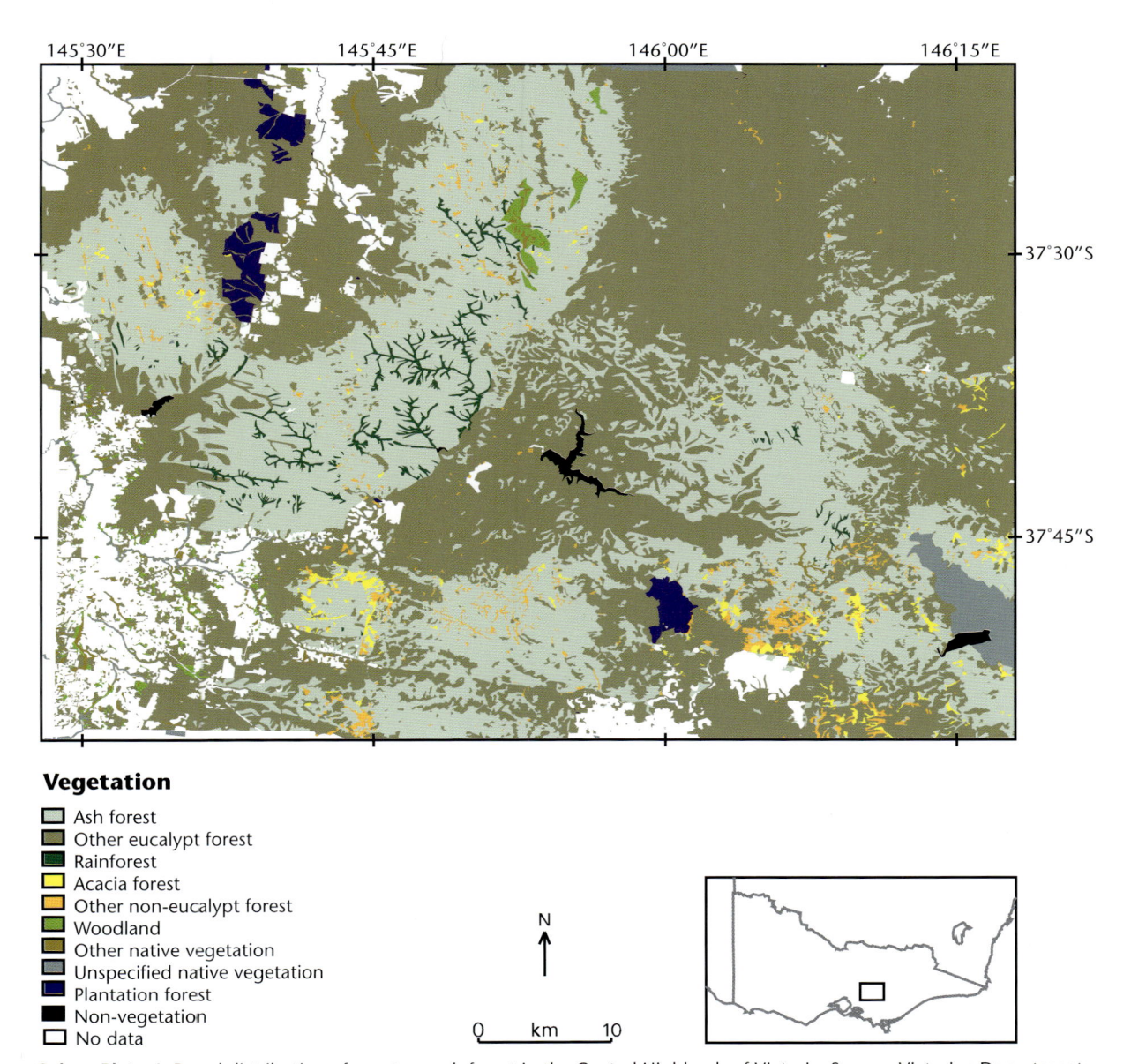

Vegetation

- Ash forest
- Other eucalypt forest
- Rainforest
- Acacia forest
- Other non-eucalypt forest
- Woodland
- Other native vegetation
- Unspecified native vegetation
- Plantation forest
- Non-vegetation
- No data

Colour Plate 4: Broad distribution of montane ash forest in the Central Highlands of Victoria. Source: Victorian Department of Sustainability and Environment.

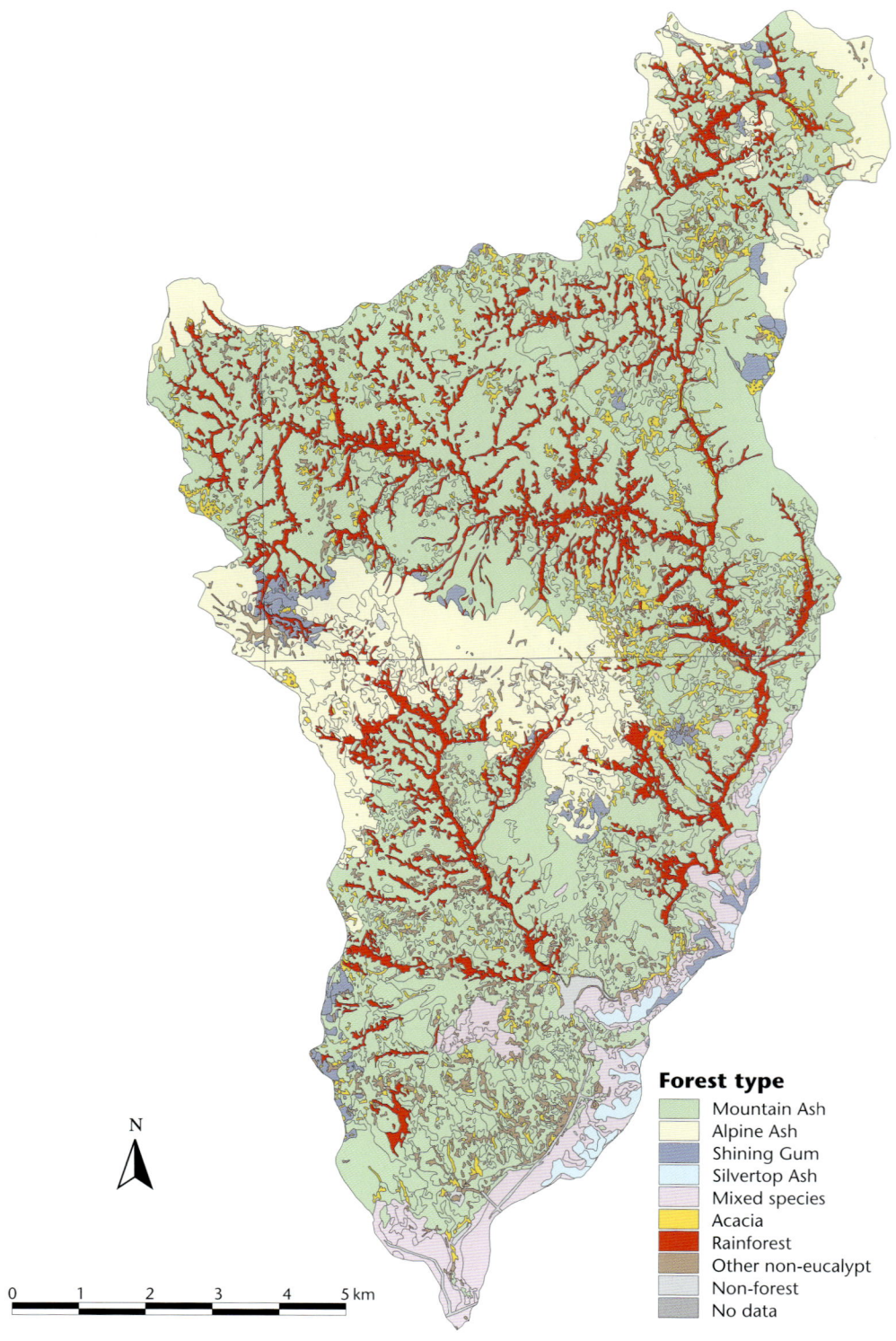

Colour Plate 5: Forest types in O'Shannassy catchment. (Redrawn from Mackey *et al.* 2002). Spatial information on the location of particular forest types was used to examine their environmental relationships (see Chapter 4).

Forest Pattern and Ecological Process | 91

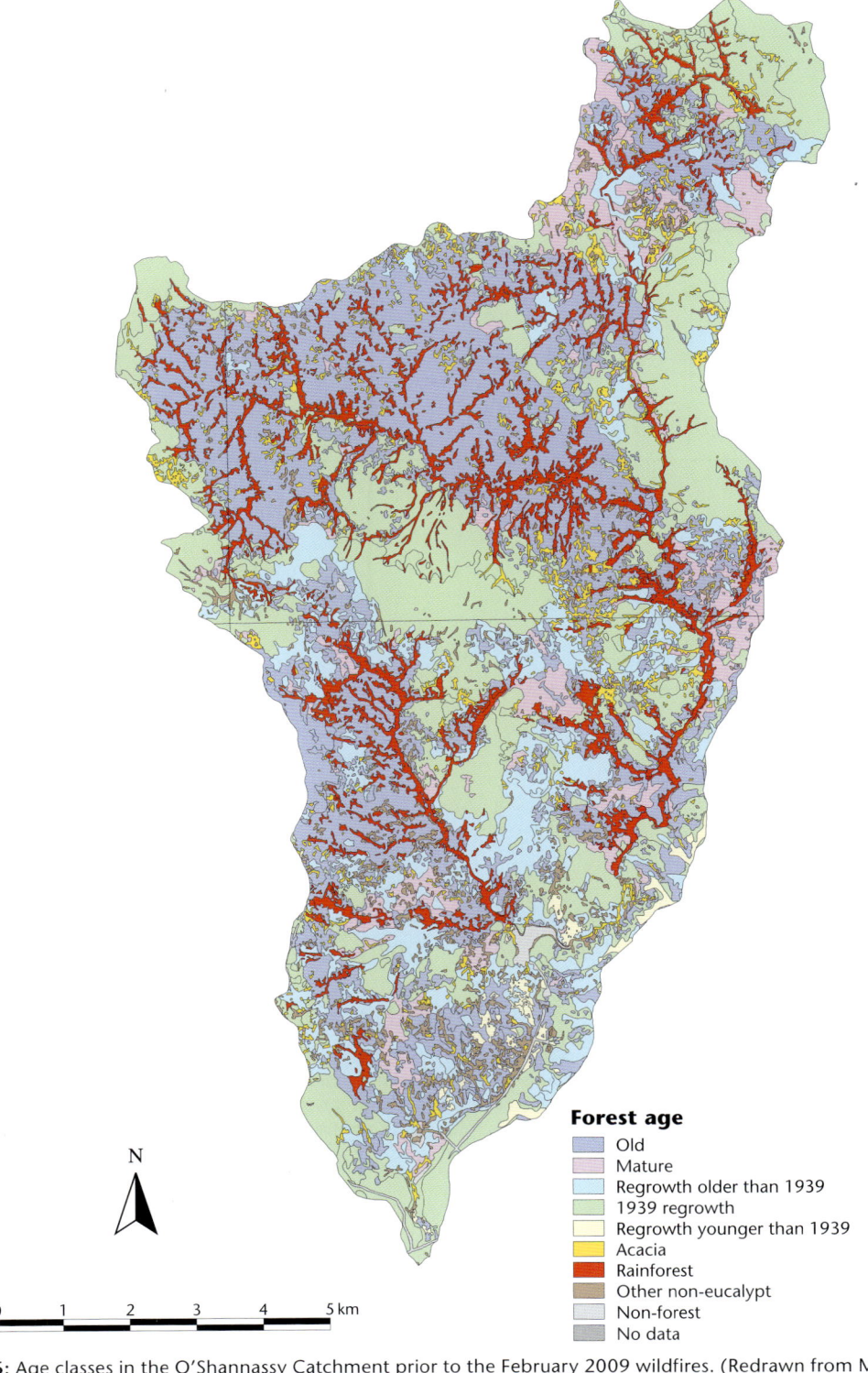

Colour Plate 6: Age classes in the O'Shannassy Catchment prior to the February 2009 wildfires. (Redrawn from Mackey *et al.* 2002).

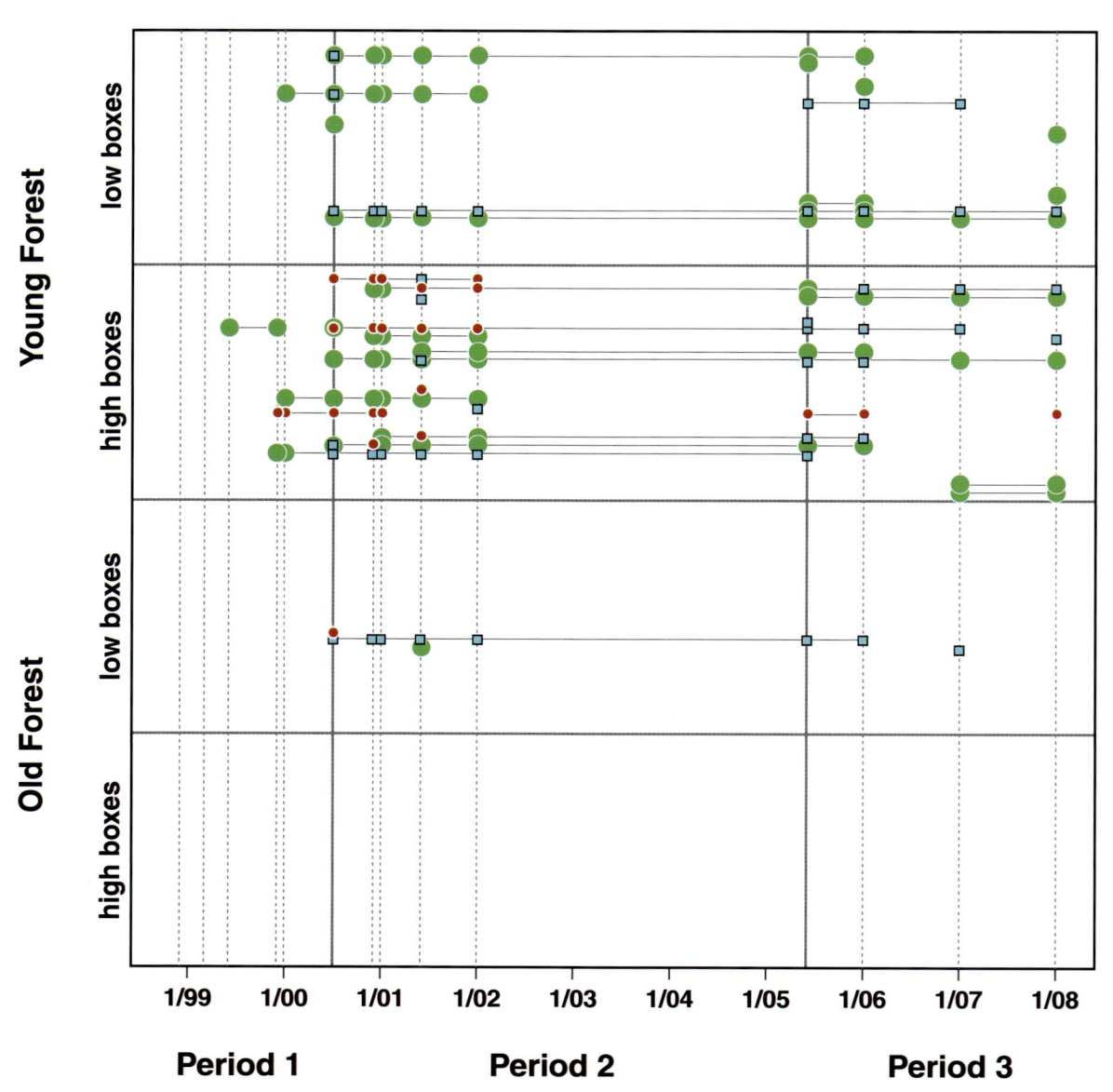

Colour Plate 7: Nest box occupancy by arboreal marsupials over 10 years. Nest boxes are represented horizontally and are grouped by nest box height within each forest age class; vertical lines indicate surveys grouped into three periods. Solid horizontal lines represent continuous evidence of occupancy by any of the three species of arboreal marsupials – Leadbeater's Possum (red), Mountain Brushtail Possum (blue) and Common Ringtail Possum (green).

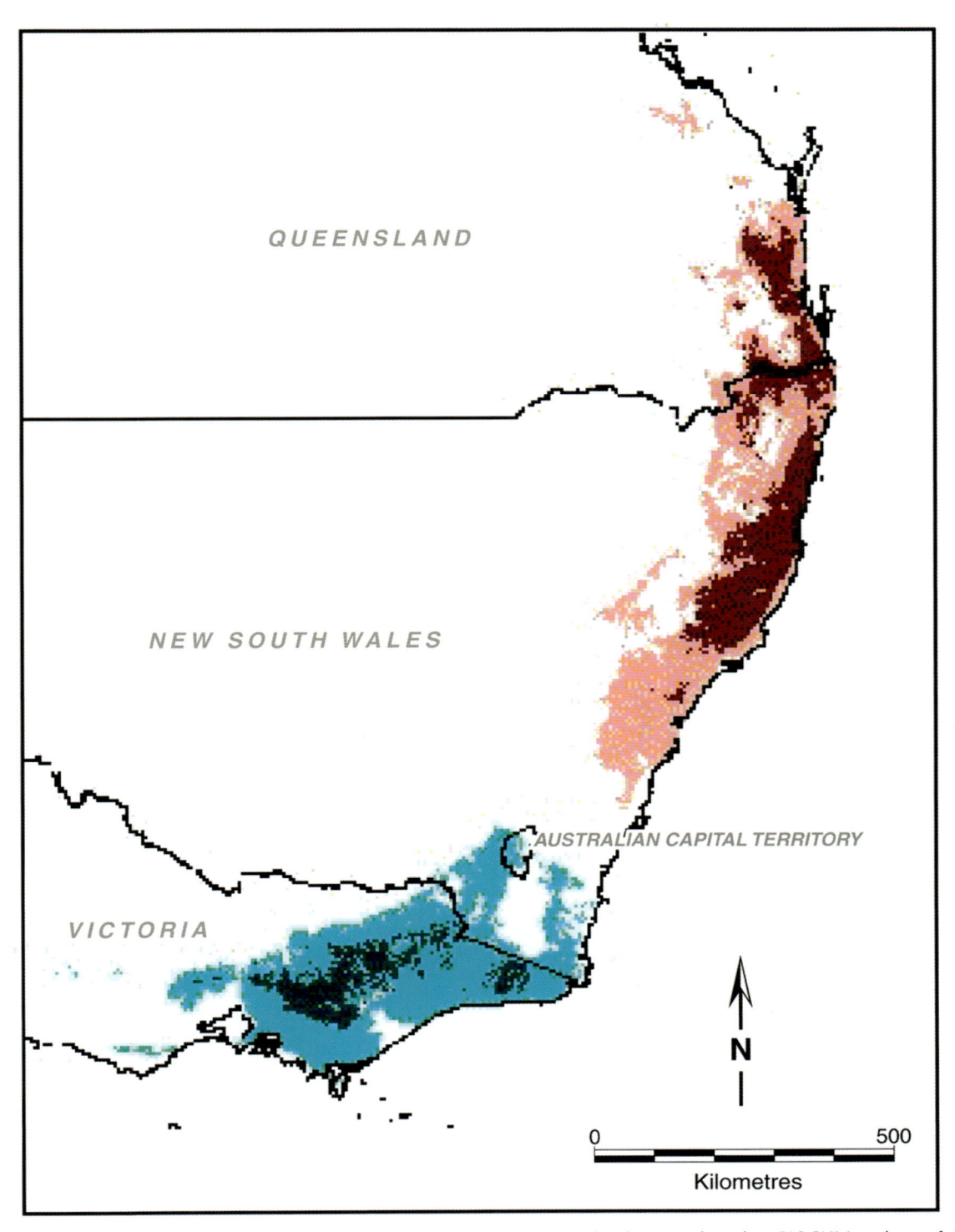

Colour Plate 8: Predicted broad-scale distribution patterns of Mountain Brushtail Possum based on BIOCLIM analyses of the species. The separation of the predicted core and range distributions underscored the likely occurrence of two distinct species. (Redrawn from Fischer *et al.* 2001).

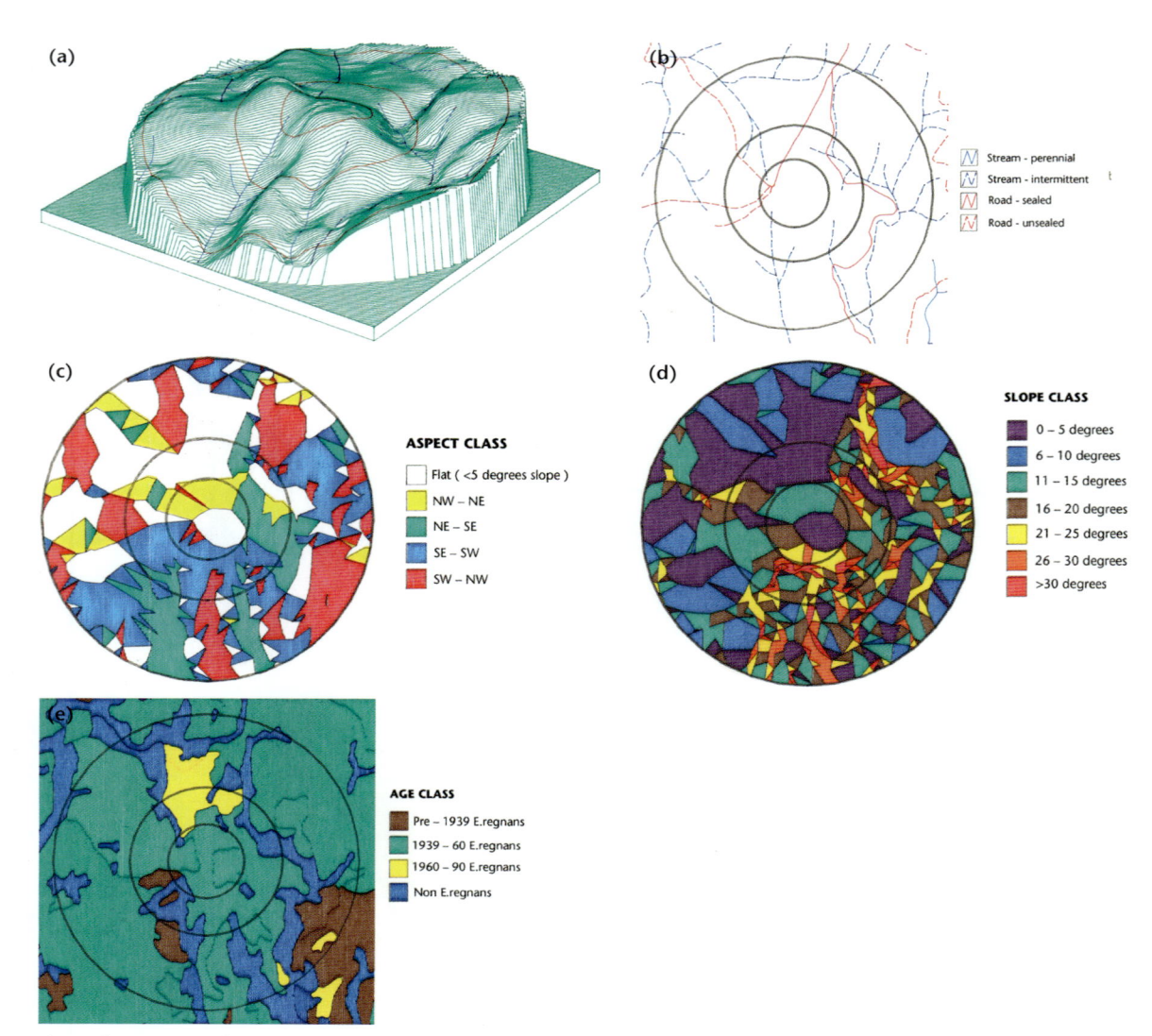

Colour Plate 9: Spatial context of a forest landscape surrounding a field site. Attributes in the polygon were used in statistical analyses of the influence of landscape composition and landscape context on the occurrence of arboreal marsupials. (a) Roads, streams and topography (3D view). (b) Roads and streams (flat view). (c) Forest aspect classes. (d) Forest slope classes. (e) Forest age classes. (Redrawn from Lindenmayer et al. 1999a).

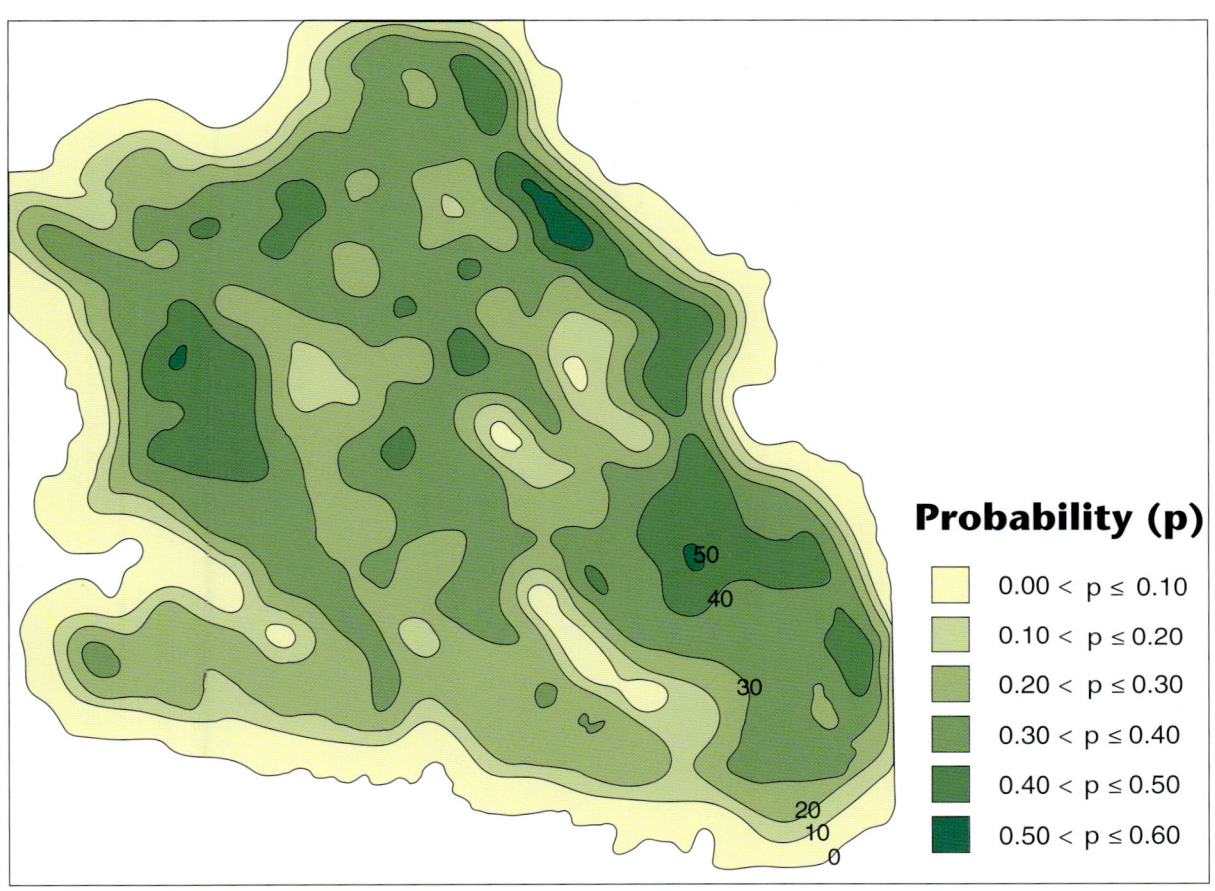

Forest Pattern and Ecological Process | 95

Colour Plate 10: Spatial prediction of the probability of occurrence of the Greater Glider in the Ada Forest Block. (Redrawn from Lindenmayer *et al.* 1995b).

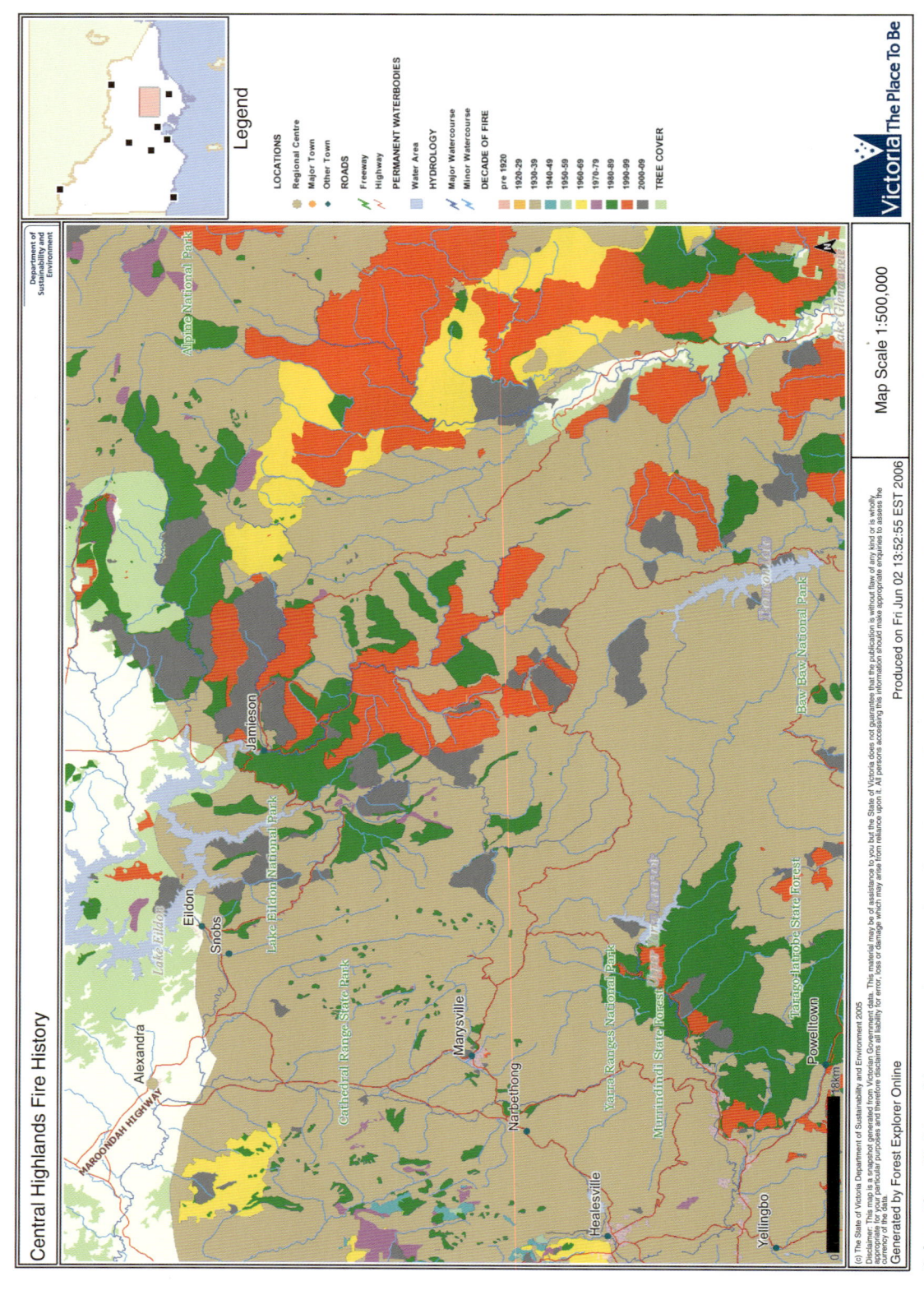

Colour Plate 11: Map of forest age classes resulting from fires in different decades highlights the prevalence of stands dating from the 1939 wildfires. (Map courtesy of Victorian Dept of Sustainability and Environment).

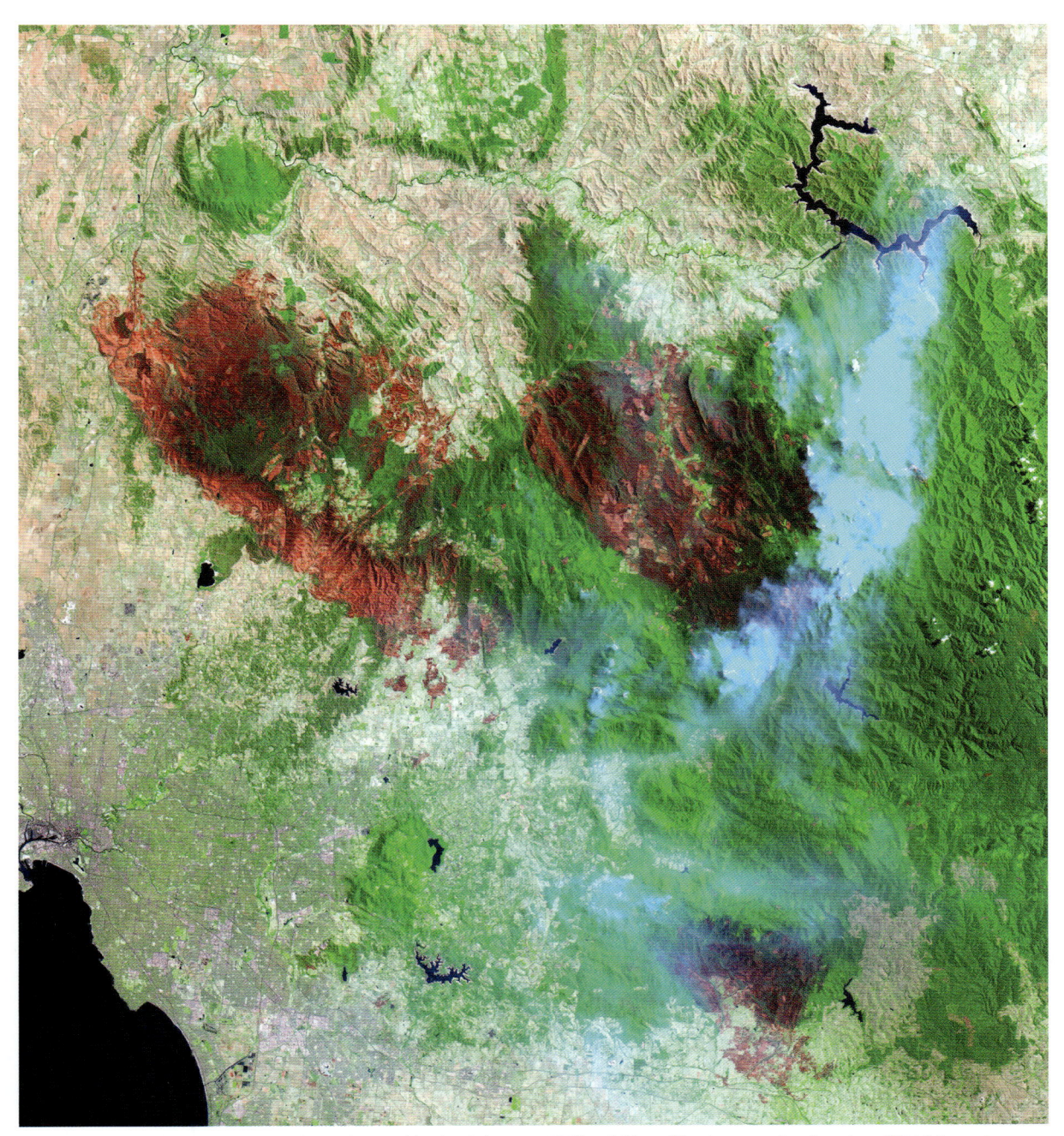

Colour Plate 12: Satellite image of area burned in the February 2009 wildfires. The imagery is a Landsat 5TM false colour composite (RGB = 7:4:3) acquired on 16 February 2009 at 10.49 am. The RGB 743 false colour composite was generated to enhance the contrast between burned and unburned features, combining the near-infrared reflectance decrease with the mid-infrared reflectance increase following wildfire (Epting *et al.* 2005).

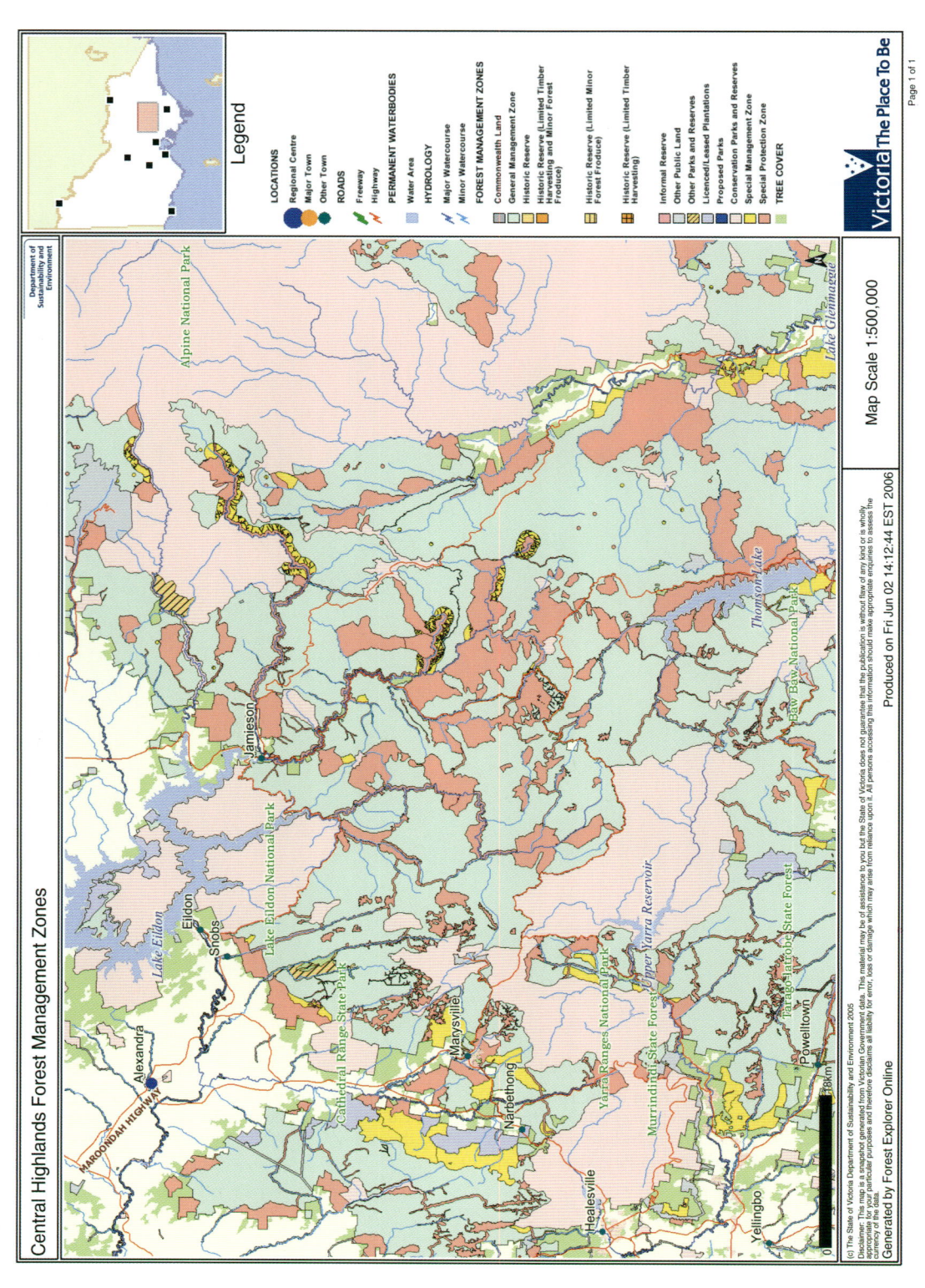

Colour Plate 13: Management zones within the Central Highlands of Victoria. Logging is precluded from special protection areas, special management zones and parks and reserves. (Map courtesy of Victorian Dept of Primary Industries).

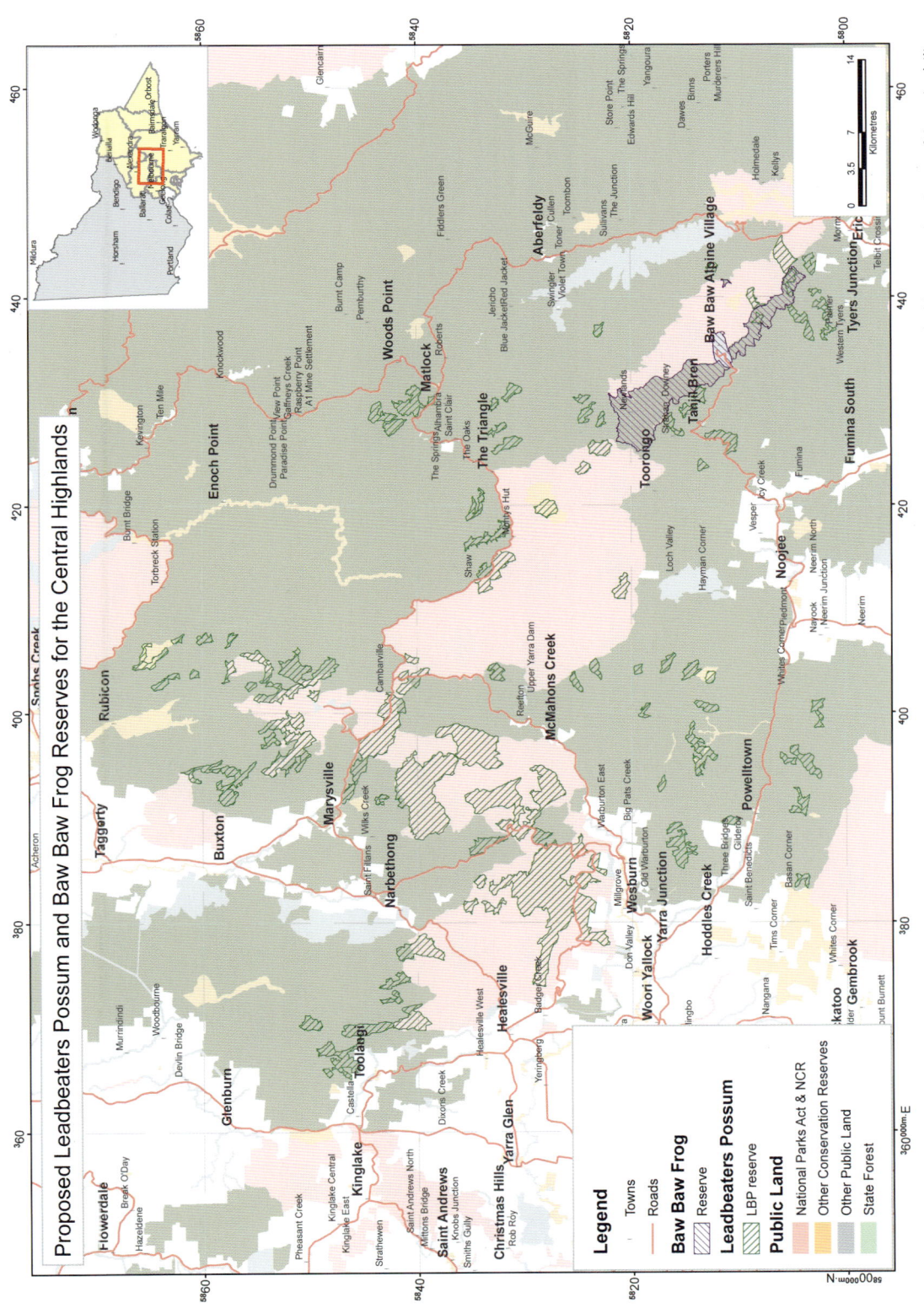

Colour Plate 14: Proposed meso-scale reserves for Leadbeater's Possum and the Baw Baw Frog. (Redrawn from maps provided by Victorian Dept of Sustainability and Environment).

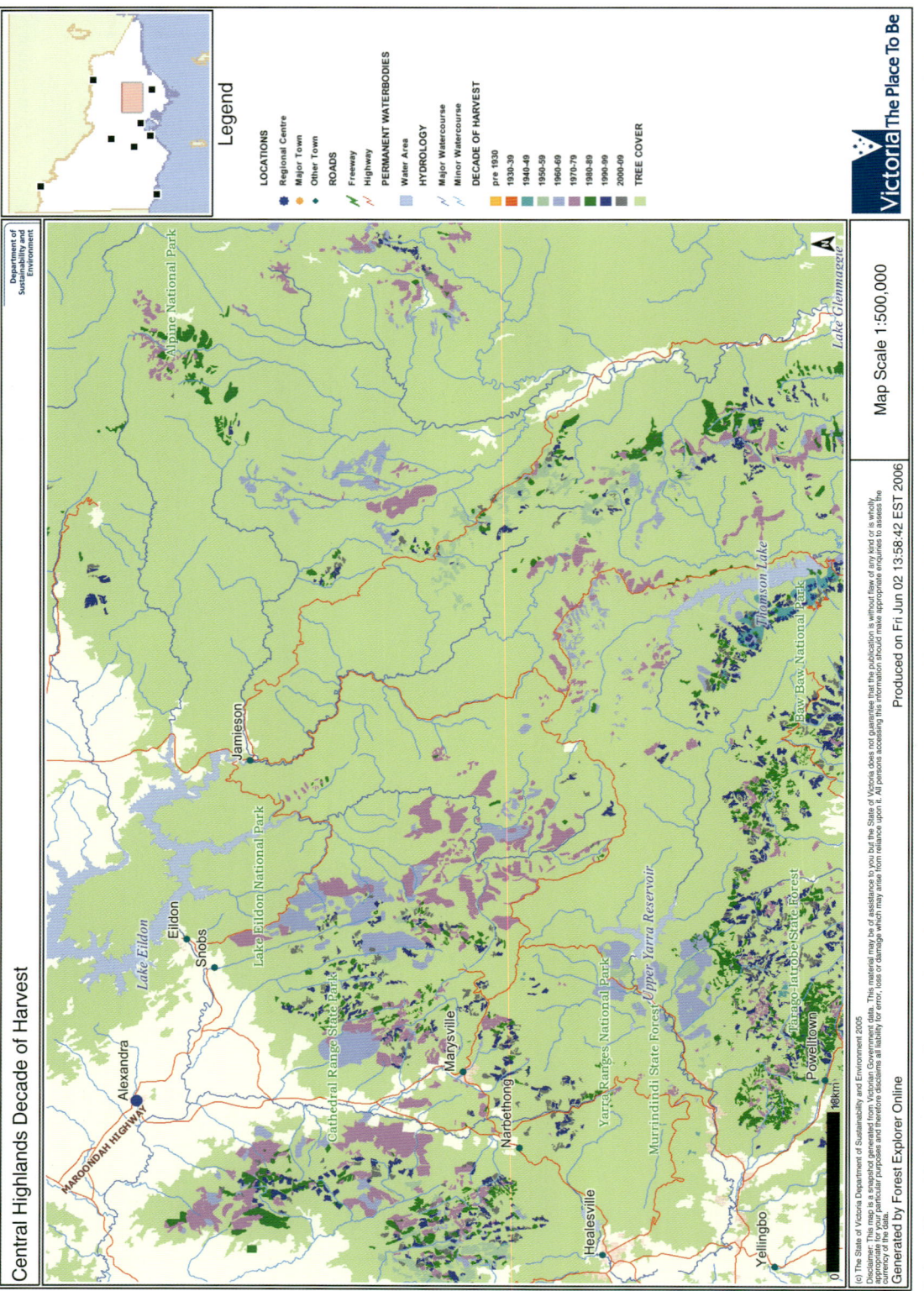

Colour Plate 15: Logged stands in wood production areas in part of the Central Highlands of Victoria. (Data provided by the Victorian Dept of Primary Industries).

The shape of a patch of forest habitat can be a third key reason why not all trees with hollows may be occupied by arboreal marsupials. This was apparent in studies of narrow strips of forest (or wildlife corridors) left uncut within logged montane ash forest (Lindenmayer *et al.* 1993a). It is likely that some species of animals, such as Leadbeater's Possum, had difficulties gathering food in linear-shaped patches of forest. This probably precluded them from inhabiting such sites, even though other habitat characteristics were often suitable (Lindenmayer *et al.* 1993a). Finally, an important factor affecting the occupancy of trees is their spacing. When several hollow trees are close together, they may all occur with the home range of a given colony or pair of animals. Territorial behaviour may exclude other animals from the area, leaving many unoccupied hollow trees (Lindenmayer *et al.* 1990a).

Long-term tree occupancy

Trees with hollows are long-lived entities and may remain standing for prolonged periods, even after they have died (Lindenmayer and Wood 2009). The animals that use them may change as the attributes, morphology and condition of trees change in response to processes of growth, decay and cavity development. In other cases, the same trees with hollows may be used for prolonged periods by the same individuals or different individuals of the same species. For example, detailed studies of radio-collared animals at Cambarville have revealed that the same trees with hollows have been occupied by colonies of Leadbeater's Possum for over 25 years (Lindenmayer 1991; Lindenmayer, unpublished data). Similarly, individuals of the Mountain Brushtail Possum used trees with hollows in 2008 that were occupied by the species in 1990 (Banks *et al.* unpublished data). Because these trees have not been monitored permanently, it is not possible to determine if they have been used on a constant basis. Given the longevity of Leadbeater's Possum and Mountain Brushtail Possum in the wild (about 5–10 years and 7–15 years, respectively), it is highly unlikely that any of the animals now using these trees were those observed many years ago.

Co-occupancy

It is rare in montane ash forests for more than two species of animals to occur in the same nest tree. Indeed, only 10 (<1%) of more than 1100 hollow trees stagwatched between 1983 and 1989 contained more than one species of possum or glider (Lindenmayer *et al.* 1990a). When co-occupancy occurred it was usually between a large and a small species. For example, rare records of co-occupancy of trees with hollows by Leadbeater's Possum were with two other species – Mountain Brushtail Possum and Greater Glider (Table 6.3) (Lindenmayer *et al.* 1990a). The entrances to nests used by co-occupying species are usually different, and typically at different heights above the ground (Lindenmayer *et al.* 1991c).

The processes of cavity development in montane ash trees result in extensive interlinked hollows that

---

> **Box 6.2: Shifting baselines and levels of tree occupancy in montane ash forests**
>
> The concept of 'shifting baselines' has begun to emerge in ecological studies, where management history and other factors have created modified environmental conditions that strongly influence human perceptions of what is 'natural'. The concept is used increasingly in marine research, for example, to highlight historical population sizes of pelagic sharks and the extent of recent declines (Baum and Myers 2004), and the magnitude of recent changes in the cover of coral reefs (Greenstein *et al.* 1998). The concept of shifting baselines is relevant in terrestrial conservation biology and forest ecology and management, including in Victorian montane ash forests. When studies of arboreal marsupials and trees with hollows commenced in 1983, typically one in every three trees was occupied by an animal. This continued until 1993, after which occupancy levels steadily declined so that now one in every five trees is occupied. It is not possible to determine whether initial levels of tree occupancy were inflated well above background or baseline levels, or if current levels of tree occupancy are below a given benchmark. However, the changes in tree occupancy quantified over time indicate that human perceptions of a 'natural' baseline have the potential to change dramatically over time.

Table 6.3: Patterns of co-occupancy among pairs of possum and glider species

Species code	LP	SG	GG	BP	FTG	YBG	RTP
LP	X	0	1	3	0	0	0
SG	–	X	0	0	0	0	0
GG	–	–	X	2	0	0	0
BP	–	–	–	X	0	0	0
FTG	–	–	–	–	X	0	0
YBG	–	–	–	–	–	X	0
RTP	–	–	–	–	–	–	X

The number in each cell represents the incidence of co-occupancy in 1125 different hollow trees.
LP = Leadbeater's Possum; SG = Sugar Glider; GG = Greater Glider; BP = Mountain Brushtail Possum; FTG = Feathertail Glider; YBG = Yellow-bellied Glider; RTP = Common Ringtail Possum. The dashes (–) represent pairs of species for which values are already given in the matrix.
Modified from Lindenmayer et al. (1990b).

connect to the outside of the main trunk at several points (Ambrose 1982). These interconnected hollows mean that animals which inhabit hollow-bearing trees have a good chance of being disturbed by other species. This may explain the comparative rarity of co-occupancy of hollow trees. Co-occupancy of trees is more common in other types of eucalypt forest (Gibbons and Lindenmayer 2002; Mackowski 1987). This may be because discrete self-contained cavities appear more common in the non-ash types of eucalypt trees.

MULTIPLE TREE HOLLOW USE BY ARBOREAL MARSUPIALS: DEN-SWAPPING

Two radio-tracking studies have been completed on the use of trees with hollows by arboreal marsupials.

Figure 6.18: Radio-tracking the Mountain Brushtail Possum at Cambarville. (Photo by Esther Beaton).

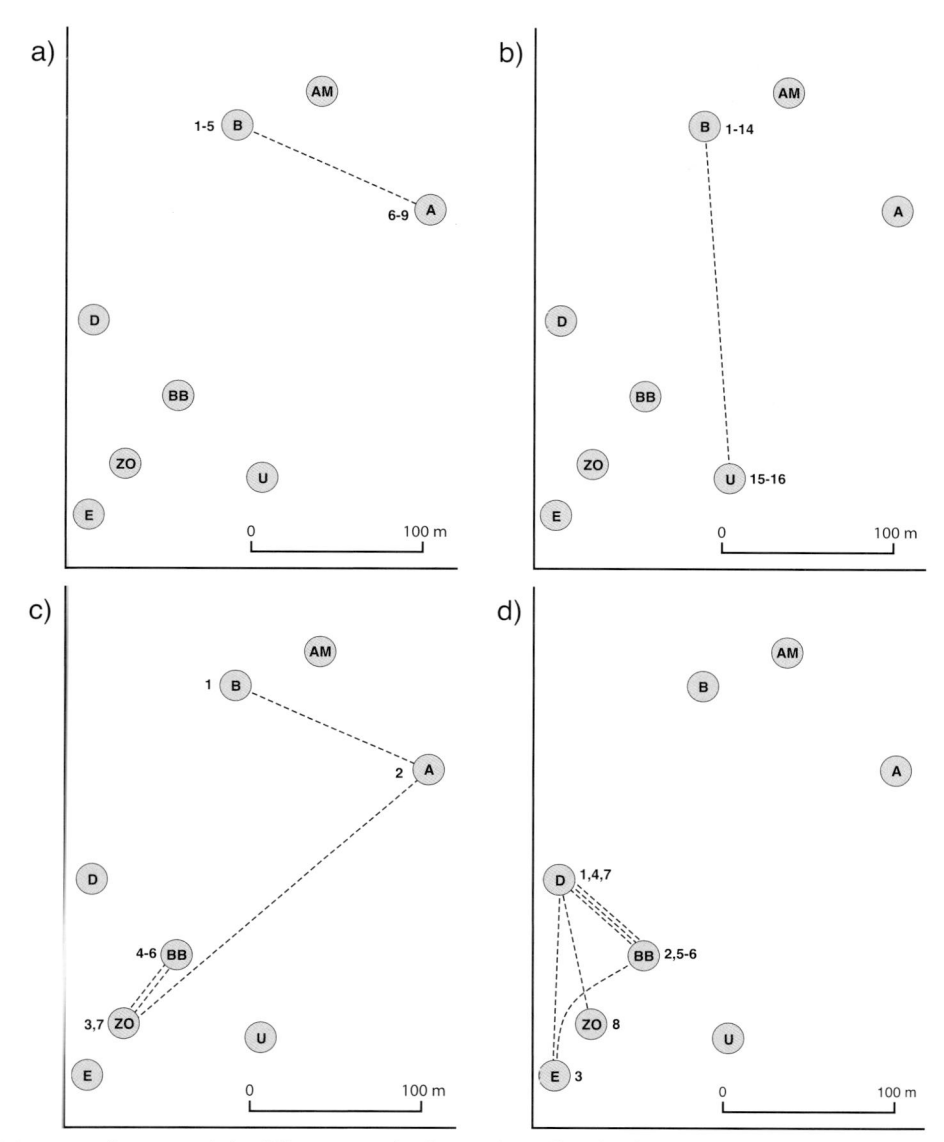

Figure 6.19: Movement between eight different trees by four radio-collared individuals of Leadbeater's Possum in late June and early July 1991. Each den site is assigned a different combination of letters. Lines show movements between different trees. The numbers show the days in which a given animal was in a particular den tree. (Redrawn from Lindenmayer 1996).

Both were undertaken in the mixed-aged forests at Cambarville and focused on the Mountain Brushtail Possum and Leadbeater's Possum, respectively. Both studies showed that animals swapped regularly between different den trees. This type of den-swapping behaviour is common in many species of nocturnal marsupials and is seen in other kinds of cavity-dependent species such as other mammals (e.g. bats and carnivorous marsupials) and birds (Gibbons and Lindenmayer 2002). In the work on Leadbeater's Possum, animals were commonly recorded moving 50–200 m between different den trees on successive nights (Figure 6.19).

The study of den-swapping in the Mountain Brushtail Possum was far longer and more comprehensive than that on Leadbeater's Possum. A total of 16 adults was tracked for 18 months. These animals used an unexpectedly large number of trees with

Table 6.4: Number of trees with hollows used by individual Mountain Brushtail Possums at Cambarville

Animal no.	Gender	Trees used	Days tracked	Trees/day tracked[a]
1	M	18	151	0.12
2	F	5	164	0.03
3	M	23	166	0.14
5	F	17	161	0.11
6	M	6	31	0.19
7	M	13	124	0.11
9	M	8	139	0.06
10	F	16	129	0.12
11	M	6	38	0.16
12	M	13	133	0.10
13	F	7	125	0.06
14	M	6	85	0.07
15	F	7	69	0.10
16	M	12	85	0.14
32	F	17	142	0.12
46	M	12	154	0.08

a Calculated by dividing column 3 by column 4
Modified from Lindenmayer *et al.* (1996c).

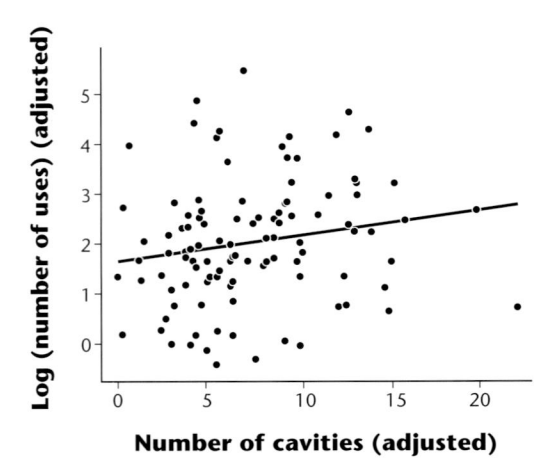

Figure 6.20: Relationships between the number of uses of a tree with hollows by a radio-collared Mountain Brushtail Possum and the number of cavities in the tree. (Redrawn from Lindenmayer *et al.* 1996d).

hollows as den sites (113 trees), with some individuals using more than 20 different den sites, moving between a number of different trees in any given week (Lindenmayer *et al.* 1996c). Despite the extensive nature of the investigation, the number of occupied trees did not reach an asymptote by the conclusion of the study (Lindenmayer *et al.* 1996c).

Although a large number of trees were occupied by the Mountain Brushtail Possum at Cambarville, each animal appeared to have a few primary den sites (typically 1–5 trees) where it spent most of its time (Lindenmayer *et al.* 1996c). Analyses of data aggregated for all individuals revealed that den trees used most frequently were characterised by particular morphological features, such as the abundance of cavities and the density of the surrounding vegetation (Lindenmayer *et al.* 1996d) (Figure 6.20). These characteristics may reflect important attributes of trees such as the internal thermal properties of stems, the size of the internal chamber of trees and the ease of access to, and movement away from the nest site to the surrounding forest (Lindenmayer *et al.* 1996d). Thus, the quality of den sites presumably influenced the extent of their use by the Mountain Brushtail Possum.

Additional analyses examined factors influencing the movements of collared individuals of Mountain Brushtail Possum between den trees on successive days (Welsh *et al.* 1998). This work revealed that inter-tree movements were individual-specific. That is, each animal exhibited different patterns of transition between trees. Movement patterns were more complex in older animals than in younger ones, possibly because of attempts to gain access to more mates.

Another surprising outcome was the prevalence of den-sharing, with instances of up to six animals per tree (Lindenmayer *et al.* 1997a). Den-sharing was common among males but rarely recorded for females (Welsh *et al.* 1998). The extent of den-sharing could be related to population density and may be more common on sites supporting large numbers of animals, as at Cambarville where densities appear to be approximately three times higher than at other places where the species has been studied (e.g. northern New South Wales) (How 1972).

Possible reasons for den-swapping

The reasons for den-swapping behaviour in arboreal marsupials are not known but some plausible explanations are as follows.

Table 6.5: Instances of co-sharing of dens by radio-collared individual Mountain Brushtail Possums

Animal no.	Gender	No. and identity of other co-sharing animals
1	M	8 (2, 5, 6, 7, 10, 11, 32, 46)
3	M	1 (14)
6	M	9 (1, 2, 5, 7, 9, 10, 11, 32, 46)
7	M	6 (1, 2, 6, 9, 32, 46)
9	M	4 (2, 6, 7, 32)
11	M	5 (1, 2, 6, 10, 12)
12	M	2 (11, 16)
14	M	1 (3)
16	M	2 (10, 12)
46	M	5 (1, 2, 6, 7, 32)
2	F	8 (1, 6, 7, 9, 10, 11, 32, 46)
5	F	3 (1, 6, 32)
10	F	5 (1, 2, 6, 11, 16)
13	F	–[a]
15	F	–[a]
32	F	7 (1, 2, 5, 6, 7, 9, 46)

[a] A dash indicates that a given animal did not share a den site with any of the other radio collared individuals tracked during the study.
Numbers in parentheses are the identities of co-sharing animals. The information includes data from pairs as well as groups of three and four animals.
Redrawn from Lindenmayer et al. (1997a)

- If an animal regularly changes nest trees, potential predators such as large forest owls may be less likely to detect its nest sites and/or learn its patterns of emergence behaviour as it leaves the hollow.
- Species such as Leadbeater's Possum and the Sugar Glider construct a nest of bark strips and leaves. Nests are often infested with large numbers of parasites such as ticks, fleas and mites (Lindenmayer et al. 1994e). Changing nest trees may reduce parasite burdens. In some situations, frequent den-swapping may provide a mechanism for transporting parasites between trees.
- Environmental conditions may vary between different trees with hollows. For example, dead trees often contain large amounts of rotten wood that produce considerable heat – as occurs in a typical backyard compost heap. Dead trees may be favoured nest sites in winter when ambient climate conditions are cold and additional heat from the 'compost' inside the stem helps animals stay warm. In summer, when animals are susceptible to overheating, they may prefer living trees that are insulated by water flowing in the conducting tissues.
- Animals often scent-mark a range of different den trees, perhaps as a signal that a set of den trees and/or a territory is occupied. This may deter individuals of the same species from transgressing territory boundaries and preclude rivals from using particular den trees.
- Trees with hollows undergo constant morphological change. Small cavities in living trees may be occluded as the wounds are filled by callus tissue produced by a tree. Conversely, the activities of termites, fungi and bacteria steadily enlarge cavities. Changes in the size of a cavity will alter its suitability for occupancy, and animals may change den sites in an effort to find the best nesting conditions.
- The food resources within an animal's territory may be depleted or their spatial location may vary between seasons. It may be more energy-efficient for a den site to be located close to available food resources, making it necessary to have several different nest sites scattered across a home range.
- Movement between different den sites may allow an animal to increase familiarity with its territory, thereby enabling it to better escape predation or detect congeners crossing territory boundaries. It may also enable an animal to border the territories of more animals, providing increased opportunities to find a mate or, in males, gain paternity of more offspring.
- Different den sites may have different functions. For example, young animals from a colony may be creched in particular trees, as is thought to occur in social species such as the Sugar Glider. Adult females of the Yellow-bellied Glider frequently return to trees where their young had been living. Similarly, a genetic study of the Greater Glider showed that animals that shared dens were often closely related, particularly females and their recent offspring.

ARTIFICIAL HOLLOWS: STUDIES OF NEST BOXES IN MONTANE ASH FORESTS

It is increasingly recognised that there is a shortage of large trees with hollows in many forests worldwide. One solution has been to install artificial cavities such as nest boxes (Beyer and Goldingay 2006; Harper *et al.* 2005; Newton 1998), which have sometimes been remarkably successful. Nest boxes added to forests in Germany throughout the 1950s resulted in a five- to 20-fold population increase in some bird species (Bruns 1960). Artificial cavities have resulted in other spectacular population recoveries of birds such as the Pied Flycatcher (*Ficedula hypoleuca*) in Finland (von Hartman 1971), three species of Bluebirds (*Sialia* spp.) and the Wood Duck (*Aix sponsa*) in North America (Haramis and Thompson 1985). Nest boxes have been added to logged forests (where trees with hollows had been removed), with significant recoveries of populations of some cavity-dependent species (Smith and Agnew 2002; Taulman *et al.* 1998).

Given the impending shortage of trees with hollows in montane ash forests, coupled with the loss of these trees on clearfelled sites, studies of the effectiveness of nest boxes have been underway for over a decade (Lindenmayer *et al.* 2003c, 2009c). Three key areas have been examined – occupancy by arboreal marsupials, infestation by pest invertebrates, and nest box attrition. Two areas – the Powelltown State Forest and Toolangi State Forest – were chosen for our study. The dominant age classes of forest at Toolangi and Powelltown are 68-year-old stands recovering after the 1939 wildfires and those 20 years or younger that are regenerating after high-intensity clearfell logging. Six sites were established in each of these two forest age classes within both the Toolangi and Powelltown State Forests. A square of 20 × 20 m was established at each site and a nest box was attached to a tree on each of the four vertices of the square. Two nest box sizes (small and large) and two location heights (3 m and 8 m above the ground) were chosen for comparison. Thus, the complete study involved eight combinations of

Figure 6.21: A large, high nest box being checked during a study of vertebrate occupancy patterns in montane ash forests. (Photo by Esther Beaton).

forest age, nest box size and location height in the design, giving a total of 2 × 6 × 8 = 96 nest boxes.

Nest box occupancy

A large proportion of nest boxes (68.8%) were never used, even by the endangered target species, Leadbeater's Possum. Nest boxes were more often occupied in younger forests (58.3% compared with 4.1% for old forest) (Colour Plate 7), but these sites were where nest box infestations by the introduced European Honeybee were greatest and the probability of nest boxes collapsing was highest. In young forest, there was a weak effect of box height; and high nest boxes had a higher probability of being occupied than those closer to the ground (Lindenmayer et al. 2009c).

All but two instances of nest box occupancy were in young forest. That is, 46 of 48 nest boxes in old forest (= 95.8%) remained unoccupied for the entire 10 years of the investigation. The most plausible explanation was the paucity of naturally occurring trees with hollows in young forest (i.e. 20-year-old logged and regenerated stands) compared with old forest (i.e. stands burnt in 1939). The average number of trees with hollows within 40 m of each nest box in younger forest was 0.33 compared with 1.25 in older forest. Large-cavity trees in these older forests were trees persisting after past fires and/or logging and were old-growth stems of earlier old-growth forests on these sites (Lindenmayer et al. 1991b). The relative paucity of trees with hollows in young forest would mean that animals had few, if any, alternative nesting sites to the nest boxes. This finding is consistent with several nest box studies conducted elsewhere in the world (reviewed by Newton 1994).

The findings suggest that it is important to carefully target the locations where nest boxes should be established. However, these places are also those where most effort is required to maintain them, either by removing pest invertebrates and/or by reattaching them to trees after they have fallen.

Effective occupancy time and nest box effectiveness

Reasonable levels of occupancy by arboreal marsupials were recorded in young forest ~2–3 years after box establishment, but high levels of nest box attrition (>51%) occurred after 8–10 years. This suggests an effective occupancy time for arboreal marsupials of only ~5 years. The relatively limited effective occupancy time suggests that many such replacements might be required if a management aim is to provide a perpetual supply of potential cavities for ~120–150+ years until sufficient naturally occurring hollows develop in Mountain Ash trees, suitable for hollow-using arboreal marsupials (Ambrose 1982; Lindenmayer et al. 1993b). Such an undertaking would clearly have substantial logistical and financial implications (McKenney and Lindenmayer 1994). The work highlights the need to weigh the ecological and economic effectiveness of nest boxes against other strategies, such as implementing alternative silvicultural systems that result in the creation and retention of more hollow-bearing trees within logged areas.

An additional nest box study

A sobering outcome of the initial nest box study was discovering the low overall rates at which they were occupied by vertebrate species. This was despite the fact that a high proportion of the nest boxes were designed specifically for an endangered species (Leadbeater's Possum), following a design that had high rates of occupancy by that species in the lowland Swamp Gum (*Eucalyptus ovata*) dominated Yellingbo Nature Reserve (Harley 2004) outside the Central Highlands of Victoria. It was hypothesised that other designs might have returned higher occupancy rates. As a consequence, in 2001 a separate study with 96 additional nest boxes was established. Results of that work are difficult to formally report as, after seven years, only two detections of Leadbeater's Possum have been recorded (Lindenmayer et al. 2009c).

SUMMARY

Trees with hollows are a fundamental part of the structure of stands of montane ash forests. These trees play many key ecological roles and are an essential habitat attribute for a wide range of species, including the high-profile and endangered Leadbeater's Possum. Many investigations of trees with hollows have been completed and several others are ongoing. Indeed, it seems plausible that trees with hollows in montane ash forests have been subject to more study than similar kinds of trees in any other forest worldwide. The work has spanned the lifecycle of these trees from the ontogeny of cavity development and recruitment of trees with hollows, through the decay patterns of trees with hollows to their eventual collapse to the forest floor. Because these are long-term processes, the study is an ongoing effort that has spanned the entire duration of the 25-year research effort in the Central Highlands of Victoria (Lindenmayer and Wood 2009).

An additional key research theme has been the quantification and testing of relationships between the occurrence of trees with hollows and the presence and abundance of cavity-dependent arboreal marsupials. Work has attempted to determine the reasons for such strong relationships, which appear more pronounced than in many other types of eucalypt forest around Australia. The primary (and often related) reasons for such relationships appear to be:

- partitioning of the nest tree resource between different species of arboreal marsupials (Lindenmayer *et al.* 1991c, 1991e);
- selection of trees with different characteristics by different species, which makes a subset of trees with hollows on a given site unsuitable for occupancy by particular taxa (Lindenmayer *et al.* 1990a);
- frequent den-swapping that results in the use of multiple den trees by a given individual animal (Lindenmayer and Meggs 1996; Lindenmayer *et al.* 1996c).

There was some evidence that the trends observed for usage patterns of trees with hollows were reflected in the usage patterns of nest boxes by arboreal marsupials. For example, nest boxes were significantly more likely to be used in young forests where naturally occurring cavities were uncommon (Lindenmayer *et al.* 2009c) and particular nest box attributes such as entrance size and height above the ground influenced the probability of occupancy (Lindenmayer *et al.* 2003c, 2009c).

LESSONS LEARNED

Many lessons have been learned and many new insights gained from the extensive body of research on trees

---

Box 6.3: Postscript – tree hollows, nest boxes and the 2009 wildfires

Wildfire is a double-edged sword for tree hollows and hollow-dependent fauna. Fire can promote the development of cavities in trees (Inions *et al.* 1989), but many trees with hollows can be destroyed by wildfire. The 2009 Black Saturday wildfires are likely to exacerbate the projected shortage of trees with hollows in montane ash forests, both reducing the population size of current trees and setting back the time until recruitment of new trees. The 2009 wildfires therefore highlight a need to re-examine relationships between animal abundance and the availability of trees with hollows on burned sites. The examination of such relationships can be based on the use of long-term monitoring data to make comparisons between pre- and post-fire populations of animals. Like trees with hollows, nest boxes are often destroyed by wildfire. For example, almost all the nest boxes established for Leadbeater's Possum at Lake Mountain were destroyed on Black Saturday (as were the animals inhabiting them). Lake Mountain was severely burned and few areas of unburned vegetation remain. Hence, there may be value in establishing new nest boxes in places at Lake Mountain where they were previously located, to provide potential nesting sites for species such as Leadbeater's Possum.

---

with hollows in montane ash forests. Perhaps the most sobering has been the recognition that these kinds of trees will become an increasingly scarce resource in future decades, with the potential for significant corresponding negative impacts on cavity-dependent biota. It is also clear that the shortage of trees with hollows in these forests will be prolonged in the extensive areas of 1939-aged regrowth forest that comprise the vast majority of stands in the Central Highlands of Victoria (Lindenmayer *et al.* 1997b; Lindenmayer and Wood 2009). Traditional forms of clearfell logging will further accelerate the loss of trees with hollows, not only those which are retained on cut blocks but in the adjacent areas of unlogged forest, most likely through exposure to altered microclimate conditions.

It seems unlikely that the provision of artificial cavities (nest boxes) will significantly alleviate the upcoming paucity of trees with hollows in 1939-regrowth forest – almost none of the nest boxes established in those areas were occupied over the decade-long period they were available for use (Lindenmayer *et al.* 2009c). The nest box program has proved a major logistical exercise that is expensive to maintain. Replacement and re-attachment were common tasks, as were clearing bees and ants from the boxes.

These costs, together with the relatively low levels of nest box occupancy, emphasise the importance of other strategies for providing resources for cavity-dependent species in forests, particularly retaining and perpetuating mature trees on logged sites (McKenney and Lindenmayer 1994). Strategies such as altered silvicultural systems and nest box programs can be complementary rather than competing approaches to promote the conservation of cavity-dependent forest wildlife (Lindenmayer *et al.* 2009c). In addition, the instigation of nest box programs can have other values such as the engagement of community groups in conservation. Nest box studies over the past decade have involved the participation of several hundred volunteers.

KNOWLEDGE GAPS

Although more work has been conducted on trees with hollows in montane ash forests than perhaps any other forest ecosystem worldwide, key knowledge gaps remain. The most prominent gap concerns the potential for accelerated cavity development in montane ash trees. This is important because it is clear that there will be a prolonged shortage of trees with hollows in the extensive areas of regrowth ash-type forest in the Central Highlands of Victoria (Lindenmayer and Wood 2009). It is also clear that nest boxes are likely to be of limited value for cavity-dependent vertebrates in regrowth forests (Lindenmayer *et al.* 2009c).

Considerable work has been conducted in North America on methods to accelerate the development of hollows in trees (Rose *et al.* 2001). For example, killing live trees to promote the creation of trees with hollows is a common silviculture practice (Hutto 1995). Another strategy is to apply pheromones to attract bark beetles and other decay-promoting insects to trees (Bull and Partridge 1986). Mechanically girdling trees at their base is another technique (Baumgartner

Figure 6.22: Topping live green trees is one method of accelerating the development of trees with hollows in the wet conifer forests of western North America. (Photo by Jerry F. Franklin).

1939; Hennon and Loopstra 1991; Moriarty and McComb 1983), although tree death may not occur for several years. Removing the entire canopy from live trees (topping) can be accomplished with explosives or chainsaws (Bull and Partridge 1981; Chambers *et al.* 1997; Sanderson 1975).

There has been no work in Australia equivalent to that on accelerated tree hollow development in North American forests. Hardwood eucalypts may differ from conifers in the ways and rates at which hollows develop. For example, Australia lacks cavity-excavating species such as woodpeckers, which are important for hollow formation on other continents (Gibbons and Lindenmayer 2002). Therefore, ways to accelerate hollow formation in eucalypts, including ash-type tree species, remains an important area of research in this country.

Another significant knowledge gap is the nature of animal responses to dwindling numbers of trees with hollows in the extensive areas of 1939-regrowth montane ash forest. Observational datasets on animal occurrence, coupled with predictions of tree fall, suggest that animal numbers should decline dramatically in the coming decades (Lindenmayer *et al.* 1997b). However, monitoring data suggest that this has not happened. The prevalence of den-swapping may have diminished in response to the reduced numbers of trees with hollows. A recently commenced radio-tracking study of the Mountain Brushtail Possum at Cambarville aims to determine if this is indeed the case (Banks *et al.* unpublished).

Finally, parts of overstorey trees other than cavities require attention from researchers. For example, very little is known about canopy invertebrates and how these assemblages may change between the pyramid-shaped crowns of regrowth trees and the more open crowns of old-growth stands.

FOREST PATTERN, ECOLOGICAL PROCESS AND LINKS TO OTHER CHAPTERS

The distribution of large trees with hollows is an important spatial pattern in montane ash forests. These trees are readily distinguished from other smaller stems that do not contain easily observable cavities. The occurrence of these trees is influenced by processes such as climatic conditions and disturbance regimes, both human and natural (Lindenmayer *et al.* 1991b, 1993b). Their pattern of occurrence strongly influences other patterns such as the occurrence of different species of arboreal marsupials (Lindenmayer *et al.* 1994b), as well as fine-scale patterns like den-swapping behaviour (Lindenmayer *et al.* 1996c) and nest tree partitioning (Lindenmayer 1997). The pattern of occurrence of trees with hollows also affects key ecological processes such as the recruitment of coarse woody debris to the forest floor (Lindenmayer *et al.* 1999), as well as the processes of carbon storage and nutrient cycling (Keith *et al.* 2009).

Links to other chapters

The preceding discussion on ecological processes and forest patterns indicates that the body of work on trees with hollows is strongly interlinked with a wide range of other research themes in montane ash forests. Trees with hollows are a key part of the habitat requirements of many species of vertebrates in montane ash forests and therefore play an important role in shaping the distribution and abundance of these taxa (Lindenmayer *et al.* 1991d, 1990c) (Chapter 9). However, other attributes of stand structure, such as the availability of understorey vegetation, can significantly influence animal occurrence (Lindenmayer *et al.* 1991d, 1990a). Therefore, the material in this chapter links strongly with Chapter 7 on understorey as a combination of structural features that create suitable habitat for some animal species. The understorey of a forest can influence trees with hollows in other ways. For example, rates of tree fall are accelerated for stems with dense surrounding understorey vegetation (Lindenmayer and Wood 2009). This is possibly because of shading, which promotes tree decay and leads in turn to an increased probability of collapse of trees with hollows (see Chapter 7).

The dynamics of populations of trees with hollows was an integral part of the parameter set used in developing computer simulation models for analyses of the viability of populations of arboreal marsupials (see Chapter 10).

Collapsed trees with hollows add to the architecture of the forest floor, contributing to the large volumes of coarse woody debris that characterise many areas of montane ash forest (Lindenmayer *et al.* 1999d) – a topic that forms part of the discussion of the logs in Chapter 8.

Logging has a major impact on the structure of stands of montane ash forest, including reducing the

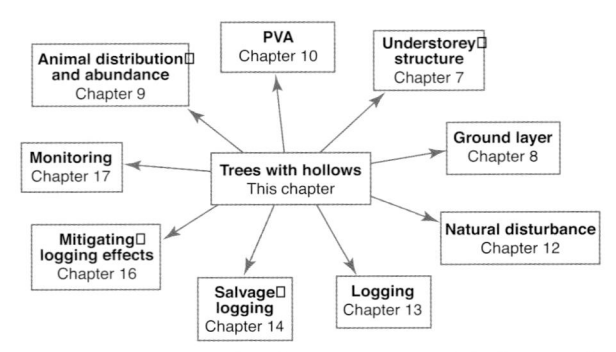

Figure 6.23: Relationships between trees with hollows, and key topics in other chapters.

abundance of trees with hollows (Lindenmayer and Franklin 1997). Such effects are a prominent topic in Chapter 13, which summarises the impacts of timber harvesting.

The abundance of trees with hollows forms a key part of the zoning system for Leadbeater's Possum in wood production montane ash forests (Lindenmayer and Cunningham 1996b; Macfarlane *et al.* 1998). This is discussed in detail in the management section of this book (Part VI, Chapter 15).

Considerable effort has been dedicated to quantifying the rate of decay and collapse of trees with hollows over the past 25 years (Lindenmayer *et al.* 1997b, 1990b; Lindenmayer and Wood 2009). Animal populations have been monitored on a subset of sites where tree fall rates have been documented on a repeated basis. Part of Chapter 17, on monitoring, discusses how the long-term work on the ecological process of tree fall has been linked with the ecological pattern of animal abundance, particularly during the past decade.

As discussed earlier in this chapter, rates of tree fall appear to have slowed in the past 15 years. A possible reason could be the considerable drying of climate conditions during that time (McAlpine *et al.* 2007). Therefore, the topic of dynamics of populations of trees with hollows is potentially linked with that of climate change, discussed in Chapter 18.

7 Key structural features: understorey trees and the shrub layer

INTRODUCTION

The understorey and shrub layers are critical vegetation layers of montane ash forests, both for biodiversity conservation and the maintenance of key ecosystem processes. Limited crown development in montane ash trees, partially as a result of crown shyness, together with the isolateral leaf form, enable high levels of light to penetrate to the forest floor. This, in turn, allows luxuriant understorey and shrub layers to grow (Jacobs 1955). These layers play many important ecological roles (see Table 7.1).

This chapter briefly summarises some of the research that has directly or indirectly examined the understorey and shrub layers of montane ash forest. Most of this work has entailed extensive and intensive vegetation surveys in which the height, diameter and species of all understorey and shrub species were carefully measured at literally thousands of vegetation

Table 7.1: Ecological roles of the understorey in montane ash forests

Role	Reference
Contributing significantly to overall species richness	Mueck (1990)
Storing large amounts of biomass carbon	Mackey *et al.* (2008)
Providing food resources and habitat for vertebrates	Seebeck *et al.* (1984); Smith (1984b)
Providing food resources and habitat for invertebrates	Woinarski and Cullen (1984)
Contributing to nutrient cycling	Ashton (1976); Adams and Attiwill (1984)
Serving as nursery sites for other plants (e.g. tree ferns)	Duncan and Isaac (1986); Ough (2002)
Acting as nesting and perching sites and movement routes for birds and other animals	Smith (1980); Loyn (1985)
Providing microhabitats for the development of fungi which are food resources for animals	Lindenmayer *et al.* (1994d)
Contributing to vertical canopy diversity, thereby increasing the range of foraging substrates for birds and bats	Lindenmayer *et al.* (unpublished data); Brown *et al.* (1997)

Figure 7.1: Well-developed understorey and shrub layers in a stand of Mountain Ash. (Photo by Esther Beaton).

plots throughout the Central Highlands region (Lindenmayer *et al.* 2000a). These datasets have underpinned many analyses. The ones which feature prominently in this chapter are studies of:

- relationships between the understorey and shrub layers and the occurrence of vertebrates;
- relationships between the understorey and shrub layers and stand age classes and disturbance regimes (wildfire and logging).

BROAD COMPOSITION OF THE UNDERSTOREY AND SHRUB LAYERS IN MONTANE ASH FORESTS

The understorey and shrub layers support several hundred species of plants (Mueck 1990) and therefore contribute significantly to the overall species richness of montane ash forests. Understorey trees in montane ash forests include Silver Wattle (*Acacia dealbata*), Mountain Hickory Wattle (*Acacia obliquinervia*) and Forest Wattle (*Acacia frigiscens*). These plants can be relatively tall, exceeding 20 m in height (Adams and Attiwill 1984). Two key species of cool temperate rainforest are also prominent in the understorey layer of montane ash forests, particularly in gullies and old-growth stands (Lindenmayer *et al.* 2000b). These are Myrtle Beech (*Nothofagus cunninghamii*) and Southern Sassafras (*Atherosperma moschatum*). Because of the importance of cool temperate rainforest from an ecological perspective (Burgman and Ferguson 1995) an entire chapter was dedicated to discussing the work conducted on it during the past 25 years (see Chapter 5). Thus, the contribution of cool temperate rainforest to the understorey of montane ash forest is given only limited attention in this chapter.

The shrub layer in montane ash forest varies from 2 m to 15 m in height and supports a diverse array of plants. These species include Soft Tree Fern, Rough Tree Fern, Musk Daisy Bush, Dusty Daisy Bush, Hazel

Table 7.2: Studies directly or indirectly examining the understorey and/or shrub layers in the montane ash forests of the Central Highlands of Victoria

Study description	Reference
Arboreal marsupial occurrence and the understorey and shrub layers	Lindenmayer et al. (1990c); Lindenmayer et al. (1991d); Lindenmayer et al. (1994d)
Small terrestrial mammal occurrence and the understorey and shrub layers	Lindenmayer et al. (1994c); Cunningham et al. (2005)
Tree ferns and animal occurrence	Lindenmayer et al. (1994c)
Bird occurrence and the understorey layer	Lindenmayer et al. (2009b)
Understorey and shrub layers in old-growth forest and stands of different age classes	Lindenmayer et al. (2000a)
Sugar and nitrogen content of gums of understorey *Acacia* spp. trees	Lindenmayer et al. (1994a)

Pomaderris, Mountain Correa, Dogwood, Tree Geebung, Stinkwood, Victorian Christmas Bush and Austral Mulberry (Costermans 1994; Mueck 1990; Ough and Ross 1992).

The understorey may be the dominant layer of vegetation in montane ash forest because the overstorey of ash-type eucalypts may be missing. Where natural disturbances have been absent for a prolonged period (exceeding several centuries), the overstorey eucalypts may be replaced by Myrtle Beech-dominated cool temperate rainforest (Gilbert 1959), although not all parts of landscapes are suitable for the development of these kinds of forests (Lindenmayer et al. 2000b) (see Chapter 5). Conversely, if there is a rapid sequence of high-severity wildfires and young regenerating trees are reburned before they reach sexual maturity, ash-type forest may be lost from a stand and replaced by a dense stocking of *Acacia* spp. trees.

BACKGROUND DATASETS

A wide range of measurements of the understorey and shrub layers were made as part of extensive vegetation surveys of more than 520 sites throughout the Central Highlands region (see Chapter 3). These sites varied in the age of their overstorey ash forest, which facilitated analyses of stand-age relationships with understorey structure and composition (Lindenmayer et al. 2000a).

A subset of the 520 sites was surveyed for arboreal marsupials and birds, which enabled animal occurrence to be related to the structure and composition of the understorey and shrub layers (Lindenmayer et al. 1991d, 2009b).

Gum nodules were collected from 57 *Acacia* spp. trees in 33 Mountain Ash stands and 24 Alpine Ash stands (Lindenmayer et al. 1994a). Gums from three tree species – Silver Wattle, Mountain Hickory Wattle and Forest Wattle – were subject to chemical analyses of sugar and nitrogen content (Lindenmayer et al. 1994a).

Two studies of small terrestrial mammals were completed to explore relationships between animal occurrence and the structure and composition of the understorey and shrub layers. The first involved combining data from hairtubing and vegetation surveys of 49 linear strips retained within clearfelled coupes (Lindenmayer et al. 1994c). The second was a trap–recapture study which explored relationships between the probability of capture of small terrestrial mammals and the structure and composition of the understorey and shrub layers (Cunningham et al. 2005).

UNDERSTOREY TREES AND MODELS OF ANIMAL OCCURRENCE
Arboreal marsupials
Several studies have used statistical modelling to link the presence and abundance of arboreal marsupials to attributes of stand structure. The extent of understorey layer features prominently in these models of the habitat requirements of arboreal marsupials. The presence and abundance of Leadbeater's Possum increases significantly with increasing amounts of

Acacia spp. basal area in stands of montane ash (Lindenmayer *et al.* 1991d) (Figure 7.2). *Acacia* spp. trees are a source of sap that is a major component of the species' diet (Smith 1984b). Many insects are associated with the foliage of *Acacia* spp. trees, particularly Silver Wattle (Woinarski and Cullen 1984); this may be another reason why the insectivorous Leadbeater's Possum (Smith 1984b) is most often found in stands characterised by large amounts of understorey *Acacia*. Finally, *Acacia* spp. trees support many interconnected lateral branches. These facilitate the movement of the non-volant Leadbeater's Possum throughout montane ash forests (Smith 1980).

Statistical modelling has shown that the Mountain Brushtail Possum is significantly more likely to occur in forest stands with high values of *Acacia* spp. basal area (Lindenmayer *et al.* 1990c). The foliage of *Acacia* spp. trees, particularly Silver Wattle, is a substantial component of the diet of the Mountain Brushtail Possum (Owen 1964; Seebeck *et al.* 1984). The species uses the interconnected branches of these trees to move through the forest. Finally, the basal area of *Acacia* spp. in the understorey influences the species richness of arboreal marsupials as well as the overall abundance of animals in this group (Lindenmayer *et al.* 1991d) (Figure 7.2).

Models of the habitat requirements of arboreal marsupials have been important for informing forest management practices and management planning (e.g. Macfarlane *et al.* 1998). Therefore, an important part of ongoing work is assessing the performance of these models. Careful analyses have demonstrated that the models perform well when tested with additional field datasets (Lindenmayer *et al.* 1994b). The relationships observed between understorey *Acacia* spp. and arboreal marsupials have remained robust over two decades of field studies and empirical analysis (Lindenmayer *et al.* 1994b) (Figure 7.3).

The statistical models described above relate understorey *Acacia* spp. to the presence and/or abundance of arboreal marsupials on 3-ha field sites. It is possible to explore understorey relationships at smaller spatial scales, such as the occupancy of individual trees with hollows and micro-sites where animal detections are made using hairtubing (see Chapter 3). Such analyses further demonstrate the

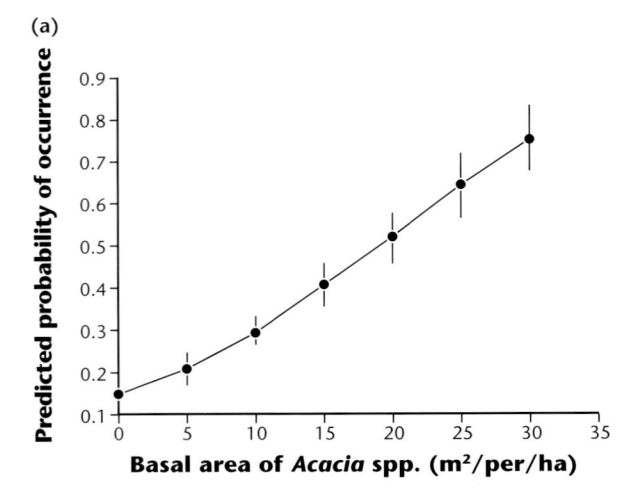

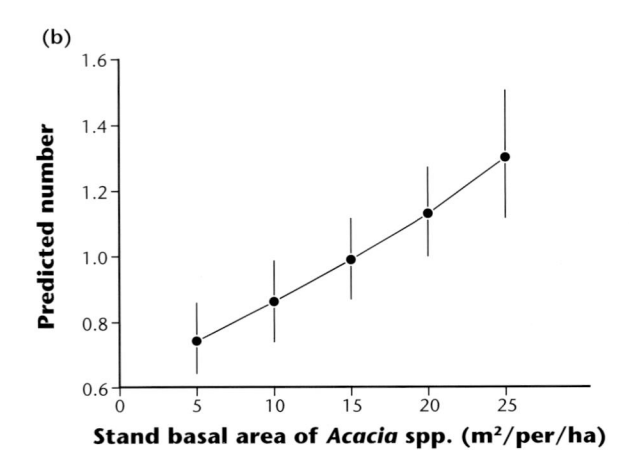

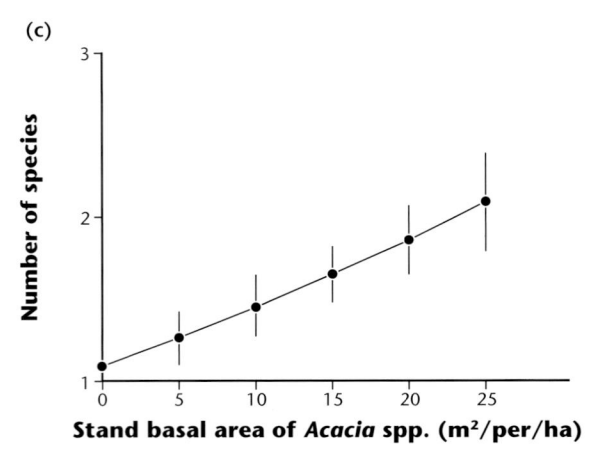

Figure 7.2: Relationships between the basal area of *Acacia* spp. and (a) the probability of occurrence of Leadbeater's Possum, (b) the abundance of the Mountain Brushtail Possum and (c) species richness of arboreal marsupials.

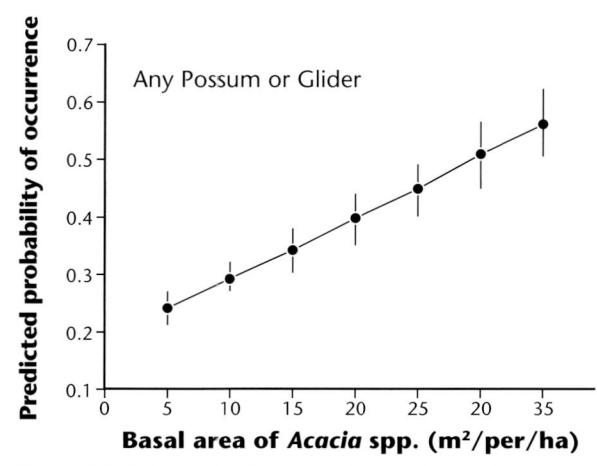

Figure 7.3: Relationship between the basal area of *Acacia* spp. and the probability of occupancy of a tree with hollows. (Modified from Lindenmayer *et al.* 1990a).

importance of the amount of *Acacia* spp. in the understorey. First, the probability that a given tree with hollows will be occupied by an arboreal marsupial was found to increase significantly with increasing amounts of *Acacia* spp. in the understorey of the surrounding stand (Lindenmayer *et al.* 1990a). Second, the probability of detection of the Mountain Brushtail Possum by hairtubing in 10 × 10 m microplots showed that the species was highest where there was a higher cover of *Acacia* spp. trees (Lindenmayer *et al.* 1994d).

Small terrestrial mammals

The importance of the understorey layer has been explored in several studies of the occurrence of small terrestrial mammals in the montane ash forests of the Central Highlands of Victoria. One species, the Agile Antechinus, shows a significant positive relationship with the presence of Mountain Hickory Wattle and Forest Wattle (Lindenmayer *et al.* 1994d). Another species, the Dusky Antechinus, is most likely to be found in stands characterised by a large number of *Acacia* spp. trees in the understorey (Cunningham *et al.* 2005). The Agile Antechinus is insectivorous (Hall 1980) and partially scansorial and *Acacia* spp. trees may provide a valuable foraging substrate for the species, particularly because of the large number of invertebrates often associated with the trees (Woinarski and Cullen 1984).

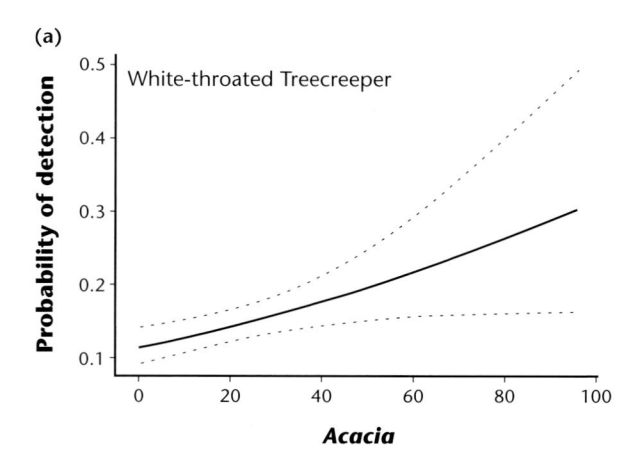

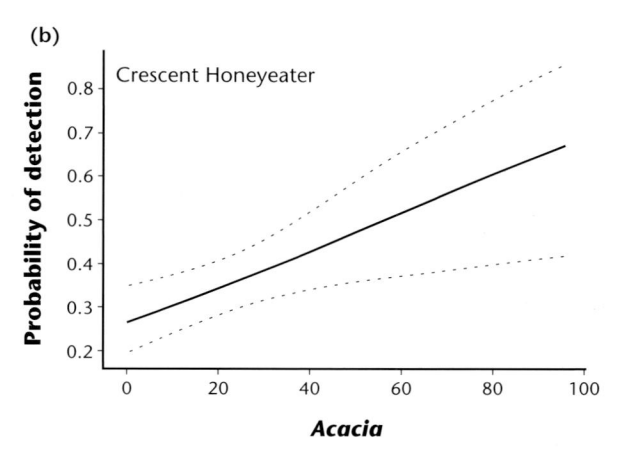

Figure 7.4: Prevalence of *Acacia* spp. in the understorey of montane ash forests and (a) the occurrence of the White-throated Treecreeper and (b) the occurrence of the Crescent Honeyeater.

Birds

Considerable effort has been dedicated to quantifying the attributes of stand structure and plant species composition that significantly influence the occurrence of birds in montane ash forests. The amount of *Acacia* spp. in the understorey of these forests has been found to be an important part of models developed for overall bird species richness. For example, the occurrence of the White-throated Treecreeper and the Crescent Honeyeater is positively associated with increasing amounts of *Acacia* spp. stems in montane ash forests. The bird assemblage in montane ash forests is strongly shaped by the prevalence of *Acacia* spp. in the understorey.

bellied Glider is occasionally observed eating *Acacia* spp. gum. It is not a readily digestible kind of food and arboreal marsupials have a range of special adaptations for eating it. The sugar and nitrogen content of the gums of these tree species could influence the distribution and abundance of the species which depend on them as a food source. A pilot study was completed of the sugar and nitrogen content of the gums of Silver Wattle, Mountain Hickory Wattle and Forest Wattle (Lindenmayer *et al.* 1994a). Two interesting questions were posed were: Do the gums produced by different species of wattles vary in nitrogen and sugar content? Are interspecific differences in nitrogen and sugar content influenced by the dominant overstorey forest type in which *Acacia* spp. trees occur?

Chemical analyses revealed that the sugar content of gum samples collected from Mountain Ash forests was significantly lower than that from Alpine Ash forests (Lindenmayer *et al.* 1994a). There were differences in the sugar content of gum samples gathered from different species of *Acacia* trees. Three variables were found to influence the nitrogen content of gum samples – the species of *Acacia* tree, tree diameter, and the forest type in which trees occurred.

The results indicate that the sugar and nitrogen content of *Acacia* spp. gums may vary within and across forest landscapes. This may partly explain why the distribution and abundance of arboreal marsupials is patchy within montane ash forests (Lindenmayer 1996) and why the basal area of *Acacia* spp. is an important explanatory variable in the presence

Figure 7.5: *Acacia* gum nodule. (Photo by David Lindenmayer).

Variation in nutrient content of wattle tree species

The gum of *Acacia* spp. is an important food source for several species of arboreal marsupial in montane ash forests (Smith 1984b). The gum produced by *Acacia* spp. trees is a primary part of the diet of Leadbeater's Possum and the Sugar Glider. The Yellow-

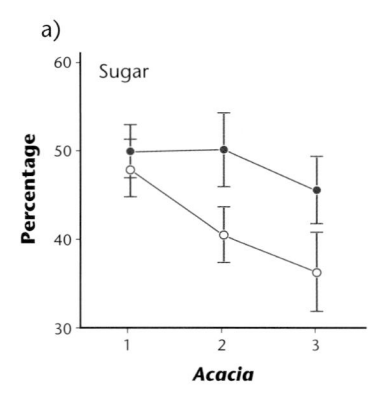

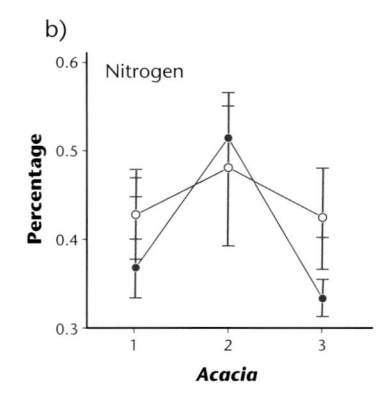

Figure 7.6: Relationships between (a) sugar content and species of *Acacia* tree in Mountain Ash and Alpine Ash forests and (b) nitrogen content and species of *Acacia* tree in Mountain Ash and Alpine Ash forests. (Redrawn from Lindenmayer *et al.* 1994a).

and abundance of gum-feeding species such as Leadbeater's Possum (Lindenmayer *et al.* 1991d).

SHRUB AND OTHER LAYERS AND MODELS OF ANIMAL OCCURRENCE

The shrub, and other layers, feature in occurrence models of arboreal marsupials and small terrestrial mammals. The abundance of Leadbeater's Possum and the Mountain Brushtail Possum is negatively related to the number of shrubs in a stand (Lindenmayer *et al.* 1991d), possibly because the presence of many shrubs may preclude the development of *Acacia* spp. Conversely, the probability of detecting the Bush Rat by hairtubing (see Chapter 3) increases significantly with the number of shrubs (Lindenmayer *et al.* 1994c). The Agile Antechinus and Bush Rat are both more likely to be detected where levels of ground cover vegetation are high (Lindenmayer *et al.* 1994c). These species often seek shelter while foraging (Predavec 1990) and ground cover may provide protection from potential predators.

Tree ferns can be a significant part of the shrub layer in montane ash forests (Ough 2002; Ough and Ross 1992). The prevalence of this vegetation component has been found to be important in several studies. For example, detections of Mountain Brushtail Possum in hairtubing surveys are highest in locations characterised by numerous tree ferns (Lindenmayer *et al.* 1994d)

Figure 7.8: Tree ferns are a key understorey component of wet eucalypt forests and rainforests throughout eastern Australia. They are a nursery site for many other plant taxa, nesting places for birds and some mammals and a food source for large vertebrates such as the Mountain Brushtail Possum. Tree ferns are sensitive to intensive harvesting practices, such as clearcutting, and special consideration needs to be given to their retention during regeneration harvest. (Photo by Esther Beaton).

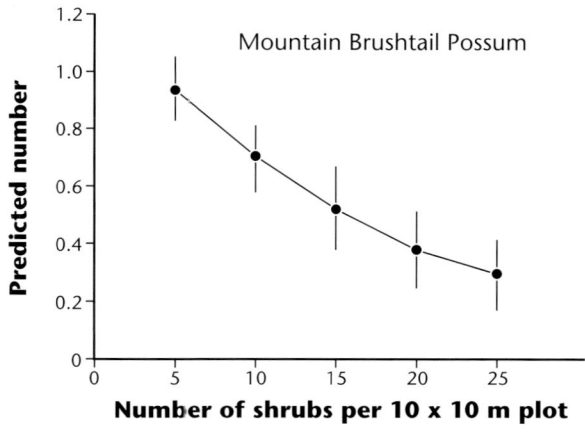

Figure 7.7: Relationship between the occurrence of Mountain Brushtail Possum and the number of shrubs in stands of montane ash forests. (Redrawn from Lindenmayer *et al.* 1990c).

(see Chapter 9). Seebeck *et al.* (1984) observed the species feeding on the fronds of these plants. They also recorded the foliage of tree ferns in the scats of the Mountain Brushtail Possum. Notably, tree ferns were used as part of trapping strategies for the species when it was hunted. A favoured method of capture was to break open the top of a tree fern trunk and fasten a wire noose to snare an animal attracted to the exposed pithy core (Edward Goldsmith, *pers. comm.*).

Other relationships between animal occurrence and tree ferns are somewhat intriguing. For example, recent work has revealed that nest boxes in stands with more tree ferns have a lower probability of infestation by the Feral Honeybee (Lindenmayer et al. 2009c). The reasons remain unclear, but the Feral Honeybee may be less likely to occur where the vegetation supports more tree ferns that are non-flowering plants and thus do not produce the nectar and pollen gathered by these invertebrates.

DISTURBANCE AND OTHER DYNAMICS OF THE UNDERSTOREY AND OTHER NON-OVERSTOREY VEGETATION LAYERS

A body of research in the montane ash forests has examined changes in the structure and composition of the vegetation with stand age, natural disturbance (wildfires) and human disturbance (logging). The response of the understorey and shrub layers has been a component of that research (Lindenmayer et al. 2000a).

Understorey *Acacia* spp. layer: succession and disturbance

Many understorey plants die as montane ash forests age (Adams and Attiwill 1984; Ashton 1976). For example, the longevity of many *Acacia* spp. (especially Silver Wattle and Mountain Hickory Wattle) rarely exceeds 60–100 years and there is a pronounced decline in the number of *Acacia* spp. with stand maturation (Ashton 1976). However, our studies of stands of different ages revealed complex effects, with spikes in the amount of *Acacia* spp. trees within old-growth and young regrowth age cohorts (Figure 7.9). Thus, there was an unexpected prevalence of *Acacia* spp. trees in the understorey of old-growth stands. Seeds from these plants may remain in the soil for a prolonged period (up to several centuries) (Gilbert 1959), and germinate following disturbance. The prevalence of these short-lived trees in old forests could indicate that one or more (fire) disturbance events rejuvenated this component of the understorey. Thus, the age of some understorey components was not the same as the overstorey in many stands of old-growth montane ash forest.

An alternative explanation for the prevalence of *Acacia* spp. trees in the understorey – canopy gaps created by ash stems – seems unlikely due to their close proximity to overstorey stems. Other data from studies in the Central Highlands of Victoria (McCarthy and Lindenmayer 1998) indicates that low- or medium-intensity fires are the most likely mechanism leading to asynchrony in the ages of the overstorey and understorey. In particular, many older stands were often not truly even-aged, but supported trees from two or more markedly different age cohorts (Lindenmayer et al. 2000a). Indeed, the probability of two or more classes occurring on a site is significantly greater with increasing stand age (Lindenmayer et al. 2000a). Such disturbances in old-growth forest appear to have stimulated the development of understorey plants but have not led to the death of all understorey trees – thus they have not been complete stand-replacing events. This is consistent with the occurrence of fire-scars on many living, old, large-diameter montane ash trees (Lindenmayer et al. 1991b, 1991g).

Tree fern layer: succession and disturbance

Analyses of extensive vegetation data gathered from many thousands of plots in stands of different ages revealed that the probability of occurrence of tree ferns was significantly higher in older stands of montane ash forests (Lindenmayer et al. 2000a) (Figure 7.10). These findings parallel studies by Ashton and Willis (1982), which demonstrated an increased development

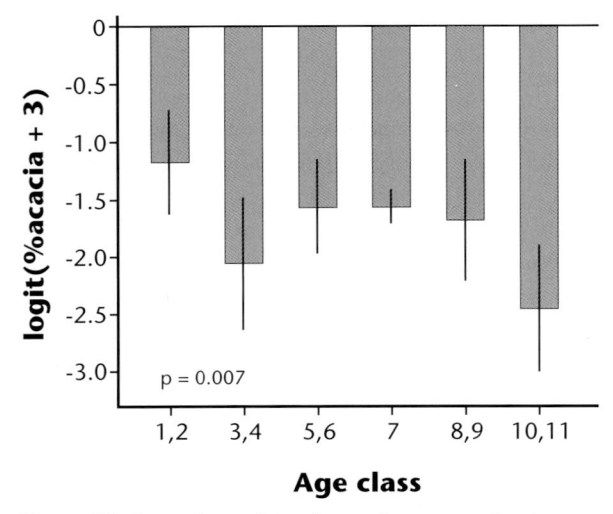

Figure 7.9: Percentage of *Acacia* spp. trees occurring in stands of different ages. (Redrawn from Lindenmayer et al. 2000a). Stands of old growth correspond to age classes 1 + 2.

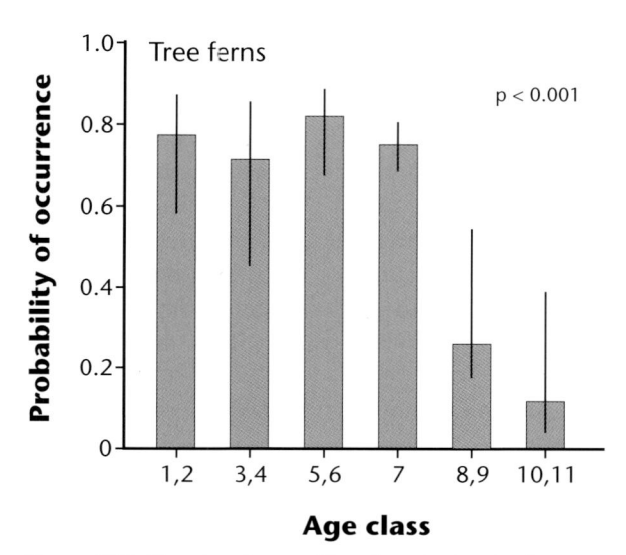

Figure 7.10: The abundance of tree ferns in stands of different ages. (Redrawn from Lindenmayer et al. 2000a). Age classes 1 + 2 correspond to old-growth forest.

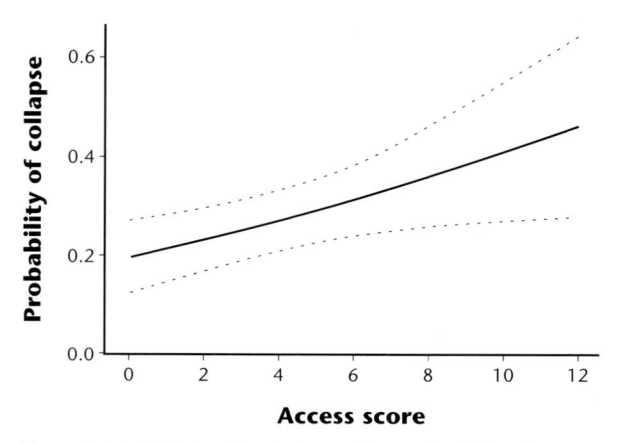

Figure 7.11: Relationships between the probability of collapse of tree with hollows and a score corresponding to the amount of dense adjacent understorey vegetation.

of the fern layer with stand succession. It is possible that the age class effects observed were related to the contrast in disturbance regimes between old and young forest. In the study by Lindenmayer et al. (2000a), younger forests were those that had been subjected to intensive clearfelling operations. Other workers have recorded significantly lower numbers of tree ferns in cutover stands of montane ash forest than in burned forest (Ough and Murphy 1996). There is evidence that tree ferns can survive the effects of wildfires in montane ash forests. In the study, tree ferns exceeding 3 m in height were often recorded in stands where the overstorey ash trees dated from the 1939 wildfires. Tree ferns of this size are likely to be considerably older than the 59-year-old overstorey trees. This finding is consistent with carbon dating work by Mueck et al. (1996) which showed that tree ferns exceeding 350 years occurred in areas where the overstorey montane ash trees dated from the 1939 fires.

Understorey layer and relationships with the collapse of overstorey trees with hollows

The work on tree fern age and overstorey tree age suggests there are interesting relationships between the ages of plants in the different layers of montane ash forest. There can be other, sometimes quite unexpected, relationships between the understorey and overstorey of these forests. As an example, recent work has examined the factors influencing the probability of collapse of large trees with hollows (Lindenmayer and Wood 2009). Trees with large amounts of nearby dense surrounding vegetation were those most likely to collapse between 1983 and 2007 (Lindenmayer and Wood 2009). It is likely that this is an outcome of the effect of adjacent vegetation on the moisture content and rate of deterioration (and ultimately the collapse) of large trees with hollows.

SUMMARY

An array of studies completed over the past 25 years has demonstrated the importance of the understorey, shrub and other (non-overstorey) vegetation layers as key habitat components for a range of vertebrate taxa in montane ash forests. The basal area of *Acacia* spp. trees, the abundance of shrubs and abundance of tree ferns are attributes that commonly emerge in statistical modelling of animal presence and/or abundance. On most occasions, these attributes make intuitive ecological sense as they describe the suitability of foraging substrates for animals, the availability of protective cover while animals are foraging or the interconnectivity of the vegetation for promoting movement through the forest. The ecological importance of these habitat relationships have often been strengthened by tests of model performance on new datasets, and by broadly similar kinds of habitat relationships emerging from analyses on datasets gathered at different spatial scales or using different field methods.

The development of the understorey, and the shrub and other layers of the vegetation, is an outcome of disturbance and post-disturbance successional processes. Analyses of the relationships between stand age and the understorey and shrub layers of montane ash forests have indicated that the structure of some stands of montane ash forest is more complex than previously recognised – some stands support rejuvenated understorey plants, have asynchronous ages of understorey and have some (older) overstorey components (Lindenmayer et al. 2000a). This also indicates that the disturbance dynamics of montane ash forests are more complex than previously recognised (see Chapter 12).

LESSONS LEARNED

Asynchrony between the age of the understorey and overstorey indicates a need to rethink what is actually meant by multi-aged stands of montane ash forest. Typically, multi-aged montane ash stands are regarded as those supporting two or more age cohorts of dominant overstorey (ash-type) trees (Ashton 1981c; Chesterfield et al. 1991). However, it is clear that where, for example, old-growth montane ash trees overtop a rejuvenated understorey or where old tree ferns (exceeding 350 years) persist in 1939-aged regrowth ash stands (Mueck et al. 1996), different ages of plants occur on the same site. It could be argued that these cases are examples of multi-aged stands. Such differences in the age of the overstorey and understorey of montane ash forests can be important because extensive field studies have demonstrated that stands with a combination of old-growth trees overtopping regrowth understorey stems provide highly suitable habitat for endangered species such as Leadbeater's Possum (Lindenmayer et al. 1991d) and other taxa such as the Mountain Brushtail Possum (Lindenmayer et al. 1990c).

KNOWLEDGE GAPS

The initial sections of this chapter focused on relationships between the understorey and other vegetation layers and the occurrence of selected species of vertebrates in montane ash forests. As a result of that work (and research by others), the habitat value of understorey and shrub and ground layers is now relatively well-known for vertebrates. There has been limited data collection for invertebrates in montane ash forests (but see Neumann 1991, 1992) and this knowledge gap needs to be closed, particularly as invertebrates are the bulk of forest biodiversity.

The potential existence of a wider range of natural disturbance pathways in montane ash forests suggests this area needs further exploration. In particular, there will be considerable value in completing studies of spatial variation in fire severity in the closed water catchments which were burned in the 2009 Black Saturday wildfires, but where human disturbances (including post-fire salvage logging) has been limited for the past 200 years.

FOREST PATTERN, ECOLOGICAL PROCESS AND LINKS TO OTHER CHAPTERS

The understorey is an integral part of stand structure in montane ash forests. Its development is a function not only of ecological processes like disturbance regimes but other processes such as climatic and topographic influences. This pattern of occurrence and structural development of the understorey influences other patterns, such as animal occurrence, through important mechanisms like the provision of food and nesting sites and facilitating animal movement. The understorey influences several important processes such as carbon storage (Keith et al. 2009) and nutrient cycling, including nitrogen fixation to facilitate the growth of overstorey trees (Adams and Attiwill 1984). The microclimatic conditions created by a dense understorey layer can influence other processes in the overstorey layer such as the collapse rate of large trees with hollows (Lindenmayer and Wood 2009).

Links to other chapters

Links between forest patterns and ecological processes emphasise the relationships between this chapter and many others in this book. As outlined above, disturbance is a major factor shaping the structure and composition of the understorey and other vegetation

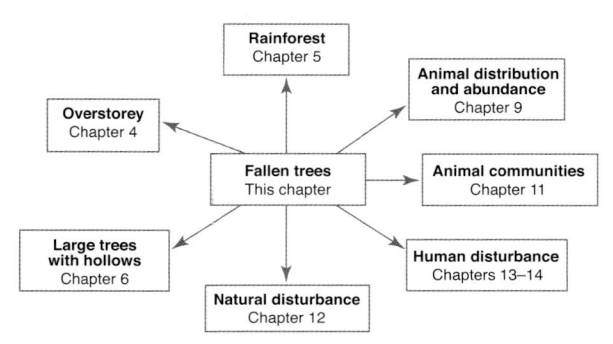

Figure 7.12: Relationships between the understorey and key topics examined in other chapters.

layers in montane ash forests. Disturbance is the key theme linking all three chapters in Part V, and its relationships with understorey and other vegetation layers features prominently in Chapters 12, 13 and 14.

Many factors influence the distribution and abundance of particular elements of the biota. Habitat suitability is a critical one, a topic addressed in detail in Chapter 9. The importance of the understorey and other vegetation layers as a key component of the habitat requirements of vertebrates in montane ash forests is part of the discussion in that chapter. The understorey can influence the species richness of groups like arboreal marsupials and the composition of the bird assemblage, and these topics feature in Chapter 11. The habitat requirements of many species in montane ash forests is a combination of understorey attributes and overstorey characteristics (e.g. forest type and the abundance of trees with hollows). Therefore, this chapter has strong links with those on the overstorey in Chapters 4 and 6. Cool temperate rainforest is a prominent part of the understorey of many areas of montane ash forest, particularly old-growth stands in gullies. Therefore, this chapter has links with Chapter 5 on cool temperate rainforest.

The structure and composition of the understorey is strongly affected by natural disturbance regimes (Chapter 12) and human disturbance (Chapters 13 and 14). For example, long-lived and disturbance-sensitive understorey species can be negatively affected by intensive forestry operations (Lindenmayer and Ough 2006; Mueck *et al.* 1996; Ough and Ross 1992). Retention strategies that focus on small 'islands of vegetation' or aggregates, which are kept free of logging and subsequent slash fires, are particularly valuable for maintaining these communities and their associated animal biota (Ough and Murphy 1998). These issues are revisited in Chapter 16, which focuses on mitigating logging impacts.

8 Key structural features: logs

INTRODUCTION

Logs are a key component of stand structure in temperate forests throughout the world (Grove and Hanula 2006; Grove et al. 2002; Harmon et al. 1986; Kaila et al. 1997; Maser and Trappe 1984; Niemela et al. 1993; Recher 1996; Sollins et al. 1987). They have many key ecological roles, which have been reviewed in detail by a number of authors (Harmon et al. 1986; Lindenmayer et al. 2002a; Sedell et al. 1988; Woldendorp and Kennan 2005). These roles are summarised in Table 8.1.

LOGS AS A KEY STRUCTURAL FEATURE OF MONTANE ASH FORESTS

Several direct or indirect kinds of studies of logs have been completed in montane ash forests over the past 25 years and are briefly summarised in this chapter.

Logs indirectly provide food resources for mycophagous (fungus-feeding) mammals. They serve as places where hypogeous (underground-fruiting) mycorrhizal fungi develop and become an important source of food for animals like the Bush Rat, Long-nosed Bandicoot and Mountain Brushtail Possum (Claridge and Lindenmayer 1993, 1998). The Mountain Brushtail Possum uses large logs as movement pathways and is often captured in traps placed on them (Lindenmayer et al. 1998; Seebeck et al. 1984). The Common Wombat and Mountain Brushtail Possum deposit scats in prominent places, such as logs, to signal territory boundaries (Triggs 1996). Small terrestrial mammals are frequently captured in traps placed close to large pieces of coarse woody debris (Cunningham et al. 2005).

In birds, the diet of species such as the Superb Lyrebird includes a wide range of prey which are strongly associated with logs and deep litter (Ashton and Bassett 1997; Lill 1996). Similarly, ground-using taxa such as the Eastern Whipbird (*Psophodes olivaceus*) forage around logs and turn over ground litter while searching for food.

Logs are very important for reptiles in montane ash forests. They provide overnight shelter sites and winter hibernacula, as well as places for reptiles to lay their eggs (Brown and Nelson 1993).

Logs are nursery sites for some species of plants and provide places for the germination and growth of ferns, mosses and liverworts in montane ash forest (Ashton 1986). Detailed vegetation surveys (Lindenmayer et al. 2000b) have shown that Myrtle Beech trees in montane ash forests often germinate on rotting logs. The water and organic matter in rotting

Table 8.1: Key ecological roles of logs

Key ecological role	Reference
Providing substrates for the germination and development of plants (nurse logs)	Harmon and Franklin (1989); Barker and Kirkpatrick (1994)
Providing habitat for rich assemblage of detritivores and decay organisms	Maser and Trappe (1984); Berg et al. (1994); Grove and Hanula (2006)
Providing shelter for a wide variety of forest-dependent vertebrates	Tallmon and Mills (1994); Wilkinson et al. (1998)
Acting as protected runways for the movement of terrestrial animals	Maser et al. (1977); McCay (2000)
Providing hunting and resting perches	Thomas (1979); Backhouse and Manning (1996)
Acting as basking sites for reptiles and mammals	Webb (1995); Wilson and Swan (2007)
Providing foraging sites for wildlife including species-rich invertebrate assemblages	Maser and Trappe (1984); Smith, et al. (1989b); Taylor (1990); New (1995); Grove and Hanula (2006)
Providing hiding cover and habitat for fish and other aquatic organisms	Harmon et al. (1986); Sedell et al. (1988); Koehn (1993); Gregory (1997); Naiman and Bilby (1998)
Influencing hydrological and geomorphological processes in streams and rivers	Harmon et al. (1986); Sedell et al. (1988); Naiman and Bilby (1998); Naiman and Turner (2000); Gippel et al. (1996)
Acting as long-term sources of energy and nutrients in forest and aquatic ecosystems	Harmon et al. (1986)
Acting a sites for nitrogen fixation	Harmon et al. (1986)
Adding to potential fire fuels	Luke and McArthur (1977); Bradstock et al. (2002)

logs then facilitates the growth of young seedlings (Howard 1973). Other understorey plant species common in montane ash forests, such as Dogwood, Musk Daisy Bush and Mountain Pepper, also germinate on rotting logs.

Table 8.2: Studies directly or indirectly examining logs and coarse woody debris in the montane ash forests of the Central Highlands of Victoria

Study description	Reference
Log volumes in different aged stands	Lindenmayer et al. (1999d)
Log decay and moss cover	Lindenmayer et al. (1999d)
Ecological roles of logs and coarse woody debris	Lindenmayer et al. (2002a)
Logs and carbon storage[a]	Mackey et al. (2002)

[a] Involved quantitative studies from forests in several parts of south-eastern Australia, including the montane ash forests of the Central Highlands of Victoria.

BACKGROUND DATASETS

Only one study has directly examined the logs and coarse woody debris of Mountain Ash forest. As part of that investigation, 984 logs were measured on 60 sites located in four distinct age classes of Mountain Ash forest:

- 20-year old regrowth;
- regrowth dating from the 1939 wildfires;
- stands dating from fires in the 1820s;
- stands exceeding 250 years old.

All logs intercepting a 100 m transect line established in each of the 60 sites were measured. The length, volume, level of decay and amount of moss cover of each log was quantified (Lindenmayer et al. 1999d). Statistical analyses were completed to quantify relationships between log attributes (number, diameter, length and moss cover) and stand age as well as the slope, aspect and topographic position of sites (Lindenmayer et al. 1999d).

LOG DIAMETER AND OVERALL LOG VOLUME

The overall volume and extent of log decay in a forest can have substantial impacts not only on habitat suitability for vertebrates, but on key ecological processes such as carbon storage and nutrient cycling (Harmon *et al.* 1986; Woldendorp and Kennan 2005). Work in the Central Highlands of Victoria revealed that the diameter of logs was significantly greater in mature and old-growth stands than in younger age cohorts. Indeed, some very large logs were measured in the older stands – the longest was 68.6 m and the greatest diameter recorded was 2.7 m (Lindenmayer *et al.* 1999d). These values were not surprising, given the height and diameter of standing living and dead trees in montane ash forests.

Log volumes were estimated using the line transect method (Van Wagner 1976; Warren and Olsen 1964) and were found to vary considerably between sites (13.5 m^3/ha up to 1026 m^3/ha). No significant age class or other effects (slope, aspect or topographic position) were found for log volume. The average value calculated from all sites was 353 m^3/ha. The log volumes exceeded 1000 m^3/ha at some sites; these estimates are similar to those recorded in other wet forests such as in Tasmania (Meggs 1996). They are somewhat higher than those calculated for many other temperate forest types around the world (Mackey *et al.* 2008).

The lack of differences between age classes for the number and overall volume of logs was surprising. It may have occurred because no attempt was made to distinguish between the trunks of collapsed trees and branches or tree crowns that had been shed and fallen to the ground. A significant portion of the logs in younger age classes may have been recruited by processes such as branch-shedding and natural self-thinning, which are prevalent in regenerating stands of Mountain Ash forest (Ashton 1975a; Cunningham 1960).

Figure 8.1: Measuring a log in a stand of montane ash forest. (Photo by David Lindenmayer).

Figure 8.2: Luxuriant moss mat. (Photo by Julie Strahan/www.flickr.com/photos/jsarcadia).

LOGS AND MOSS COVER

Moss cover on logs was significantly higher in the older stands (mature and old-growth forests) than young regrowth and 1939-aged regrowth stands. The dominant moss species found on Mountain Ash logs are *Wijkia extenuata* and *Lepidozia ulothrix* (Ashton 1986). Ashton estimated that it took 15–30 years for thick mats of moss to develop over 80–100% of logs. These age estimates may account for the lower amount of moss found on logs in the younger age classes, particularly those dating from the 1970s. There may be delayed establishment of luxuriant moss cover in stands of 1970s regeneration due to changes in solar radiation and temperature regimes in recently cutover areas (Parry 1997). Moreover, Ashton (1986) described different niches on logs for different assemblages of bryophytes. Mature and old-growth forests may provide a greater range of suitable substrates for moss development and thus facilitate the development of more luxuriant moss cover.

LOGS AS BIOLOGICAL LEGACIES IN STANDS OF MONTANE ASH FOREST

The paucity of stand-age differences in log volumes may have been related to the influence of stand conditions prior to disturbance at a site. At the younger sites, logs larger than the average present stand diameter at breast height (DBH) were often recorded. These large logs would have been left following timber harvesting. Large logs at the 1939-regrowth sites were likely to be biological legacies (Franklin *et al.* 2000) from the pre-existing forest burned in the 1939 fires.

Biological legacies are defined as 'organisms, organically-derived structures, and organically-produced patterns that survive from the pre-disturbance system' (Franklin *et al.* 2000). Logs are among the most important biological legacies in both naturally disturbed stands and forests disturbed by humans (Franklin *et al.* 2000). Various studies from forests around the world have indicated that biological legacies have many ecological roles:

- adding to the structural complexity of forest floor conditions in the regenerating stand (Lindenmayer *et al.* 1999d);
- facilitating survival of plants and animals in disturbed stands (Kotliar *et al.* 2007; Whelan *et al.* 2002);
- providing habitat for species that eventually recolonise a disturbed site (Morrison *et al.* 2006);
- influencing patterns of recolonisation in disturbed areas (Turner *et al.* 2003; Whelan *et al.* 2002);
- providing a source of energy and nutrients for other organisms (Amaranthus and Perry 1994);
- modifying or stabilising environmental conditions in a recovering stand (Perry 1994);
- shielding young plants from over-browsing by large herbivorous mammals (Turner *et al.* 2003).

It is likely that large logs play most, if not all, of these key biological legacy roles in montane ash forests (Lindenmayer *et al.* 2002a).

SUMMARY

Logs have an array of key ecological roles in montane ash forests. They are a critical part of the structure and effective functioning of these ecosystems. The extent of log decay and, in turn, cover of mosses, liverworts and lichens varies significantly along the length of any given log. Montane ash forests are characterised by extremely high log volumes, among the highest yet recorded for any forests worldwide (Lindenmayer *et al.* 1999d).

LESSONS LEARNED

As in the case of so many aspects of forest ecology and management, a scientific understanding of the world of fallen trees requires multi-scaled temporal and spatial perspectives. The size of individual logs and the overall volume of logs in a stand are intimately linked to the characteristics of the current overstorey trees. They may also be affected by the characteristics of the previous stand, because of the influence of past complete or partial stand-replacing disturbance events on the structural complexity of montane ash stands. At smaller spatial and temporal scales, individual logs can vary markedly in size, length and extent of decay. Decay and other conditions can vary along the length of a given log. This can influence the suitability of a log or parts of a log for log-associated biota.

The importance of large logs as key components of stand structural complexity have been known for many decades in forests overseas (Franklin *et al.* 1981; Harmon *et al.* 1986; Maser and Trappe 1984) but recognised relatively recently in Australian forests (Woldendorp and Kennan 2005), including areas dominated by stands of montane ash trees (Lindenmayer *et al.* 1999d).

The volume of logs quantified for montane ash forests was unanticipated, but perhaps should have been obvious given the size of the overstorey trees and the difficulty of walking through old-growth stands where the forest floor is characterised by many fallen trees! This, in turn, suggests that the importance of montane ash forests as a carbon store may have been substantially underestimated in the past (Mackey *et al.* 2008).

KNOWLEDGE GAPS

Remarkably few studies have been completed of the coarse woody debris of montane ash forests. Much remains to be learned. Relationships between logs and invertebrate biota have not been studied although work on this topic in the wet forests of Tasmania suggests that it needs urgent attention. From a management perspective, nothing is known about the amount, condition and spatial distribution of coarse woody debris that needs to be retained within logged forest to avoid replicating problems associated with the paucity of this key attribute of stand structure in other countries (Berg *et al.* 1994; Sverdrup-Thygeson and Lindenmayer 2003; Lindenmayer *et al.* 2002a).

FOREST PATTERN, ECOLOGICAL PROCESS AND LINKS TO OTHER CHAPTERS

The introduction to this chapter summarised the many ecological roles of logs in montane ash forests, ranging from influencing carbon storage and nutrient cycling (Mackey *et al.* 2008) to affecting microclimatic conditions and natural disturbance regimes. The spatial patterns of coarse woody debris influence many other spatial patterns ranging from the occurrence and luxuriance of bryophyte communities (Ashton 1986) to the provision of habitat and foraging resources for ground-dwelling animals (Ashton and Bassett 1997). Therefore, logs have a twin role – they are both a pattern and crucial contributor to key ecological processes. This is underscored by the extent to which this chapter is connected to many others in this book.

Links to other chapters

Logs are a vital habitat resource for many forest-dependent species, and their role in influencing the distribution and abundance of some of the species in montane ash forests is discussed in Chapter 9.

The cycle of tree growth, tree maturation and tree decay and collapse is critical for forest function. Large fallen logs are the next stage of decay of large standing dead trees – a topic examined in detail in Chapter 6. Logs are critical germination points for many understorey plants including cool temperate rainforest trees; aspects of the biology and ecology of rainforest are discussed in Chapter 5.

Large fallen trees can sometimes be fuel in a wildfire, although at other times they can act as micro-firebreaks. There are, therefore, important relationships between fallen trees and natural disturbance regimes in montane ash forests (see Chapter 12). A key management issue is the maintenance and long-term recruitment of logs in native forests subject to recurrent harvesting. Issues of the effects of human distur-

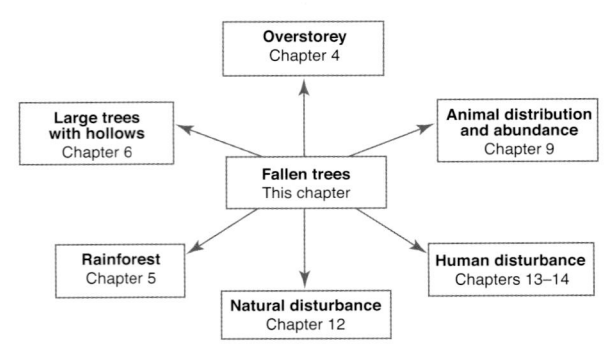

Figure 8.3: Relationships between fallen trees and key topics examined in other chapters.

bances in montane ash forests feature in Chapters 13 and 14. Failure to take account of this will ultimately lead to the simplification of stand structure – with a range of potentially negative outcomes for biodiversity and key ecological processes (Lindenmayer *et al.* 2006). Issues associated with the development of prescriptions for the maintenance of the types, numbers and overall volumes of fallen trees in wood production forests is examined in Chapters 14 and 16.

The rapid rates of tree growth in montane ash forests (Ashton 1975a), coupled with the apparent slow rates of decomposition, has meant that these areas support some of the world's highest known carbon stores in the coarse woody debris component of a forest (Mackey *et al.* 2008; Keith *et al.* 2009). This has major implications for strategies for sequestering and storing carbon as part of climate change mitigation (Mackey *et al.* 2008). The potential contribution of montane ash forest in tackling rapid climate change is discussed in the final chapter of this book (Chapter 18).

PART IV

ANIMAL OCCURRENCE

Understanding why species occur where they do is a fundamental part of ecology and has been since the field emerged as a discipline in its own right (Elton 1927). It is also a fundamental part of management and conservation. This is because it is essential to know where management actions need to take place and how to mitigate the effects of management practices where they are found to have negative impacts.

The first chapter in Part IV (Chapter 9) sets out the basis for understanding the distribution and abundance of particular species in the montane ash forests, starting with arboreal marsupials, followed by terrestrial mammals and birds. The importance of scale is paramount to the discussions in that chapter. Factors at multiple scales are critical – from whole landscapes through to patches, stands and individual trees.

The material in Chapter 9 sets out the factors which influence the occurrence of individual species in approximate hierarchical order from broad-scale climatic factors to the attributes of individual trees.

The viability of populations of an individual species can have profound impacts on the occurrence of that and other species within individual habitat patches and across ensembles of patches across a landscape. This is an immense topic and all of Chapter 10 is dedicated to providing an overview of more than a decade of research on the viability of populations of arboreal marsupials in montane ash forests. Species do not occur in isolation from one another; the third and final chapter in Part IV (Chapter 11) summarises the outcomes of various studies of different assemblages of taxa.

9 Distribution and abundance of individual species

INTRODUCTION

The distribution and abundance of all kinds of life on Earth is non-uniform. Distribution patterns may be clustered, random or some other pattern, but never uniform. Therefore, for the vast majority of species, there are places where they are comparatively common and others where they are rare or absent. This is clear from even a very rapid perusal of species distribution maps in field guides for any kind of living organism anywhere in the world. These maps usually depict the broad-scale limits of the distribution of a given species. Within these broad limits there will be places where a species will be more or less common than it is elsewhere. Analysing distribution patterns and identifying the factors which influence why species occur where they do, has been a key topic in ecology for over 80 years (Elton 1927; Krebs 2008). The animals inhabiting the montane ash forests of the Central Highlands of Victoria are no exception, and considerable effort has been expended on studying distribution patterns.

The distribution patterns of organisms can be shaped by many factors that operate at a range of different spatial and temporal scales. In the case of animals, some of these include:

- historical factors, such as the long-term biogeographic history of a region or landscape;
- climatic conditions and their relationships with the physiological tolerances of species;
- habitat suitability, including access to suitable nest sites and foraging substrates;
- the spatial arrangement of suitable remnant habitat throughout landscapes and regions;
- patterns of natural disturbance, such as wildfires;
- the impacts of human disturbance, such as land use and forestry practices;
- the prevalence of competitors, predators, diseases and parasites;
- patterns of social behaviour, such as the spacing of animals that occurs because of territorial and antagonistic interactions;
- access to mates.

Developing a detailed understanding of the factors influencing patterns of distribution and abundance is important for many reasons. It ensures that management actions to promote species conservation can be focused on areas where the target species occurs or is most likely to occur. It can help quantify the most critical resources needed by a species, how those resources might be altered by management practices and how the negative effects of human resource management practices might be mitigated.

This chapter summarises the large number of studies in the past 25 years that have examined and quantified the factors influencing the distribution and abundance of individual species of vertebrate animals within montane ash forests. By far the most work has been completed on arboreal marsupials. The work has been conducted at different spatial scales, from broad-scale climate analyses to microhabitat levels. In all cases, factors at multiple spatial scales have been found to be important. Given this, this chapter has been set out first by group – arboreal marsupials, then small terrestrial mammals and finally birds. Then, for each group, studies are presented in order of largest- to smallest-scale. In several cases, sets of species within a group have been targeted for community or assemblage-level analyses; the results of that research are summarised in Chapter 11.

Very limited work has been completed on other vertebrate groups in montane ash forests such as bats, reptiles and amphibians (Brown and Nelson 1993; Brown *et al.* 1997) as well as on invertebrates (Neumann 1991, 1992). Studies of those groups are not discussed in this chapter.

A lot of work has involved examining the factors influencing the distribution of the individual tree species which comprise the dominant overstorey and understorey layers of montane ash forests. Those topics were explored in Chapters 4 and 5 as well as in Part III, on forest structure and composition.

BACKGROUND DATASETS

Several major datasets have been gathered during studies of the distribution and abundance of arboreal marsupials in montane ash forests. First, for analyses of broad-scale climatic factors, geocoded location records for individual species were gathered from a range of museum and other sources (Fischer *et al.* 2001; Lindenmayer *et al.* 1991f). Second, analyses of landscape-level responses involved coupling distribution records of arboreal marsupials gathered during surveys of 207 field sites (Lindenmayer *et al.* 1994b) with spatial data on the amount of forest in different age, slope and aspect classes (Lindenmayer *et al.* 1999a). Third, patch-level analysis of arboreal marsupials involved linking animal occurrence in 49 retained linear strips with measured attributes of those sites (Lindenmayer *et al.* 1993a). Analyses of the occurrence of arboreal marsupials at the stand level involved establishing statistical relationships between detections of animals at 207 field sites with measurements of the vegetation structure and plant species composition at those sites (Lindenmayer *et al.* 1991d). More than 2000 trees with hollows were measured on the 207 field sites and statistical analyses were used to identify the attributes of trees which significantly influenced tree occupancy (Lindenmayer *et al.* 1991e). Finally, data on detections of arboreal marsupials by hairtubing and trap–recapture methods (see Chapter 3) were linked with measured attributes of the vegetation at these plots; this approach was used to quantify microhabitat requirements (Banks *et al.* 2008; Lindenmayer *et al.* 1994d).

Trapping and hairtubing data were the basis of studies of the factors influencing the occurrence of small terrestrial mammals in montane ash forests. At the patch level, detections of small terrestrial mammals involved linking animal occurrence in 49 retained linear strips with measured attributes of those sites (Lindenmayer *et al.* 1994c). Surveying animals in these sites involved plot-level hairtubing methods, which enabled microhabitat analyses. Extensive trap–recapture studies of small terrestrial mammals, coupled with measurements of stands and plot points where traps were set, helped quantify the factors influencing the occurrence of these animals at both the stand and microhabitat levels (Cunningham *et al.* 2005).

Work on the factors influencing the occurrence of birds in montane ash forests involved linking counts of individual species on 81 field sites with data on the measured vegetation attributes on those sites, as well as the location of the sites in forest landscapes (Lindenmayer *et al.* 2009a).

FACTORS INFLUENCING THE DISTRIBUTION AND ABUNDANCE OF ARBOREAL MARSUPIALS

The initial brief for work in montane ash forests over 25 years ago was to establish the factors influencing

Table 9.1: Studies directly or indirectly examining the distribution and abundance of arboreal marsupials in the montane ash forests of the Central Highlands of Victoria

Study description	Reference
Broad-scale climatic factors	
BIOCLIM analysis of Leadbeater's Possum	Lindenmayer et al. (1991e)
BIOCLIM analysis of Mountain Brushtail Possum	Fischer et al. (2001)
Landscape-level factors	
Landscape composition analyses of Yellow-bellied Glider	Lindenmayer et al. (1999a)
Patch-level factors	
Occurrence of individual species of arboreal marsupials in retained linear strips	Lindenmayer et al. (1993a)
Stand-level factors	
Habitat requirements of Mountain Brushtail Possum and Greater Glider	Lindenmayer et al. (1990c)
Habitat requirements of Leadbeater's Possum	Lindenmayer et al. (1991c)
Tests of the performance of habitat models for various species of arboreal marsupials	Lindenmayer et al. (1994b)
Tree-level factors	
Nest-tree models of arboreal marsupials	Lindenmayer et al. (1991d)
Tests of the performance of nest models for various species of arboreal marsupials	Lindenmayer et al. (1994b)
Microhabitat factors	
Microhabitat factors influencing the occurrence of Mountain Brushtail Possum	Lindenmayer et al. (1994d)
Relationships between microhabitat attributes and the breeding biology of Mountain Brushtail Possum	Banks et al. (2008)

the distribution and abundance of Leadbeater's Possum (Smith et al. 1985). Considerable work on Leadbeater's Possum and a range of other species has developed since that initial study. The work has encompassed studies of the factors influencing not only Leadbeater's Possum but also the Yellow-bellied Glider, Greater Glider, Mountain Brushtail Possum and Sugar Glider (Table 9.1). The work has included studies of broad-scale climatic influences and factors at the landscape, patch, stand, tree and microhabitat levels (Figure 9.1).

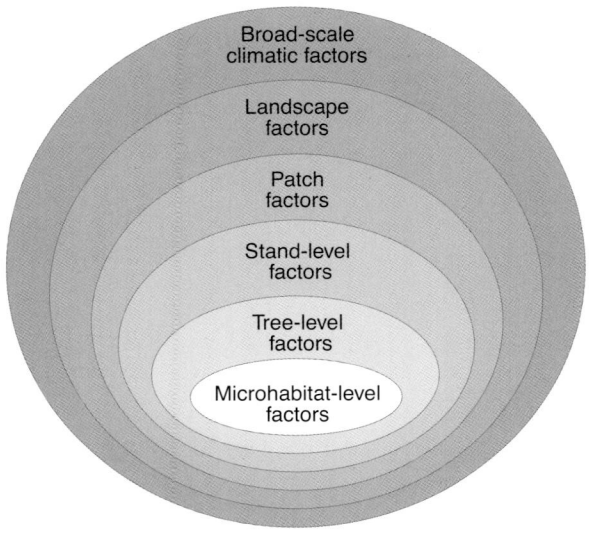

Figure 9.1: Nested hierarchy of factors at different spatial scales influencing the distribution and abundance of arboreal marsupials in montane ash forests.

Broad-scale climatic factors: BIOCLIM analyses

Climatic conditions set the broad-scale distribution limits within which a species may occur. Detailed analyses of the climate envelope were completed for two species of arboreal marsupials – Leadbeater's Possum and the Mountain Brushtail Possum. In both cases, the computer-based climate projection program BIOLCLIM (Nix 1986) was used. The sequence of major steps involved in bioclimatic analysis was the same for both taxa. The first step was to create a file

Table 9.2: Bioclimatic profile of Mountain Brushtail Possum

	Parameter	Mean	SD	Minimum	5%	10%	50%	90%	95%	Max.
1	Annual mean temperature	11	2.62	5.1	7.3	8.2	10.6	14.2	17	20.3
2	Mean diurnal range (mean(monthly maximum−minimum))	10.6	1.32	7.1	8.3	8.7	10.8	12.1	12.5	14.5
3	Isothermality (2/7)	0.48	0.03	0.36	0.42	0.43	0.48	0.51	0.52	0.54
4	Temperature seasonality (CV)	1.5	0.13	1.07	1.28	1.34	1.5	1.7	1.76	1.89
5	Maximum temperature of warmest month	23.3	2.29	16.7	19.8	20.4	23.3	26.4	27.6	31
6	Minimum temperature of coldest month	1.2	2.22	−3.6	−2	−1.6	1	4.1	5.5	10.4
7	Temperature annual range (5–6)	22.1	1.79	16.3	19.2	19.8	22.2	24.6	25.3	28.8
8	Mean temperature of wettest quarter	8.7	5.46	−0.7	2.3	2.9	7	18.2	21.3	24.5
9	Mean temperature of driest quarter	14.8	2.48	3.6	10.2	12.1	15	17.8	18.3	20.5
10	Mean temperature of warmest quarter	16.2	2.41	11	12.8	13.4	15.9	19.1	21.6	24.5
11	Mean temperature of coldest quarter	5.7	2.72	−0.7	2.3	2.8	5.3	9.3	11.8	15.6
12	Annual precipitation	1324	299.89	596	847	906	1305	1707	1831	2838
13	Precipitation of wettest month	154	46.17	63	80	97	150	211	225	451
14	Precipitation of driest month	67	11.99	29	44	50	68	79	87	112
15	Precipitation seasonality (CV)	25	10.49	10	11	12	25	39	44	62
16	Precipitation of wettest quarter	432	130.4	182	221	260	420	595	621	1271
17	Precipitation of driest quarter	222	40.84	108	149	167	224	271	301	350
18	Precipitation of warmest quarter	264	106.38	138	149	160	247	385	532	1067
19	Precipitation of coldest quarter	382	130.58	121	184	214	368	571	599	839
20	Annual mean radiation	16	0.91	14.9	15	15.1	15.7	17.4	18.4	20.2
21	Highest month radiation	25.2	0.5	23.3	24.5	24.7	25.2	26.1	26.3	26.7
22	Lowest month radiation	7	1.44	6.1	6.1	6.2	6.5	9.2	11.2	13.5
23	Radiation seasonality (CV)	43	5.85	21	26	31	46	46	46	46
24	Radiation of wettest quarter	13.4	5.46	6.9	7.2	7.5	10.8	21.9	22.5	24.5
25	Radiation of driest quarter	20.8	3.04	7.2	13.8	15.9	21.2	23.1	23.5	25.6

	Parameter	Mean	SD	Minimum	5%	10%	50%	90%	95%	Max.
26	Radiation of warmest quarter	22.8	1.29	20.7	21.2	21.3	22.7	24.6	25.1	25.7
27	Radiation of coldest quarter	8.3	1.58	7.3	7.4	7.4	7.8	11.2	12.8	15.3
28	Annual mean moisture index	0.9	0.08	0.39	0.75	0.79	0.92	0.99	0.99	1
29	Highest month moisture index	0.99	0.04	0.58	0.96	0.96	0.98	1	1	1
30	Lowest month moisture index	0.65	0.17	0.23	0.36	0.41	0.66	0.88	0.91	1
31	Moisture index seasonality (CV)	14	8.65	0	2	4	13	27	30	42
32	Mean moisture index of high quarter (MI)	0.99	0.04	0.55	0.95	0.96	0.98	1	1	1
33	Mean moisture index of low quarter (MI)	0.71	0.16	0.29	0.44	0.48	0.73	0.91	0.93	1
34	Mean moisture index of warm quarter (MI)	0.79	0.13	0.31	0.54	0.6	0.82	0.93	0.96	1
35	Mean moisture index of cold quarter (MI)	0.98	0.05	0.55	0.91	0.95	0.97	1	1	1

The 5, 10, 50, 90, and 95 percentile values for each bioclimatic variable are given in the bioclimatic profile.
SD = standard deviation, CV = coefficient of variation, MI = moisture index.
Redrawn from Fischer et al. (2001).

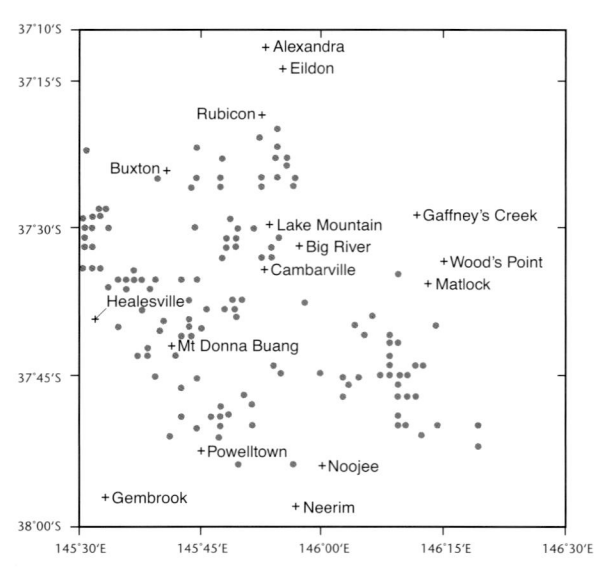

Figure 9.2: Distribution records of Leadbeater's Possum within the Central Highlands of Victoria, used in BIOCLIM modelling of the broad-scale distribution patterns of the species. (Redrawn from Lindenmayer et al. 1991f).

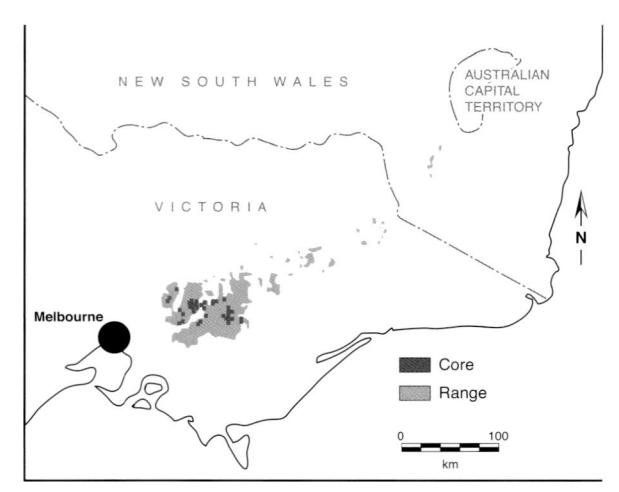

Figure 9.3: Predicted broad-scale distribution patterns of Leadbeater's Possum based on BIOCLIM analyses. The predicted range is based on the minimum and maximum levels in the species' bioclimatic profile. The predicted core distribution is based on matching values for the 10–90% percentiles in the bioclimatic profile. (Redrawn from Lindenmayer et al. 1991f).

containing location data encompassing the known distribution of the target species (Figure 9.2). Estimates of various bioclimatic parameters were then generated using BIOCLIM. A statistical summary was produced of the range of bioclimatic conditions sampled from location records. This summary is called the bioclimatic profile (Table 9.2).

The next step in BIOCLIM analysis was to compare the bioclimatic profile of the target species with a regular grid of bioclimatic indices. The result was a file of grid points where climatic conditions matched those summarised in the bioclimatic profile. Maps were generated of the matched cells, showing the potential bioclimatic domain of the target species (Figure 9.3).

BIOCLIM analyses showed that the climatic conditions most suitable for Leadbeater's Possum were virtually confined to a subsection of the Central Highlands of Victoria. Forests outside the region, including those in north-eastern Victoria where Leadbeater's Possum occurred at the turn of the 20th century (Lindenmayer and Dixon 1992), were predicted by BIOCLIM to support only marginally suitable bioclimatic conditions. Extensive field surveys at these marginal sites failed to detect Leadbeater's Possum (Lindenmayer et al. 1991f). Within the Central Highlands, statistical modelling of the relationships between the presence/absence of Leadbeater's Possum and fine-scaled BIOCLIM-derived climate values revealed that the animal is most likely to occur in ash-type forests characterised by warm and wet climatic conditions.

BIOCLIM analyses of the Mountain Brushtail Possum (Fischer et al. 2001) were based on extensive location records gathered from Melbourne to Gladstone in south-central Queensland. This highlighted the existence of two distinct bioclimatic domains in the southern and northern part of the species' range (Colour Plate 7). It suggested the existence of two separate taxa, which was later confirmed by morphological and genetic data (Lindenmayer et al. 2002d). Southern populations (from Victoria) have a significantly larger ear conch, longer pes and shorter tail than do northern populations (from New South Wales and Queensland). North–south dimorphism was strongly supported by patterns in genetic data, which show genetic distances of 2.7–3% between the southern and northern populations. This is more than twice the level of genetic differentiation between humans and other modern primates such as the Chimpanzee.

The northern form was renamed the Short-eared Possum (*Trichosurus caninus*); the southern form (which retained the common name of Mountain Brushtail Possum) was given the new Latin name *Trichosurus cunninghami* and named in honour of Ross Cunningham, the statistician who discovered the major morphological differences between the northern and southern species (Lindenmayer *et al.* 1995c).

Landscape factors

The attributes of a landscape may have a significant effect on the occurrence of a species in that landscape. For example, the spatial arrangement of particular kinds of patches may make areas suitable (or unsuitable) for a given species (Bennett *et al.* 2006; Forman 1995). For instance, a patch of old-growth forest might be more likely to be occupied by a particular species it if is surrounded by extensive areas of adjacent old growth than if it is embedded within recently clearfelled forest (Harris 1984). On this basis, an analysis examined the relationships between the occurrence of arboreal marsupials at 166 3-ha field sites and the characteristics of the forest in the area surrounding each site – the landscape context of the sites.

Data were extracted from the GIS database for the surface area of forest (measured in m^2) within 20-ha and 80-ha circles surrounding each site for different slope, aspect and age class categories. The area of forest in three age class categories of montane ash forest was calculated: stands older than 1939, forests dating from 1939 to 1960, and stands aged from 1960 to 1990. Non-ash forest was assigned to a fourth category. The surface area of forest (m^2) within each circle was assigned to one of five slope classes: 0–5°, 5–10°, 10–15°, 15–20°, 20–25°, 25–30° and >30°. The length of streams and roads (m) within each circle was recorded. The distance (m) from a given 3-ha survey site to the nearest patch of old-growth montane ash forest was calculated. An example of the terrain surrounding a site is shown in Colour Plate 8.

Analyses of relationships between landscape composition data and the occurrence of arboreal marsupials identified significant explanatory variables for only one of four species examined – the Yellow-bellied Glider (Lindenmayer *et al.* 1999a). This species was more likely to be recorded from landscapes dominated by substantial amounts of old-growth montane ash forest, particularly on steep or flat terrain and on northerly and westerly aspects (Lindenmayer *et al.* 1999a). These results were congruent with those of subsequent studies of arboreal marsupials (Incoll *et al.* 2000). No relationships with landscape composition variables were identified for Leadbeater's Possum, the Mountain Brushtail Possum or the Greater Glider (Lindenmayer *et al.* 1999a).

Patch-level factors

A major study was completed of the relationships between the occurrence of individual species of arboreal marsupials and the size, shape and stand structural attributes of 49 linear strips retained between recently clearfelled montane ash forests. The strips ranged from 30 m to 265 m wide, 125 m to 760 m long and 0.8 ha to 14.6 ha in size (Lindenmayer *et al.* 1993a). All strips were surrounded by 1939-regrowth forest that had been clearfelled up to three years earlier (Lindenmayer *et al.* 1993a).

The work highlighted the effects of patch-level variables on the presence and abundance of particular species. First, Leadbeater's Possum was found to only rarely occupy linear habitat strips even when the areas supported structural features and plant species composition required by the species based on stand-level analysis (see below) (Lindenmayer *et al.* 1993a). Leadbeater's Possum was recorded at only one of 17 strips predicted to be suitable for it, based on measures of habitat suitability (Lindenmayer *et al.* 1993a). The narrow linear configuration of the strips appears to disadvantage colonial species such as Leadbeater's Possum, which has a complex social system and forages on food that is widely dispersed (Lindenmayer *et al.* 1993a).

The Greater Glider and Mountain Brushtail Possum occurred at levels of abundance and particular expected sites, based on habitat suitability shown in stand-level analysis (see below). However, particular attributes of retained strips had a significant effect on the occurrence of both species in those areas. The Greater Glider was less likely to occur in areas that supported a watercourse, possibly because such areas are often characterised by stands of cool temperate rainforest (see Chapter 5) which are not good sources

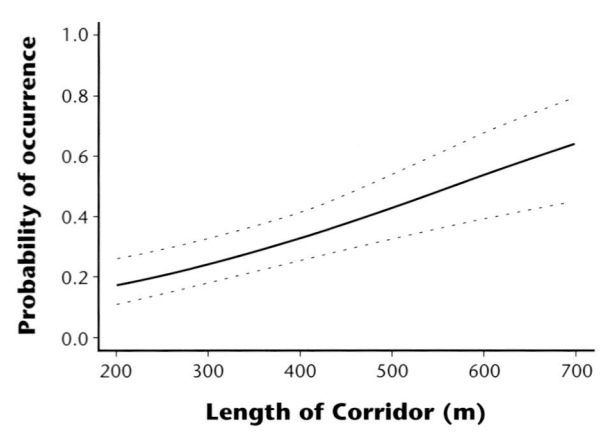

Figure 9.4: Relationships between the probability of occurrence of Mountain Brushtail Possum and the length of a retained linear strip. (Redrawn from Lindenmayer et al. 1993a).

Table 9.3: Models of factors influencing presence and abundance of arboreal marsupials

Logit models for animal presence or absence	
Presence of Leadbeater's Possum	−3.368 + 0.876 log HBT + 0.091 BAA
Presence of Mountain Brushtail Possum	−2.578 + 1.066 log HBT − 0.056 BAA - 2.348 aspect 1
Presence of Greater Glider	−0.993 + 0.554 log HBT − 1.106 AGE 2
Presence of Sugar Glider	−5.19 + 0.909 log HBT + 0.133 BARK
Poisson models for animal abundance	
Mountain Brushtail Possum	−0.408 + 0.502 log HBT
Leadbeater's Possum	1.166 + 0.215 log HBT + 0.034 BARK - 0.085 SHRUBS − 0.025 SLOPE
Greater Glider	1-068 - 0.652 AGE 2

Explanation of habitat variables given in models (see Lindenmayer et al. (1991c) for further details).
HBT = trees with hollows; BAA = basal area of *Acacia* spp.; BARK = index of decorticating bark; SHRUBS = abundance of shrubs; SLOPE = inclination of site; ASPECT = site aspect: divided into two classes – ASPECT 1 = NW–SW, ASPECT 2 = SW–NW; AGE = stand age: divided into two classes – AGE1 = forests older than 1900, AGE2 = forests younger than 1900.

of eucalypt food. The Mountain Brushtail Possum was significantly more often detected in retained linear strips that were 150 m or more in length than in shorter sites (Lindenmayer et al. 1993a) (Figure 9.4). In both cases, the number of trees with hollows in the retained linear strips was important – an attribute in common with stand-level analyses of sites that were not confined to a narrow linear shape (see below).

Stand-level factors

Studies at the stand level encompassed three key steps:

- counting arboreal marsupials at a large number of 3-ha sites;
- collecting various measures of environmental, structure and composition attributes of the vegetation at those sites;
- using statistical analysis to identify relationships between the presence or abundance of a species and measured characteristics of the forest (Lindenmayer et al. 1991d, 1990c).

A major outcome was the production of statistical equations relating the presence or abundance of a given species of arboreal marsupial to a subset of forest measurements. Two relationships were developed: a logistic regression model for the presence or absence of a species and a Poisson regression model for the abundance of a species (Table 9.3). Both types of relationship have been termed habitat requirements models. The performance of each has been tested on a new set of data collected in additional surveys in montane ash forests. Lindenmayer et al. (1994b) showed the models were robust and performed well when rigorously assessed in this way.

The significant explanatory variables in the models in Table 9.3 show differences in the characteristics of the forest where each species is most likely to occur or will be most abundant. The variables typically reflect the suitability of the forest for foraging and the availability of nesting sites for a given species (Lindenmayer et al. 1991d, 1990c). For example, the basal area of *Acacia* spp. probably reflects the availability of leaves for the Mountain Brushtail Possum (Seebeck et al. 1984) and saps for Leadbeater's Possum (Smith 1984). Loose bark hanging on the branches of montane ash trees is an important microhabitat for invertebrates consumed by Leadbeater's Possum and the Sugar Glider. The abundance of trees with hollows was common to all models. Almost all species of arboreal marsupials are obligate cavity-dependent animals and cannot survive without access to den and nest sites in large trees with hollows (see Chapter 6).

Box 9.1: Spatial prediction of an arboreal marsupial species in montane ash forest

The distribution of the Greater Glider is positively influenced by two variables – the age of the forest and the abundance of trees with hollows. Data on these two key variables have been mapped in extensive parts of montane ash forests, using high-quality aerial photographs and expert interpretation of the images. It was possible to use spatial data on forest age and the abundance of trees with hollows, and combine those data with the habitat requirements model developed for the Greater Glider to predict its spatial distribution for the Ada Forest Block – a 6700-ha area in the Powelltown region of the Central Highlands of Victoria (Lindenmayer et al. 1995b) (Colour Plate 9). The resulting map shows areas where the probability of occurrence of the species was very high. These are typically old-growth stands with numerous trees with hollows. Lightly coloured areas are young regrowth forest with few trees with hollows (Lindenmayer et al. 1993a).

Tree level

Several studies in montane ash forests have examined the use of hollow trees as nests or dens by arboreal marsupials (Lindenmayer et al. 1991c, 1991e, 1996c; Lindenmayer and Meggs 1996; Smith et al. 1982) (see Chapter 6). Investigations of the use of trees with hollows for nesting involved three key steps:

- scanning trees with hollows at dusk to determine if they were occupied by an arboreal marsupial;
- measuring the height, diameter and number of types of cavities in each tree;
- using statistical analyses to develop logistic regression models of the relationships between the probability of trees with hollows being occupied by a given species and measured characteristics of those trees (Lindenmayer et al. 1991e).

The performance of the model developed for each species was tested on a new set of data collected in additional surveys of trees within montane ash forests. Lindenmayer et al. (1994b) showed the models were robust and performed well when rigorously assessed in this way.

Characteristics of trees such as the number of types of cavities, overall size and shape, and quantity of surrounding vegetation significantly influenced their likelihood of occupancy. The model for each species contained a different set of significant variables (Table 9.4).

Differences in the characteristics of the occupied trees may be related to factors such as body size of arboreal marsupials (Chapter 6). A change in the size, number and depth of cavities in montane ash trees occurs as they progress through the stages of senescence, death, decay and collapse (Ambrose 1982; Lindenmayer et al. 1993b). Temporal changes in the attributes of trees mean a change occurs in the identity of arboreal marsupials likely to occupy them. For example, small cavities are available in the relatively early stages of tree growth (Ambrose 1982) which favours their use by smaller species and precludes use by larger animals.

Microhabitat-level factors

A substantial field study was used to examine the microhabitat relationships of one species – the Mountain Brushtail Possum. Factors influencing detections of the species by hairtubing at 10 ×10 m plots were quantified. The work demonstrated that the Mountain Brushtail Possum was most likely to be detected by hairtubing in plots where there were numerous tree ferns and understorey trees such as Silver Wattle and Forest Wattle (Lindenmayer et al. 1994d). The

Table 9.4: Logit models of nest tree requirements of arboreal marsupials

Any species of arboreal marsupial	−1.205 − 0.724 shape + 0.089 holes
Leadbeater's Possum	−4.239 − 0.603 size − 0.853 shape + 0.239 holes + 0.137 access
Sugar Glider	−3.275 + 0.436 fissures
Greater Glider	−1.816 + 0.974 size
Mountain Brushtail Possum	−0.080 − 0.694 size − 0.458 shape − 0.196 holes

Shape = transformed variable derived from diameter and height; Size = transformed variable derived from diameter and height; Holes = number of holes in a tree; Fissures = number of fissures in a tree; Access = amount of dense vegetation surrounding a tree.

Table 9.5: Studies directly or indirectly examining distribution and abundance of small terrestrial mammals in the montane ash forests of the Central Highlands of Victoria

Study description	Reference
Patch-level factors	
Small terrestrial mammals in retained linear strips and contiguous forest	Lindenmayer et al. (1994c)
Microhabitat factors	
Microhabitat factors influencing occurrence of small terrestrial mammals in retained linear strips	Lindenmayer et al. (1994c)
Capture probabilities of small terrestrial mammals and microhabitat attributes	Cunningham et al. (2005)

relationship with the presence of the two species of *Acacia* seems likely to be related to dietary requirements. Studies by Owen (1964) and Seebeck et al. (1984) have demonstrated that the Mountain Brushtail Possum consumes the leaves of these *Acacia* species. Studies of the species at the stand level have indicated that the basal area of *Acacia* spp. was an important factor influencing its presence and abundance (Lindenmayer et al. 1990c). These findings emphasise the role of *Acacia* spp. as a key habitat component for wildlife in montane ash forests. The positive relationship between the detection of Mountain Brushtail Possum in hairtubes and the number of tree ferns was an interesting finding and most probably reflects the importance of these plants in the species' diet (Seebeck et al. 1984).

FACTORS INFLUENCING THE DISTRIBUTION AND ABUNDANCE OF SMALL TERRESTRIAL MAMMALS

Three species of small terrestrial mammal – the Bush Rat, Dusky Antechinus and Agile Antechinus – have been the focus of work on factors influencing animal occurrence. Unlike studies of individual species of arboreal marsupials, where up to six spatial scales of work have been completed, investigations of small terrestrial mammals have been far more modest (cf Tables 9.1 and 9.5). Work at two spatial scales is summarised below – the patch level and the microhabitat or plot level.

Patch-level factors
A major study was completed on the relationships between the occurrence of two species of small terrestrial mammals detected by hairtubing (see Chapter 3) and the attributes of 49 linear strips retained between recently clearfelled montane ash forests. The Agile Antechinus was more likely to be detected in:

- sites within contiguous forest than in linear strips of retained forest;
- retained linear strips that spanned several parts of the topography compared with those confined solely to a gully, a midslope or a ridge;
- retained linear strips characterised by few vegetation gaps created by roads (Lindenmayer et al. 1994c).

The Bush Rat was more likely to be detected in sites within contiguous forest than in linear strips of retained forest, but within retained linear strips was more likely to be detected in gullies than elsewhere in the landscape (Lindenmayer et al. 1994c).

Microhabitat-level factors
Data from animal detections from hairtubing studies of retained linear strips and areas of contiguous forest were analysed at the plot level. This was done by creating statistical models of the relationships between animal occurrence and the measured vegetation and other attributes of the 10 ×10 m plots that were the survey unit for the study.

For the Agile Antechinus, plots where the species was most likely to be detected were characterised by large amounts of *Acacia* spp. in the understorey, nearby large trees with hollows, and large amounts of ground cover vegetation. The final statistical model for the Bush Rat also included a variable for the amount of ground cover as well as measures reflecting the density of vegetation at less than 1 m above the ground, and the number of shrubs. The Bush Rat was more likely to

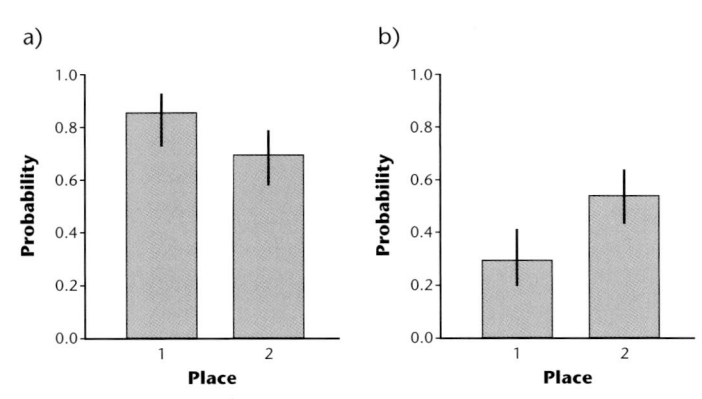

Figure 9.5: Relationships between the variation in topography in retained linear strips and the proportions of hairtubing plots where (a) the Agile Antechinus and (b) the Bush Rat were detected. The categorical variable 'Place' has two levels. Place 1 corresponds to strips that linked two or more parts of a forest landscape (e.g. gully–midslope). Sites in Place 2 were confined to one topographic position. (Redrawn from Lindenmayer et al. 1994c).

be recorded in hairtubes in plots close to gullies (Lindenmayer et al. 1994c). An interesting outcome was that the variables in the final models for both species of small mammal often differed between sites in retained linear strips and contiguous forest, reflecting an interaction between patch-level responses and plot-level responses (Lindenmayer et al. 1994c).

A second study involving trap–recaptures of the Agile Antechinus, Dusky Antechinus and Bush Rat and fine-scale vegetation measures provided data to further explore the influence of microhabitat attributes on occurrence (Cunningham et al. 2005). For example, captures of Bush Rat males were significantly less likely to occur where cover of Myrtle Beech was high (see Chapter 5). This may be due to the fact that the dense canopies of these trees are often associated with an open ground layer supporting few understorey, shrub layer or ground layer plants. Male and female Dusky Antechinus were less likely to be caught in traps where the surrounding area was characterised by many rocks, a high litter cover and numerous overstorey trees. As in the model for Bush Rat males, these measures may be acting as a surrogate for limited ground layer cover. Several other studies have demonstrated that small mammal populations often respond negatively to the absence of localised microhabitat features such as shrub and ground cover (Knight and Fox 2000). These microhabitat features might be important in contributing to providing food resources such as fungi (for the Bush Rat) and invertebrates (for Antechinus spp.). They may also be critical for providing protection from predators.

FACTORS INFLUENCING THE DISTRIBUTION AND ABUNDANCE OF BIRDS

Considerably less work has been conducted on birds in montane ash forests than on arboreal marsupials and small terrestrial mammals. The results of the single investigation to date are therefore not subdivided by the spatial scale of response. This single study was based on repeated points counts (Pyke and Recher, 1983) (see Chapter 3) on 81 sites, and detailed measurements of the vegetation cover at those sites. Statistical modelling procedures have been used to explore the datasets for relationships between the occurrence of a dozen species of birds and environmental attributes, as well as attributes of plant species composition and vegetation structure (Lindenmayer et al. 2009a). These results of statistical modelling are summarised in Table 9.6.

The detailed analyses for individual species of birds revealed some interesting new insights. First, there were different combinations of significant explanatory variables in the final model for each species, with the combinations often including an overstorey tree (forest type) effect and an understorey covariate (e.g. the amount of Acacia spp). Second, there were major differences in bird responses between wet ash forest types

Table 9.6: Statistically significant (P < 0.001) explanatory variables in models of the 12 species of birds recorded in repeated surveys of montane ash forest, 2004–07

Species	Detections	Most important predictors		
White-browed Scrubwren	951	Age (fewest detections in old forest)	Aspect (fewest detections on easterly facing sites)	Tenure (fewest detections in forests managed by Parks Victoria/Melbourne Water)
Brown Thornbill	738	Forest type (AA < SG < MA)		
Grey Fantail	664	Forest type (AA < [SG, MA])	Year (2007 < [2004, 2005])	Aspect (fewest detections on south-facing sites)
Golden Whistler	590	Forest type (AA < SG < MA)	Elevation (most detections on low elevation sites)	
Crescent Honeyeater	541	Forest type (AA < MA < SG)	Number of *Acacia* spp. stems (most detections on sites with high numbers of *Acacia* spp. stems)	Interaction (MA < SG < AA) (response to increasing *Acacia* spp. stems strongest in Alpine Ash forests)
Silvereye	522	Year (2004, 2005) < 2007		
Striated Thornbill	511	Number of vegetation strata (most detections on sites with more vegetation strata)	Forest type (SG < AA < MA)	Tenure (fewest detections in forests managed by Parks Victoria/Melbourne Water)
Eastern Spinebill	371	Elevation (most detections on low elevation sites)		
Crimson Rosella	295	Stocking level of Myrtle Beech stems (most detections on sites with many Myrtle Beech stems)	Elevation (most detections on low elevation sites)	Stumps (most detections on sites with few cut stumps)
Striated Pardalote	290	Age (most detections in old forest)	Year (2005 < 2007 < 2004)	Topographic position ([ridges, midslopes] < gullies)
Eastern Yellow Robin	259	Forest type (AA < SG < MA)	Topographic position (ridges < [midslopes, gullies])	Stocking level of Myrtle Beech stems (fewest detections on sites with many Myrtle Beech stems)
Grey Shrike-thrush	238	Forest type ([AA, SG] < MA)	Forest tenure (fewest detections in forests managed by Parks Victoria/Melbourne Water)	Slope (fewest detections on steep slopes)
White-throated Treecreeper	215	Year ([2005, 2007] < 2004)	Number of *Acacia* spp. stems (most detections on sites with high numbers of *Acacia* spp. stems)	Elevation (most detections on low elevation sites)

Species listed in order of decreasing numbers of detections. Significant explanatory variables are followed by a brief description of response.
MA = stands dominated by Mountain Ash; AA = Alpine Ash; SG = Shining Gum.

such as Mountain Ash and Alpine Ash, even though stands of these tree species can appear superficially similar in structure and plant species composition. Third, many individual species showed a highly significant negative response to elevation. That is, they were less likely to be detected on sites at higher altitudes. The influence of elevation on birds might be associated with temperature effects on bird physiology as well as on food resources such as prey availability, nectar and pollen. Fourth, there were large-scale tenure effects. For example, the Striated Thornbill, White-browed Scrubwren and Grey Shrike-thrush were least likely to be detected in areas managed by Parks Victoria and Melbourne Water. The reasons for these findings are unclear, but they could be associated with landscape-level differences in the management regimes in areas of different tenure. For example, extensive stands of dense young regrowth regenerating after disturbances such as logging are rare in forests managed by Parks Victoria and Melbourne Water, and these appear to be the kinds of stands favoured by species such as the White-browed Scrubwren. However, other factors are likely to contribute to the tenure effects because stand age effects did not characterise the final statistical models for the Striated Thornbill and Grey Shrike-thrush. A fifth insight derived from the detailed analyses of data for individual species of birds was the paucity of stand age effects. The models for only two bird species – the Striated Pardalote and White-browed Scrubwren – included the variable 'stand age'. This was surprising, as earlier, largely qualitative or descriptive studies suggested differences in bird responses to chronological stand age (Loyn 1985).

A final and very important outcome of the study was that the kinds of significant explanatory variables influencing the occurrence of individual bird species were often markedly different from those in models developed in extensive studies of arboreal marsupials. For example, the abundance of trees with hollows, in particular, has been a recurrent significant covariate in models of arboreal marsupials. This variable did not appear in any of the models developed for birds. This was not surprising for most of these taxa, because they do not depend on trees with hollows for nesting. However, although the Crimson Rosella and the White-throated Treecreeper are both hollow-users, the abundance of trees with hollows did not feature in models for either.

SUMMARY

Studies of the factors influencing the distribution and abundance of animals, particularly arboreal marsupials, have been at the core of the research program in montane ash forests for 25 years. The work has generated valuable new insights. First, it is clear that factors at multiple spatial scales are important for individual species. Many workers and resource managers are acutely aware that different species (e.g. owls and lizards) have different spatial requirements. However, many are less aware that the occurrence of any individual species can be influenced by a suite of different factors that operate at different spatial scales (Lindenmayer 2000). This is an important insight because it means that strategies for conservation management will need to be implemented at multiple spatial scales (Lindenmayer and Franklin 2002) as discussed in considerable detail in Part VI.

A second important insight from work on patterns of distribution and abundance has been that, irrespective of the scale studied, different combinations of significant explanatory variables appear in the final model for each species within a given group. This was true for arboreal marsupials, small terrestrial mammals and birds. This provides strong evidence of interspecific partitioning of habitat and hence the resources associated with those habitats. In the case of small terrestrial mammals, evidence was found of intraspecific resource partitioning: different factors characterised the models for the probability of capture of males and females of the Agile Antechinus, Dusky Antechinus and Bush Rat (Cunningham *et al.* 2005).

A third insight has been the substantial effect of the dominant forest type on many species. The occurrence of many species of arboreal marsupials, small terrestrial mammals and birds varied markedly between Mountain Ash and Alpine Ash forests. Therefore, despite these forests often appearing superficially similar to the human eye, they are different enough in structure, plant species composition and environmental conditions that major differences in

> **Box 9.2: Postscript – animal distribution and abundance following the 2009 wildfires**
>
> Natural disturbances such as wildfires can be major ecological processes shaping the distribution and abundance of many elements of the biota (Turner *et al.* 2003), either directly, through killing animals and plants, or indirectly through modifying habitats upon which plants and animals depend (Keith *et al.* 2002). The February 2009 wildfires have undoubtedly had major direct and indirect influences on the biota of montane ash forests. For example, the severity of the fire and the number of biological legacies remaining after the fire are very likely to strongly affect the rate of post-disturbance recovery of biodiversity, as has been observed in other studies in south-eastern Australia such as at Jervis Bay (Lindenmayer *et al.* 2008b) and Tumut (Lindenmayer *et al.* 2005). This will strongly influence the spatial pattern of animal and plant distribution across montane ash forest landscapes. On this basis, a key aim of work following the Black Saturday wildfires will be to contrast the trajectory of post-disturbance ecological recovery in areas that were subject to low-severity fire and where many biological legacies remain, with those where the fire was extremely severe and numbers and types of biological legacies are limited.

vertebrate occurrence have developed. An entire chapter has been dedicated to overstorey forest composition (Chapter 4).

LESSONS LEARNED

Many lessons have been learned as part of an improved understanding of the factors influencing the distribution and abundance of individual animal species in montane ash forests. The first is that identifying separate spatial scales to complete analyses has been an approach of convenience. In reality, many ecological phenomena operate across a continuum of ecological scales (Jarvis and McNaughton 1986; Pickett 1989). However, it is logistically impossible and statistically intractable to work on continua.

It was decided very early in the research program to consider the six scales of analysis summarised in Figure 9.1 as nodal points along a spatial continuum (Mackey and Lindenmayer 2001).

A second important lesson has been that the approach of linking animal occurrence and vegetation attributes required deep thought about the appropriate covariates to be gathered. These needed to be ecologically meaningful for the species targeted for study, and measured at a scale that was appropriate. For example, in the case of the wide-ranging Yellow-bellied Glider that has a home range of 30–60 ha, analyses of forest composition necessarily involved gathering landscape attributes from areas of 20–80 ha around the survey sites (Lindenmayer *et al.* 1999a). Scales of 1–3 ha were appropriate for all other species of arboreal marsupials, which have home ranges broadly similar to the area covered by the field study sites. Building statistical models for the occurrence of species also required careful statistical planning and considerable expertise from professional statisticians. In one case, that of the rare Leadbeater's Possum, a new application of the statistical approach of zero-inflated Poisson regression modelling was developed to deal with the fact that the data collected included many zeros because the species was absent from many survey sites (Welsh *et al.* 1996). That is, the zero class was inflated and the distribution of the data made traditional classes of statistical models difficult to implement (Welsh *et al.* 1996). This innovative approach has many potential applications in studies of other rare species where datasets will be similar to those collected for Leadbeater's Possum (Cunningham and Lindenmayer 2005).

Models of the occurrence of arboreal marsupials were important for many aspects of forest management in the Central Highlands of Victoria. When models may be applied in a management context, it is important to be confident that they are ecologically and statistically robust. On this basis, a further innovation associated with several studies of species occurrence was not to use cross-validation methods, but go to the considerable extra effort of gathering a new dataset to formally test the performance of the statistical models (Lindenmayer *et al.* 1994b).

Factors found to be important for arboreal marsupials, such as stand age and the extent of multi-agedness, either did not appear or appeared only rarely in the final models for birds. Such differences between models for arboreal marsupials and for birds, together with the different models for different species of birds, strongly suggest that neither group, nor a particular taxon within a group, would be a robust surrogate or biodiversity indicator for other species or groups. This topic is examined in more detail in Chapter 11.

A final and somewhat sobering lesson from 25 years of work has been that an understanding of the factors influencing the occurrence of a given species may not extrapolate well between different kinds of forest. That is, the models may perform well in montane ash forests, but poorly in other forest types. The Greater Glider provides a useful example. The occurrence of the species is strongly associated with the age of ash-type forests and the abundance of trees with hollows (Lindenmayer et al. 1995b). However, neither stand age nor the abundance of trees with hollows were significant explanatory variables in models of the species' occurrence in the Tumut region of southern New South Wales. In that region, forest type is a significant explanatory variable and stands dominated by Narrow-leaved Peppermint (*Eucalyptus radiata*) and Manna Gum (*Eucalyptus viminalis*) are where the species is most likely to occur (Lindenmayer et al. 1999c). Other factors such as slope, which are not important in models of the occurrence of the Greater Glider in ash-type forests, are prominent in equivalent models for the species at Tumut (Lindenmayer et al. 1999a). Similar outcomes have resulted from work on the same species of small terrestrial mammals in the montane ash forests in Victoria (Lindenmayer et al. 1994c) and at Tumut (Lindenmayer et al. 1999c), as well as for birds that inhabit both environments (Lindenmayer et al. 2002c, 2009a).

KNOWLEDGE GAPS

The focus of this chapter and much of the animal occurrence work in montane ash forests has been on arboreal marsupials and, to a lesser extent, on small terrestrial mammals and birds. Other groups have received much less attention. Bats, for example, have not been well studied even though they are a species-rich group of mammals in montane ash forests. Other researchers have examined issues such as foraging patterns (Brown et al. 1997), but only limited additional work has extended beyond that investigation (e.g. Lumsden et al. 1991). There are good reasons for this. Bats move large distances in a single night (Lumsden et al. 1994) and data gathered from methods such as trapping of individuals are difficult to interpret. It is hard to know if the point of capture reflects the place where animals are foraging, nesting or moving between the two.

Reptiles inhabiting montane ash forests have received very limited attention from researchers (Brown and Nelson 1993). Again, there are good reasons for this – primarily because the reptile assemblage in montane ash forests is not particularly species-rich and the abundance levels of the vast majority of individual species appear to be very low. This makes traditional kinds of statistical analyses of occurrence patterns not particularly tractable.

As well as these issues with studies of such groups as bats and reptiles, there are important gaps in knowledge about their response to natural and human disturbance and, in turn, the effectiveness of management practices designed to conserve these elements of biodiversity in montane ash forests.

Other groups in montane ash forests have received little attention from researchers. These include the invertebrates – the groups which comprise the vast proportion of biodiversity. Some early studies investigated these kinds of animals (Neumann 1991, 1992) and others looked at pest taxa (Neumann and Marks 1976). However, much remains to be done on invertebrates in montane ash forests. In 2009, a new set of initiatives began examining the response of particular groups, such as beetles, to natural disturbance and contrasting it with response to disturbances from logging operations. It is hoped that this new work will build into a consolidated research program in the coming years. The importance of this work has become particularly important in the context of the 2009 wildfires and ecological recovery following them.

Finally, the body of work summarised in this chapter has focused on the pattern of animal occurrence in the landscape. Much less work has explored the ecological processes underlying these distribution patterns. For example, levels of reproduction within a site may not only influence the persistence of a species at that site, but the dispersal of offspring from that area can influence levels of patch occupancy and distribution in other ensembles of patches and hence across broader landscapes (Hanski 1998). This kind of phenomenon underlies concepts such as source-sink dynamics (Pulliam et al. 1992), in which source populations produce many offspring that maintain populations in other areas characterised by lower rates of reproduction. Thus, there are distinct cross-scale effects where site-level processes influence site, patch and landscape patterns of occurrence. These links between the ecological processes of reproduction, dispersal and site occupancy and, in turn, larger-scaled distribution patterns, are notoriously difficult to study. However, long-term demographic and genetic datasets gathered for the Mountain Brushtail Possum (Lindenmayer et al. 1998; Viggers and Lindenmayer 2004) are enabling the quantification of exciting new relationships between population dynamics, dispersal and distribution patterns within montane ash forests (Banks et al. 2009, 2008).

FOREST PATTERN, ECOLOGICAL PROCESS AND LINKS TO OTHER CHAPTERS

Animal distribution is a forest pattern. The distribution and abundance of a species are influenced by many factors at a range of spatial scales (Mackey and Lindenmayer 2001). These include ecological processes such as climate regimes and disturbance regimes. They also include other forest patterns such as the occurrence of particular forest types and individual plant species (e.g. *Acacia* spp trees in the understorey), the occurrence of other individual species (e.g. predators and competitors) and the occurrence of particular structural attributes of stands or landscapes (e.g. large trees with hollows).

Links to other chapters

Because the distribution of particular species of animals is influenced by a large number of other patterns and ecological processes, this chapter is strongly linked with almost every other chapter in this book (Figure 9.6). This highlights the value of studying a range of topics in the one ecosystem because, collectively, the body of knowledge gained substantially exceeds the sum of its parts.

Animal occurrence at a site is a function of the viability of populations in landscapes and regions – a topic examined in Chapter 10. This chapter has examined the occurrence of individual species, but each taxon is influenced by the activities of other taxa. Chapter 11 focuses on aggregate species responses and community-level topics.

The distribution of many individual species is strongly related to the composition of the overstorey ash forest (Chapter 4) and rainforest (Chapter 5). The various models for individual species highlight the importance of stand structural attributes and all of

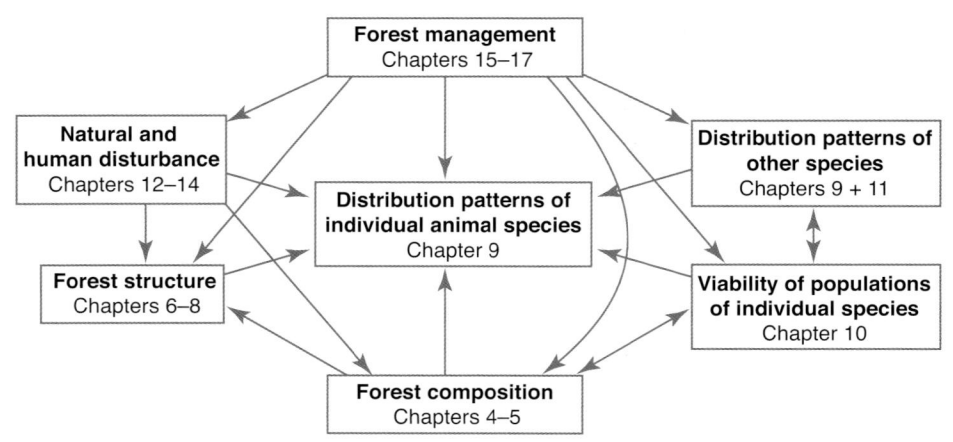

Figure 9.6: Ways in which species distribution patterns are directly or indirectly influenced by a range of key factors examined in other chapters.

Part III is dedicated to that topic – the structure of the overstorey in Chapter 6, the understorey in Chapter 7 and the ground layer in Chapter 8. The composition and structure of the overstorey, understorey and ground cover and, in turn, the distribution and abundance of individual species are heavily influenced by natural disturbance (particularly wildfires) and human disturbance (primarily logging). These topics are the focus of Part V (Chapters 12–14). The ways in which montane ash forests are managed has a major influence on the composition and structure of the overstorey, understorey and ground cover and therefore on the distribution and abundance of the species associated with these key forest attributes. Hence, this chapter is intimately linked with the three chapters on forest management in Part VI.

Following the 2009 Black Saturday wildfires, post-fire ecological recovery will be a critical factor influencing the distribution and abundance of biota in montane ash forests. Continuation of the long-term monitoring described in Chapter 17 will be vital in quantifying the recovery process and, in turn, better understanding patterns of animal distribution and abundance in montane ash forests.

10 Viability of populations of individual species

INTRODUCTION

The previous chapter focused on the wide range of studies that have examined the factors influencing the distribution of particular species in montane ash forests. Habitat suitability is a key factor, but in some circumstances, although suitable habitat may be available, a species may not occupy it. For example, local populations of a species may not be viable, leading to local extinctions and hence unoccupied suitable habitat. Thus, population viability can be an important factor influencing the distribution and abundance of many species (Possingham et al. 2001).

A substantial body of work in the Victorian montane ash forests has involved modelling the viability of populations of arboreal marsupials using generic models for population viability analysis (PVA). The research was some of the first to link life history data for arboreal marsupials, spatial and temporal data on habitat availability for these animals, and information on disturbance regimes (particularly fire and logging). Extensive work using PVA models was completed during the mid 1990s and insights from that research were used to guide management planning, particularly in the recovery plan for Leadbeater's Possum (Lindenmayer and Possingham 1995c).

This chapter contains three sections. The first is a short overview of some background to the application of PVA models. The second summarises key results for species that were the focus of PVA modelling in the Central Highlands of Victoria. The final section comprises a general synthesis of modelling outcomes.

BACKGROUND: PVA

Extinction risk assessment tools such as PVA explore issues associated with the viability of populations (Fieberg and Ellner 2001; Possingham et al. 2001). PVA is critical because the assessment of species extinction risk lies at the heart of conservation biology (Burgman et al. 1993; Fagan et al. 2001). The objective of PVA is to provide insights into how management can influence the probability of extinction (Boyce 1992; Possingham et al. 2001). It provides a basis on which to evaluate data and anticipate the likelihood that a population will persist. More generally, PVA may be seen as a systematic attempt to understand the processes that make a population vulnerable to decline or extinction (Gilpin and Soulé 1986; Shaffer 1990). In practice, PVA usually refers to building computer-based models of the likely fate of a population.

Probabilities of extinction are estimated by Monte Carlo simulation. The most appropriate model structure depends on the availability of data, the essential features of the species' ecology and the kinds of questions that managers need to answer (Burgman *et al.* 1993; Starfield and Bleloch 1992).

PVA has been used widely around the world – there are hundreds of examples of its use in studies on a wide range of species (Beissinger and McCullough 2002), particularly on rare, declining and threatened taxa (Lindenmayer and Burgman 2005). There are several examples where predictions from models and actual dynamics of populations have compared favourably, such as the Puerto Rican Parrot (*Amazona vittata*), Whooping Crane (*Grus americana*) and Lord Howe Island Woodhen (*Tricholimnas slyvestris*) (Brook *et al.* 2000; Lacy *et al.* 1989; Mirande *et al.* 1991). However, in general, PVAs have produced very variable predictions (Coulson *et al.* 2001; Ellner *et al.* 2002).

Since Brook *et al.* (2000) assessed the predictive accuracy of PVA for 21 populations (eight bird, 11 mammal, one reptile and one fish species), discussion of the predictive accuracy of PVAs has increased (Beissinger and McCullough 2002; Boyce 1992; Brook 2000; Brook *et al.* 2000; Fagan *et al.* 2001; Fieberg and Ellner 2001; O'Grady *et al.* 2004; Reed *et al.* 2002). Data availability often limits the predictive accuracy of most PVAs (Boyce 1992; Burgman *et al.* 1993; Caughley 1994; Ellner *et al.* 2002; Taylor 1995). Even the simplest models require more parameters than are usually available. Even so, PVAs can be valuable in several ways. For example, the use of the models can help organise information for subsequent empirical tests (Walters 1986), especially if they summarise available data consistently and transparently (Brook *et al.* 2000). Their use can help identify knowledge gaps (Burgman *et al.* 1993), highlight problems for which pre-emptive action may be beneficial, and promote the understanding of complex ecological processes that might otherwise be overlooked in fieldwork (Bender *et al.* 2003; Gilpin and Soulé 1986; Temple and Cary 1988). In addition, through sensitivity analyses, PVA models allow the identification of which model structures and parameters are most important (Lindenmayer and McCarthy 2006; Possingham *et al.* 2001).

PVA MODELLING OF ARBOREAL MARSUPIALS IN MONTANE ASH FOREST

Three species of arboreal marsupials have been targeted for extensive analysis using PVA models. These are Leadbeater's Possum, Greater Glider and Mountain Brushtail Possum. Leadbeater's Possum has received by far the greatest amount of attention because forest and wildlife management strategies have been based, in part, on the extensive risk assessments and other related modelling studies completed on that species (Table 10.1). Three widely available models were used in simulations of Leadbeater's Possum:

- VORTEX (Lacy 2000);
- ALEX (Possingham and Davies 1995);
- RAMAS (Akçakaya and Ferson 1990).

The three models are structured quite differently and simulate population processes in markedly different ways (Lindenmayer *et al.* 1995a). For example, VORTEX contains detailed subprograms for modelling inbreeding depression and losses of genetic variability in small populations. ALEX is a more general simulation package focusing on the dynamics of a single (population-limiting) sex – typically females. A detailed description of the models is beyond the scope of this book; a comprehensive discussion of each is presented in papers by the architects of each model (Akçakaya and Ferson 1990; Lacy 2000; Possingham and Davies 1995).

A key principle underpinning the use of generic models for PVA is that different models have different strengths, limitations and underlying assumptions. Thus, a given model may be better suited to a particular problem than another model is (Lindenmayer *et al.* 1995a). Even for the same species, different problems may be best addressed using different packages. For example, in the case of Leadbeater's Possum, ALEX was the best program for comparing and ranking management options in a dynamic and complex landscape mosaic where timber harvesting was also permitted (Lindenmayer and Possingham 1995c). Conversely, an examination of the behaviour of small metapopulations of Leadbeater's Possum in response to an interplay of genetic and demographic processes was best completed using VORTEX (Lindenmayer *et al.* 1995a). Finally, Burgman *et al.* (1995) used one of

Table 10.1: Studies of arboreal marsupials using PVA

Species	General objective	Model used	Reference
Leadbeater's Possum	Viability of single populations	VORTEX	Lindenmayer et al. (1993c)
	Group sizes for reintroduction	RAMAS	Burgman et al. (1995)
	Inter-patch dispersal in a metapopulation	VORTEX	Lindenmayer and Lacy (1995c)
	Wildfire effects	ALEX	Lindenmayer and Possingham (1995b)
	Logging effects	ALEX	Lindenmayer and Possingham (1996d)
	Optimal reserve size	ALEX	Lindenmayer and Possingham (1996d)
	Importance of habitat patches	ALEX	Lindenmayer and Possingham (1995a)
	Inter-patch dispersal and connectivity	ALEX	Lindenmayer and Possingham (1996a)
	Tests of model predictions	ALEX	Lindenmayer and McCarthy (2006)
Mountain Brushtail Possum	Inter-patch dispersal in a metapopulation	VORTEX	Lindenmayer and Lacy (1995b); Lindenmayer and Lacy (1995a)
	Genetic effects of inter-patch dispersal	VORTEX	Lacy and Lindenmayer (1995)
Greater Glider	Persistence in a network of patches	ALEX	Possingham et al. (1994)
	Inter-patch dispersal in a metapopulation	VORTEX	Lindenmayer and Lacy (1995a)
	Reserve design and old-growth patch management	ALEX	McCarthy and Lindenmayer (1999)
	Tests of model predictions	ALEX	Lindenmayer and McCarthy (2006)

the packages in the library of RAMAS programs to predict the size and social structure of groups of Leadbeater's Possum for release, to maximise the success of a reintroduction program.

The broad range of studies completed using PVA models is summarised in Table 10.1. A more detailed overview of the key results of these various investigations is presented below.

BACKGROUND DATASETS

One of the strengths of using PVA for modelling studies is that the approach provides a framework for drawing together relevant information for a particular species or a particular problem (Burgman et al. 1993). Many kinds of data were used in work involving PVA models in montane ash forests. They included:

- life history attributes of target species such as fecundity, mortality, social organisation and carrying capacity. Information was used from populations from within the Central Highlands of Victoria when available, and from nearby populations when it was not;
- information about natural and human disturbance regimes in montane ash forests (see Part V), particularly the frequency and intensity of wildfires and logging and their effects on the suitability of the forest as habitat for animals (see Chapter 9);
- spatial data on the current and predicted future location of suitable habitat patches for arboreal marsupials. Much of this information was derived from high-resolution aerial photography done for the Victorian Department of Sustainability and Environment.

PVA MODELLING OF LEADBEATER'S POSSUM

By far the largest body of PVA work has been on Leadbeater's Possum, examining a wide range of issues. Indeed, Leadbeater's Possum has undergone more PVA modelling than possibly any other species worldwide.

Viability of single populations

The first PVA work conducted on Leadbeater's Possum involved using VORTEX to explore relationships between the size of single populations and their susceptibility to extinction (Lindenmayer et al. 1993c). The work suggested that single isolated populations of fewer than 50–100 animals were highly susceptible to extinction over a 100-year period and were likely to suffer significant levels of inbreeding depression (Lindenmayer et al. 1993c). Increasing levels of environmental variation over that time were predicted to contribute to increased risks of extinction and losses of expected heterozygosity (Figure 10.1). In contrast, populations larger than 200 individuals were predicted to be relatively stable, both demographically and genetically (Lindenmayer et al. 1993c).

This modelling study largely ignored the impacts of other factors such as logging, wildfire and decline in habitat suitability due to the collapse of trees with hollows (see Chapter 6). Notably, when the last factor was modelled, declining habitat suitability resulted in the rapid extinction of small and localised populations (Lindenmayer et al. 1993c).

Group sizes for reintroduction

In the 1980s and 1990s there was a large captive population of Leadbeater's Possum (Myroniuk 1995), and the possibility of the species being reintroduced to areas of unoccupied forest was contemplated. As a result, a small-scale nest box study was trialled with the release of captive-bred animals (M. Macfarlane, unpublished data). A key issue associated with the success of reintroduction programs is the size of the groups released (McCallum et al. 1995). General reviews of reintroduction and translocation programs indicate that they are more likely to be successful when large numbers of animals are released (Fischer and Lindenmayer 2000).

Burgman et al. (1995) used the PVA program RAMAS to examine the relationships between the size of initially released populations of Leadbeater's Possum and extinction risk. They also explored the effects on extinction risk of the composition of groups of released animals (adults, subadults) and group behaviours such as Allee effects (Allee et al. 1949). The work predicted that releases of larger groups had an increased chance of success, particularly groups above ~20 animals (Burgman et al. 1995) (Figure 10.3), where success was defined as the persistence of animals after 20 years. Burgman et al. (1993) also found that adults were more valuable than subadults in terms of contributing to a reduced risk of failure in a reintroduction program. The effectiveness of group behaviours such as mobbing to drive off predators such as owls (Smith 1980) influenced the success or failure of attempts to re-establish populations of Leadbeater's Possum (Burgman et al. 1993).

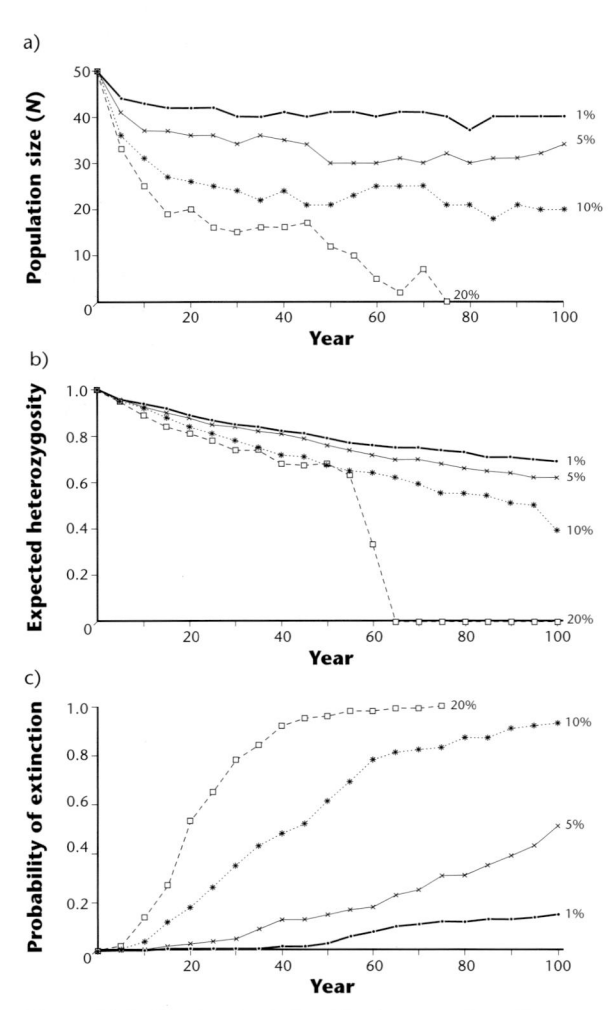

Figure 10.1: Simulated relationships between founding populations of 50 individuals of Leadbeater's Possum and expected heterozygosity and extinction probability for varying levels of environmental variation (higher values lead to increased numbers of breeding failures. (Redrawn from Lindenmayer et al. 1993c).

Figure 10.2: A captive Leadbeater's Possum in Healesville Sanctuary in 1991. No populations of the species now remain in captivity, although there is a single adult female at the Toronto Zoo in Canada. (Photo by David Lindenmayer).

Interpatch dispersal

In this study, the computer program VORTEX was used to estimate the viability of populations of Leadbeater's Possum. The study simulated population dynamics and genetic variability in multiple habitat patches of varying number, size and carrying capacity. The impacts of different rates of migration of dispersing individuals between subpopulations were also examined (Lindenmayer and Lacy 1995c).

Computer simulations of populations of 20 or fewer animals were characterised by very rapid rates of extinction, and most populations typically failed to persist for longer than 50 years. Increasing either the rate of migration or the number of small subpopulations exacerbated the demographic instability of metapopulations. This was reflected by lower rates of population growth and depressed probabilities of population persistence. These effects appeared to be associated with the substantial impacts of demographic stochasticity on very small populations, together with the dispersal of animals into empty patches or functionally extinct (i.e. single-sex) subpopulations. Lindenmayer and Lacy (1995c) identified an extinction vortex that resulted from interacting demographic and stochastic processes that they believed was likely to threaten small populations and loosely interlinked subpopulations of Leadbeater's Possum (Figure 10.4).

The study revealed significant differences between the metapopulation dynamics of 40 animals and those comprising 20 or fewer individuals. Increased migration and the addition of subpopulations of 40 animals resulted in higher rates of population growth,

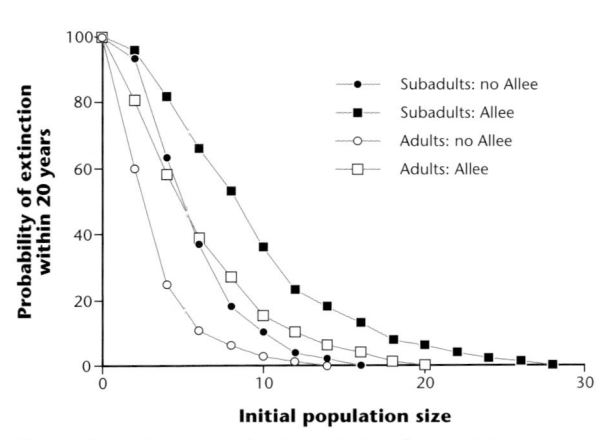

Figure 10.3: Risks of extinction of a Leadbeater's Possum population, given the release of a fixed number of subadults or adults into suitable vacant habitat. The models assumes equal numbers of males and females are released, and that the release takes place only in the first year of the simulation without any subsequent additions to the population. Simulations were run with and without Allee effects. Each point is the result of 500 replications. (Redrawn from Burgman *et al.* 1995.)

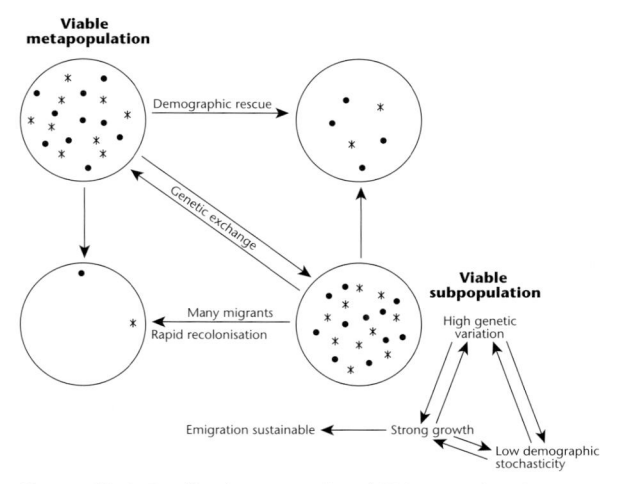

Figure 10.4: Feedback among destabilising stochastic processes that combine to create a metapopulation extinction vortex in Leadbeater's Possum. Metapopulation instability is amplified as increasing numbers of unstable subpopulations (undergoing local extinction vortices) are connected by increasing rates of migration. Symbols (dots and stars) inside the large circles represent females and males within subpopulations. (Redrawn from Lindenmayer and Lacy 1995c).

lower probability of extinction and longer persistence times. Extinctions in these scenarios were more likely to be reversed through recolonisation by dispersing individuals. At the highest rates of migration, subpopulations of 40 individuals were essentially panmictic and behaved as a single larger population. Increased numbers of subpopulations and accelerated rates of migration slowed the loss of expected heterozygosity in all scenarios. However, there was a significant (>10%) loss in expected heterozygosity over a 100-year period, even at the highest rates of migration among five subpopulations of 40 animals (Lindenmayer and Lacy 1995c).

The results of that modelling study predicted that while demographic stability may occur in populations of 200 individuals of Leadbeater's Possum, considerably more individuals may be required to avoid a significant decline in genetic variability during a 100-year period.

The outcomes suggested that, over the next 100 years, populations of Leadbeater's Possum may be lost from extensive parts of its present range. An intriguing outcome was that there may be some metapopulation structures where the addition of subpopulations and increased migration may have a negative effect on population persistence (Lindenmayer and Lacy 1995c).

Importance of ensembles of habitat patches

The population simulation computer program ALEX was used to estimate the probability of extinction of Leadbeater's Possum in complex networks of remnant habitat patches within four large areas of forest characterised by different amounts of old-growth forest and other kinds of habitat suitable for the species (Lindenmayer and Possingham 1995d). These blocks were the O'Shannassy Water Catchment and the Murrindindi, Ada and Steavenson Forest Blocks. Forest inventory data on the spatial location and size of potentially suitable habitat patches (Figure 10.5) was combined with data on the life history attributes of the Leadbeater's Possum, as well as submodels that tracked the dynamics of key habitat attributes (Figure 5.7), to simulate the behaviour of metapopulations of the species over the next 150 years.

This study involved simulated populations of animals in hypothetical isolated patches of old-growth

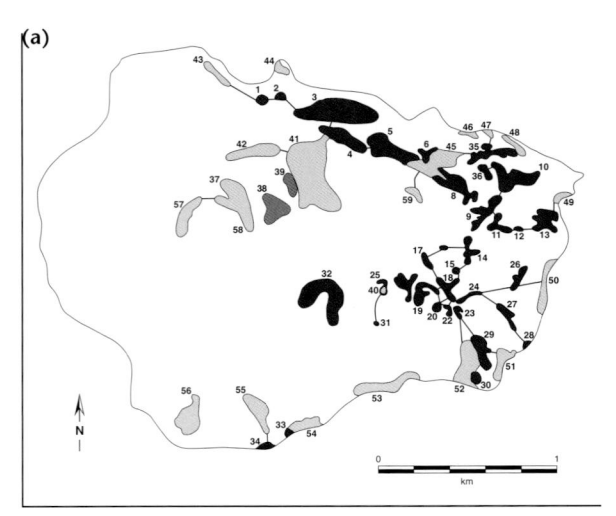

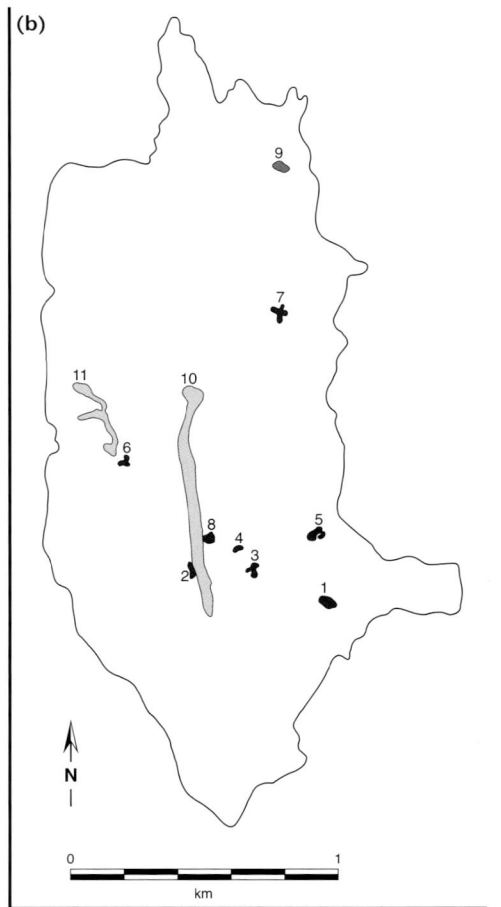

Figure 10.5: Patch structures used in simulating populations of Leadbeater's Possum. Two forest blocks are shown. (a) The 5500-ha Steavenson Forest Block. (b) The 3500-ha Murrundindi Forest Block. (Redrawn from Lindenmayer and Possingham 1995d).

a)

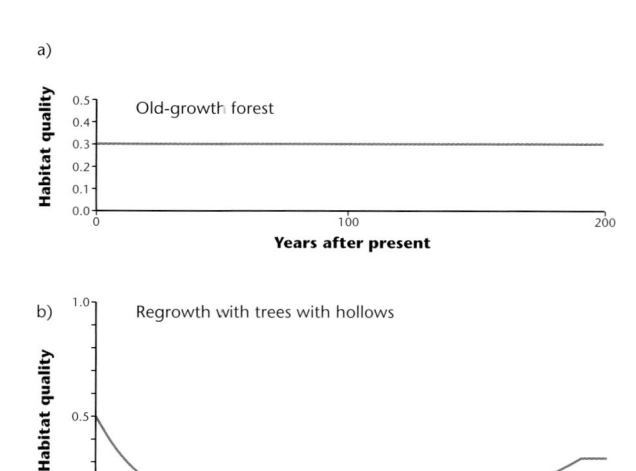

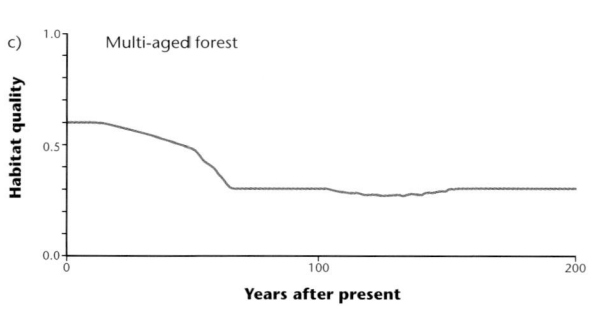

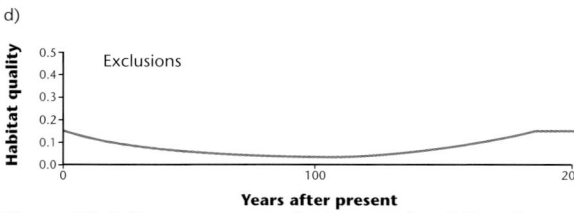

b)

c)

d)

Figure 10.6: Response curves for temporal variations in habitat suitability in four kinds of forest patches modelled for the dynamics of Leadbeater's Possum. The types of areas were (a) old-growth forest, (b) regrowth stands with numerous dead trees with hollows, (c) multi-aged forests where there was a combination of dead and living trees with hollows and (d) areas excluded from logging (steep slopes and streamside reserves). The line in each submodel represents the habitat quality value and corresponds to the maximum number of breeding females per ha that can occur in a given type of habitat. (Redrawn from Lindenmayer and Possingham 1995d).

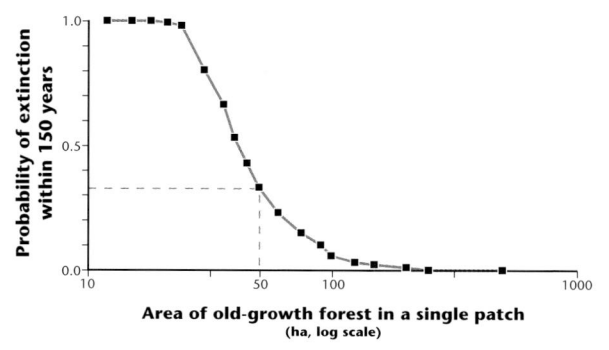

Figure 10.7: Predicted relationship between the size of a single hypothetical patch of old-growth forest and the predicted probability of extinction. (Redrawn from Lindenmayer and Possingham 1995d).

forest of varying size, and assemblages of real patches of potentially suitable habitat within each of the four study areas. In addition, a detailed investigation was completed for the Ada Forest Block, in which habitat areas of different size and spatial isolation were deleted from the patch structure.

The analyses predicted that single isolated populations inhabiting patches smaller than 20 ha were highly susceptible to extinction. For areas larger than that, the probability of survival increased with patch size and approached 1% for patches that exceeded 150 ha (Lindenmayer and Possingham 1995d).

Analyses of complex ensembles of patches indicated that Leadbeater's Possum was highly vulnerable to extinction during the next 150 years within the 3500-ha Murrindindi Forest Block, where only 25 ha of old-growth forest remain and the largest single patch is only ~3 ha in size. The species was predicted to have better prospects of survival in the Steavenson Forest Block and the O'Shannassy Water Catchment, where there are relatively large remaining areas of old-growth forest (approximately 250 ha and 6420 ha respectively) (Lindenmayer and Possingham 1995d).

The retention of the few remaining relatively large habitat patches (exceeding 12 ha) in the Ada Forest Block was predicted to be critical for the persistence of Leadbeater's Possum in the area. The removal of relatively small (<3 ha) and/or isolated patches of old growth had a minor impact on the predicted probability of extinction.

This study did not examine the impacts of major disturbances such as logging and wildfire but those factors were a focus of subsequent modelling studies, outlined below.

Table 10.2: Estimated probability of extinction (%) of populations of Leadbeater's Possum in response to wildfire in four forest management areas

Study area	Predicted probability of extinction after 150 years	Predicted probability of extinction after 300 years	Median time to extinction (years)
Murrindindi Forest Block			
No fire	100	100	10
With fire	100	100	10
Ada Forest Block			
No fire	34	48	–
With fire	78	98	87
Steavenson Forest Block			
No fire	2	5	–
With fire	48	84	162
O'Shannassy Water Catchment			
No fire	0	0	–
With fire	4	15	–

a   The annual probability of a fire was 1%; 75% of patches were burnt in a given fire event.
For those scenarios where the impacts of wildfires were modelled, there was no added impact of post-fire salvage logging operations incorporated in the analysis.

Wildfire effects

ALEX was used to simulate the impacts of wildfires on the persistence of metapopulations of Leadbeater's Possum in the O'Shannassy Water Catchment and the Murrindindi, Ada and Steavenson Forest Blocks (Lindenmayer and Possingham 1995a). Wildfires resulted in an increase in the predicted probability of metapopulation extinction in the four areas targeted for study, particularly in the Murrindindi and Ada Forest Blocks where there are lower total areas of, and significantly smaller, suitable habitat patches (Lindenmayer and Possingham 1995a).

The probability of extinction of Leadbeater's Possum was predicted to be lowest when there were frequent fires that burned relatively small areas of a given forest block. The analyses suggested that populations are vulnerable to infrequent but intensive conflagrations that burned a large proportion of the forest.

An additional study of fire effects on extinction risk in Leadbeater's Possum was completed by McCarthy and Lindenmayer (2000). This work used a reconfigured version of ALEX to model spatial correlation in wildfires, where correlations between the incidence of fires declined with distance. It showed that the predicted risk of extinction grew with an increasing variance in number of fires per year and increasing mean fire interval.

Mitigating logging effects

ALEX was used to model the effectiveness of forest management strategies to enhance the persistence of Leadbeater's Possum in wood production areas (Lindenmayer and Possingham 1995b, 1996c). The work focused on the relationships between the risk of

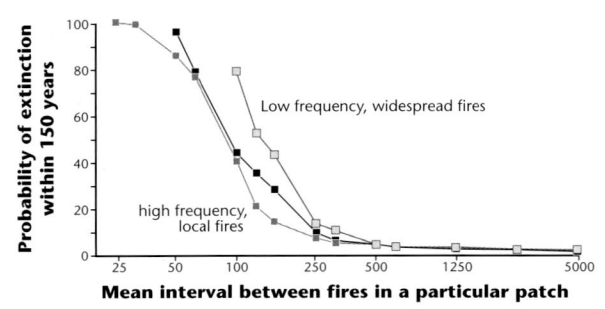

Figure 10.8: Factors affecting the probability of extinction – relationships between fire frequency, wildfire extent and interval between fire in a habitat patch. (Redrawn from Lindenmayer and Possingham 1995a).

metapopulation extinction in the Murrindindi and Steavenson blocks, the number and spatial arrangement of 50-ha logging areas that could be reserved from timber harvesting and the impacts of post-fire salvage logging in reserved areas.

The analyses showed that the exclusion of salvage logging operations from burned old-growth forests significantly improved the species' prospects of survival in both the short and long term. Withdrawal of timber harvesting from some proposed logging coupes made a significant positive long-term contribution to metapopulation persistence. Given this, the effectiveness was examined of different designs for setting aside a total reserved area of 300 ha. These ranged from a single 300-ha reserve to twelve 25-ha reserves. Populations in smaller reserves were vulnerable to extinction from demographic stochasticity and environmental variability. Conversely, a small number of larger reserves were susceptible to being destroyed in a single catastrophic wildfire, highlighting a need for several dispersed reserves.

Analyses of the sensitivity of various management options to variations in fire frequency and extent, movement capability and a wide range of other factors indicated that the conservation strategy offering the best relative outcome for Leadbeater's Possum was involved setting aside several 50–100-ha reserves in every forest block and precluding post-fire salvage logging operations from those areas if they are burned in wildfire (Figure 10.9) (Lindenmayer and Possingham 1996c).

Management recommendations for the conservation of Leadbeater's Possum

The extensive PVA modelling studies resulted in a suite of recommended management strategies for Leadbeater's Possum (Lindenmayer and Possingham 1995c), based on a two-tiered reserve system. This included a single large national park (Yarra Range National Park) and an array of smaller reserves in every wood production forest block. The design principles for reserves in each forest block were:

- reserves should be located where they have a lower chance of being burned, for example in forests on southerly slopes;
- reserves should not be salvage logged if they are burned in a wildfire;
- reserves should not be isolated from each other. Forests which connect reserves may assist animals to move between habitat patches. This could be valuable where dispersing animals may help re-establish populations in patches where localised extinctions have occurred. However, it is important to ensure that these links do not facilitate the spread of wildfires between reserves;
- the size of reserves should be approximately 50–100 ha because, in the long term, populations in smaller reserves could be vulnerable to problems such as loss of genetic variability;
- a total of 600–1000 ha of existing regrowth forest should be set aside as reserves for Leadbeater's Possum in every forest block. This corresponds to 12–20 50-ha reserves or 6–10 100-ha reserves in each block. Larger forest blocks (>4000 ha) would contain about 1000 ha of old-growth reserves. There would be approximately 600 ha of reserves in the smaller blocks (<4000 ha);
- some reserves should be located near existing patches of old-growth forest. If populations of Leadbeater's Possum manage to persist in these areas, dispersing offspring may more readily colonise the surrounding areas as suitable habitat develops. Such a strategy will eventually increase

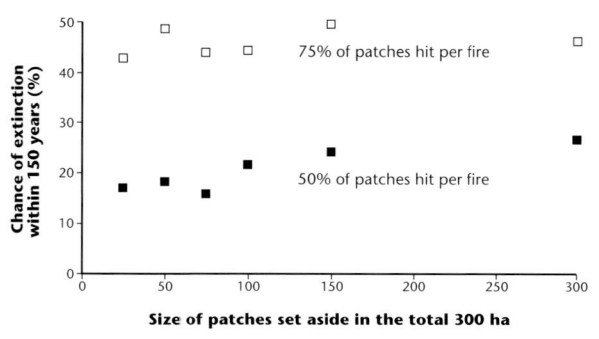

Figure 10.9: Relationships between the probability of extinction of Leadbeater's Possum in the Steavenson Forest Block, wildfire and various configurations of 300-ha reserve areas. The fire regime was an annual fire probability of 1% with 50% and 75% burned in a given conflagration. (Redrawn from Lindenmayer and Possingham 1995c).

the total size of patches of old-growth forest. This will be particularly important in some wood production forest blocks where there are limited areas of old-growth forest;
- attempts to suppress wildfires should be maintained as even the largest remaining areas of old-growth forest may be susceptible to burning by repeated widespread wildfires that could result in localised and/or global extinction of the species.

Although large and small reserves are an integral component of forest management for Leadbeater's Possum, other strategies will also be important. These include the continued use of a management zoning system, the maintenance of networks of wildlife corridors, the exclusion from logging of forest on steep terrain and streamside reserves, and the retention of trees on logged sites (Lindenmayer and Franklin 2002; Macfarlane *et al*. 1998).

Issues associated with the reserve system and the management of montane ash forests are discussed in more detail in Chapters 14 and 15.

PVA MODELLING OF OTHER ARBOREAL MARSUPIALS
Importance of habitat patches for the Greater Glider

A detailed study of the metapopulation viability of the Greater Glider in the 6600-ha Ada Forest Block was undertaken using ALEX (Possingham *et al*. 1994). As with work on Leadbeater's Possum, the model involved information on the spatial location, structure and composition of different habitat patches within the study area, estimates of the species' life history attributes and knowledge of disturbance regimes such as fire and logging (Possingham *et al*. 1994).

This work produced very pessimistic results and suggested that the vast majority of remaining patches of old-growth forest within the Ada Block had a very high probability of not supporting the Greater Glider in the coming 100–200 years (Table 10.3). That is, the risks of local extinction in these patches were extremely high, especially for those patches 20 ha or smaller – which are the typical old-growth patch sizes within the Ada Block (Possingham *et al*. 1994).

Table 10.3: Patch size, initial population size of the Greater Glider and probability of being empty in the baseline simulation for the 20 patches of old-growth forest, Ada Forest Block

Patch no.	Patch size (ha)	Mean initial population size (females only)	Probability of being empty (%)
1	45	30	11
2	3	2	73
3	3	2	69
4	3	2	72
5	3	2	84
6	3	2	81
7	3	2	80
8	3	2	80
9	3	2	78
10	12	8	47
11	6	4	65
12	3	2	79
13	24	16	25
14	6	4	62
15	9	6	76
17	3	2	76
18	3	2	76
19	3	2	75
20	21	14	29
Totals	162	108	

Not surprisingly, given the extent of localised patch extinctions, the overall risk of extinction of the Greater Glider across the Ada Forest Block was very high over the scenario period of 200 years (Figure 10.10). This high value increased even further if wild-

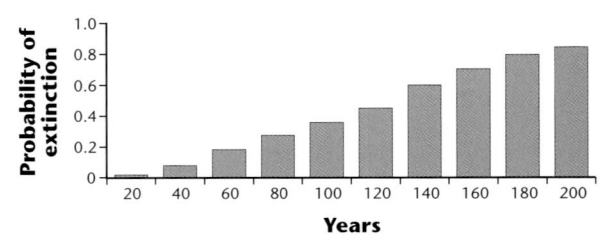

Figure 10.10: Cumulative probability of extinction of the Greater Glider in the Ada Forest Block in absence of wildfire. (Redrawn from Possingham *et al*. 1994).

fire effects were added to the model (Possingham *et al.* 1994).

McCarthy and Lindenmayer (1999) extended that initial work on the Greater Glider in the Ada Forest Block to include more complex spatially correlated simulations of wildfires. The work further emphasised the importance of old-growth forest as vital to long-term persistence of the species. McCarthy and Lindenmayer (1999) recommended a substantial increase in the area of old-growth forest reserved in each forest block to reduce the risk of localised extinction.

Population subdivision effects on the Mountain Brushtail Possum

VORTEX was used to model populations of the Mountain Brushtail Possum (Lindenmayer and Lacy 1995a, 1995b). Simulated populations of 100, 200 and 400 animals were partitioned into one to 10 subpopulations, linked by varying rates of interpatch migration. These levels of population subdivision may be typical in extensive areas of wood production montane ash forest where there are limited and generally very small patches of old-growth forest (<10 ha) that will become increasingly important refugia for many species of hollow-dependent fauna such as the Mountain Brushtail Possum.

The results demonstrated that the impacts of population subdivision on population demography were usually negative. In almost all the modelling scenarios, a single population was more stable than an ensemble of subpopulations of comparable initial size, irrespective of the rate of interpatch migration. There were marked differences in the behaviour of subpopulations initially comprising 10 and 20 individuals and those comprising 40 or more animals. Small subpopulations were predicted to exhibit high inherent levels of population instability, probably as a result of factors such as demographic stochasticity and inbreeding depression. They were characterised by high rates of extinction, small or negative values for subpopulation and metapopulation growth, and large fluctuations in population size.

Modelling the effects of animal dispersal between subpopulations on their persistence generated complex outcomes related to the final size of the populations that resulted from the process of subdivision. Among

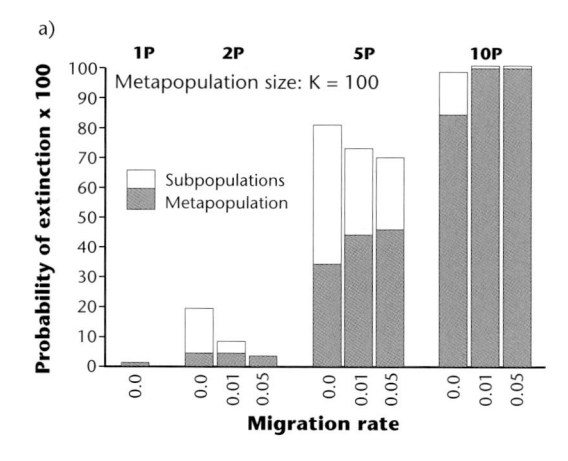

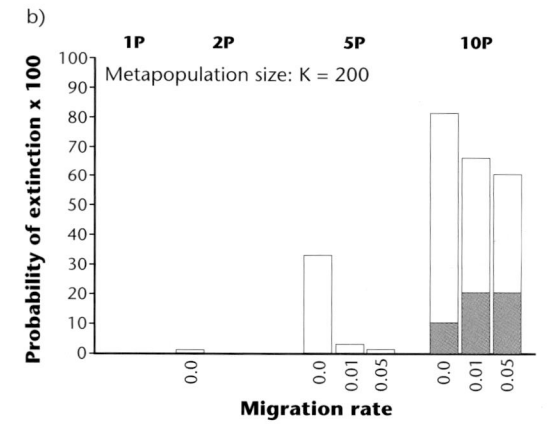

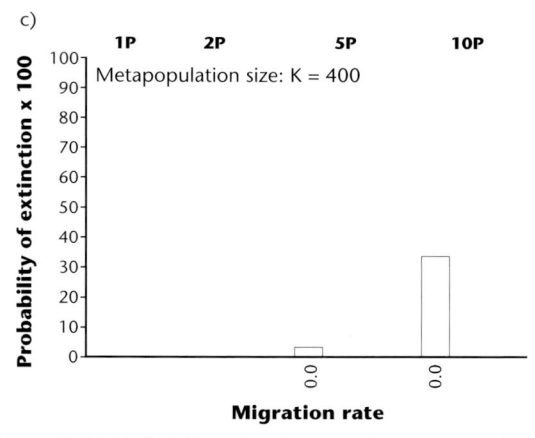

Figure 10.11: Probability of extinction of metapopulations of 100, 200 and 400 individuals of Mountain Brushtail Possum, subjected to varying levels of population subdivision (1 patch, 2 patches, 5 patches and 10 patches) and varying levels of interpatch dispersal. Values for dispersal of 0, 0.01 and 0.05 are linked with the size and structure of pairs of populations and overall population size. (Redrawn from Lindenmayer and Lacy 1995b).

ensembles of subpopulations which initially comprised 40 or more animals, increasing the rate of dispersal had a range of positive effects on subpopulation and metapopulation dynamics. Such effects included higher rates of, and smaller fluctuations in, subpopulation and metapopulation growth, and lower probabilities of subpopulation and metapopulation extinction. Many of these trends were reversed when the metapopulation structure initially comprised small subpopulations of 10–20 animals. In such cases, accelerated dispersal rates were predicted to have a detrimental impact on most measures of population demography, including the rate of extinction. A second phase of the simulation study on the Mountain Brushtail Possum highlighted how subdivision can significantly erode levels of genetic variability in metapopulations of the species, even when rates of interpatch dispersal are relatively high (Lacy and Lindenmayer 1995).

Comparisons of extinction risks of arboreal marsupials

Given extensive computer simulations of the different species of arboreal marsupials, a logical additional study was an interspecific comparison of the extinction risk. The three target species were the Greater Glider, Mountain Brushtail Possum and Leadbeater's Possum. VORTEX was used to model simulated population dynamics and genetic variability in multiple habitat patches of varying number, size and carrying capacity. Impacts of different migration rates of dispersing individuals between subpopulations were also examined (Lindenmayer and Lacy 1995a).

The study indicated for all three species there may be some metapopulation structures in which increased migration between habitat patches has a detrimental effect on demographic stability and, in turn, reduces population persistence. These negative impacts were most pronounced in metapopulations comprising an ensemble of very small subpopulations of four, 10 and 20 animals. Many of these patterns of population behaviour were accentuated both with increased rates of interpatch migration and the addition of further small subpopulations of individuals (Lindenmayer and Lacy 1995a). These trends in population demography were reversed in scenarios in which larger subpopulations of 40 animals were modelled. In such cases,

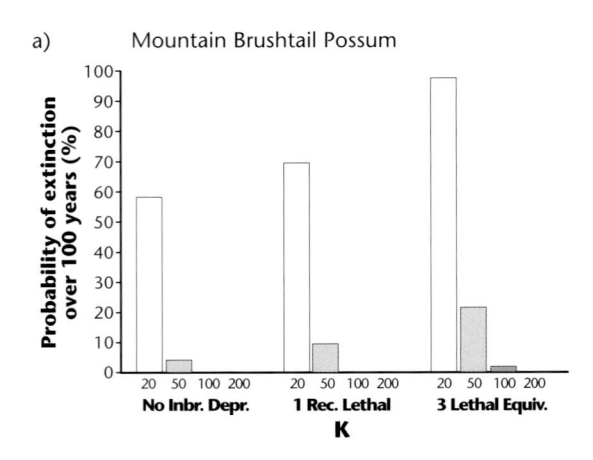

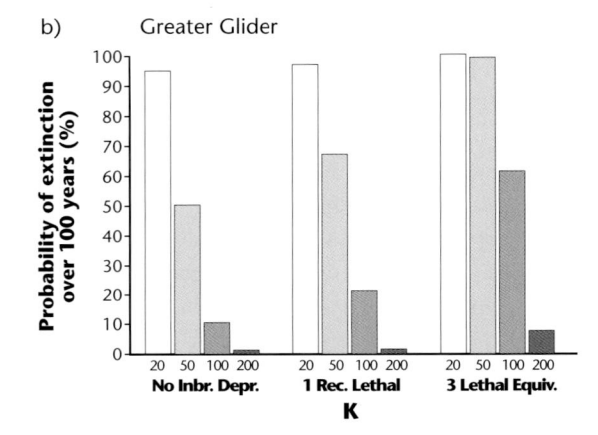

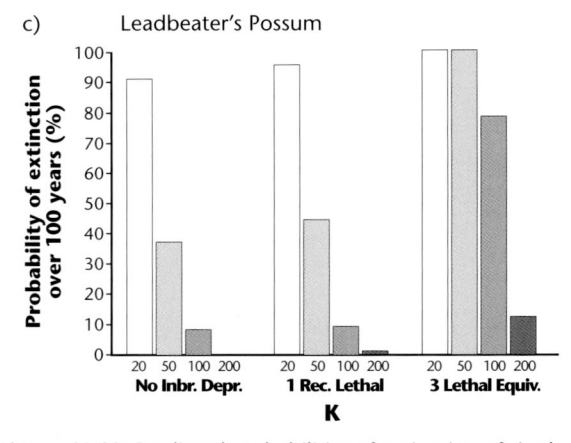

Figure 10.12: Predicted probabilities of extinction of single isolated populations of the Mountain Brushtail Possum, Greater Glider and Leadbeater's Possum over a 100-year period under different models of inbreeding depression (no inbreeding depression, one unique recessive lethal allele per initial individual, three lethal equivalents per individual) and four carrying capacities for patches (20, 50, 100 and 200 individuals). (Redrawn from Lindenmayer and Lacy 1995a).

the impacts of dispersal and additional habitat patches on subpopulation and metapopulation dynamics were generally positive. In contrast to the results for various demographic parameters, increased migration and added subpopulations had a beneficial effect on levels of genetic variability among all types of metapopulation structures examined. However, metapopulation structures with five subpopulations of 40 or fewer individuals were typically characterised by a significant (>10%) loss in expected heterozygosity over 100 years (Lindenmayer and Lacy 1995a).

The results revealed differences in the behaviour of metapopulations of the three species of arboreal marsupials. Many of these differences were attributed to variations in key life history parameters such as birth and death rates, but the trends in metapopulation viability did not correlate simply with intrinsic rates of population growth. The three species of arboreal marsupials differed in the size and structure of metapopulation required for demographic and genetic stability. In addition, the rank order of their susceptibility to demographic and genetic stochasticity was not consistent. These findings indicate that different species vary in vulnerability to disturbance and environmental perturbation and that even currently common and widespread species such as the Greater Glider and Mountain Brushtail Possum are susceptible to extinction in the medium to long term (50–100 years) in some areas of montane ash forest (Lindenmayer and Lacy 1995a). The results emphasise the need to understand the structure and composition of metapopulations when developing strategies for the conservation of wildlife at the landscape and regional scales (Lindenmayer and Lacy 1995a).

MODEL TESTING: LINKING MODELLING AND MONITORING DATA

Initial parameters of population models for Leadbeater's Possum (Lindenmayer 1995; McCarthy and Lindenmayer 2000) and the Greater Glider (McCarthy and Lindenmayer 1999; Possingham *et al.* 1994) accommodated the known information about those species.

One of the features of PVA is that models and their predictions can be updated as new information

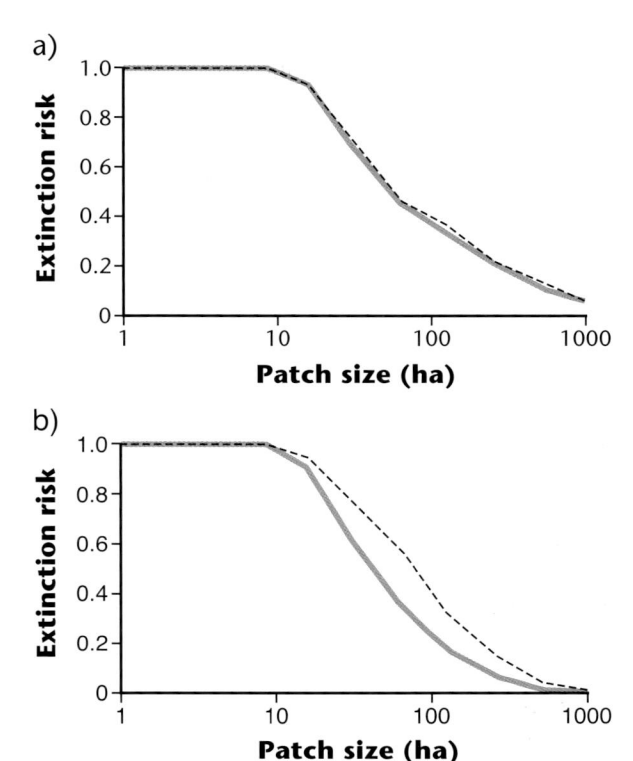

Figure 10.13: Risks of extinction within 300 years as a function of patch size for (a) Leadbeater's Possum and (b) the Greater Glider. For each species, earlier modelling (dashed line) is compared with the result for a new model updated with monitoring data (solid line). (Redrawn from Lindenmayer and McCarthy 2006).

becomes available. Updating can be conducted in an explicit and repeatable manner to ensure that the models remain internally consistent (Burgman 1993). In this way, PVA models form a part of a cycle of prediction, monitoring, refinement and further prediction (Possingham *et al.* 2001).

Lindenmayer and McCarthy (2006) used data from seven years of monitoring (1997–2003) conducted at 161 sites in montane ash forests (Lindenmayer *et al.* 2003a) (see Chapter 17) to assess population models of Leadbeater's Possum and the Greater Glider. Importantly, past conclusions about the optimal sizes of patches to reserve for Leadbeater's Possum (Lindenmayer and Possingham 1995c, 1996c) and the Greater Glider (McCarthy and Lindenmayer 1999) were robust to changes in the model (Figure 10.13). However, the results indicated that the importance of food resources for both species had been underestimated. Despite this, the modified models that included increased importance of food availability did not substantially change the predicted risks of decline, particularly for Leadbeater's Possum (Lindenmayer and McCarthy 2006).

SUMMARY

Studies using PVA models were a major focus of work in the montane ash forests for over 15 years (1991–2006). Leadbeater's Possum has been the subject of more, and a greater diversity of, studies than possibly any other species around the world. PVA modelling has provided an excellent opportunity to pull together much of what is known about a given species and the environments in which it lives. In many cases, the work has provided an opportunity to simulate the potential cumulative effects of multiple stressors (e.g. logging, fire and the paucity of suitable habitat) on arboreal marsupials (Lindenmayer and Possingham 1995c). These kinds of cumulative impacts can be extremely difficult to examine in other ways.

From a practical perspective, the results of PVA have underpinned many of the single-species conservation strategies developed for Leadbeater's Possum (Lindenmayer and Possingham 1995c). Indeed, in late 2008, the recommendation by Lindenmayer and Possingham (1995c) to set aside 600–100 ha of each forest block as a mid-spatial-scale reserve system for Leadbeater's Possum was finally ratified by the Government of Victoria (Department of Sustainability and Environment 2008).

LESSONS LEARNED

The integration of modelling and field data discussed here highlighted several new insights of management significance, even though the analysed species has been the focus of intensive research in the Central Highlands of Victoria for more than 20 years (Lindenmayer 2000). In particular, the analyses highlighted the critical importance of food availability for both the Greater Glider and Leadbeater's Possum (Lindenmayer and McCarthy 2006), although much of the past management concern for these (and other) hol-

> **Box 10.1: Postscript – population viability and the 2009 wildfires**
>
> Many of the scenarios run as part of PVA suggested that wildfire could have a significant negative impact on the viability of populations of arboreal marsupials, particularly in landscapes already heavily modified by logging operations and with few stands of old-growth forest. An important follow-up to the 2009 Black Saturday wildfires would be to rerun PVA models for populations of Leadbeater's Possum and the Greater Glider in areas which were extensively burned, such as those in the Murrundindi and Steavenson forest blocks and the O'Shannassy Water Catchment. Such analyses may help elucidate additional conservation strategies, to promote the persistence of remaining populations of these species in fire-damaged areas.

low-dependent species has been on the abundance of large trees with hollows as nest sites (Chapter 9). In the case of the endangered Leadbeater's Possum, this finding underscores the importance of new management initiatives in montane ash forests to modify current clearfelling practices to better create key habitat components in logged areas (Lindenmayer 2007). This involves the retention of patches or islands of unlogged forest within otherwise clearfelled areas to create an appropriate spatial juxtaposition of tree hollows and regenerating understorey *Acacia* spp. trees (Lindenmayer 2007). The aim is to provide a combination of suitable nesting and foraging habitat within the area that approximates the size of a home range of a colony of Leadbeater's Possum and a pair of Greater Gliders. The actual on-ground effectiveness of such altered silvicultural systems as a habitat creation strategy for Leadbeater's Possum is unknown, although early results appear promising (Lindenmayer 2007). This initiative of altered silvicultural systems is discussed further in Chapter 16, on mitigating the effects of logging.

All models simplify the systems they attempt to portray (Burgman *et al.* 1993, 2005). Metapopulation models are no different. While PVA models are less than perfect (Lindenmayer *et al.* 2003b) they nevertheless remain useful tools, particularly in the absence of alternative approaches (Ball *et al.* 2003). Improvements to PVA models need to be motivated by the cycle of testing and continuous upgrading (Possingham *et al.* 1993). This has been done during work in the montane ash forests – PVA has been characterised by a cycle of prediction, monitoring, refinement and further prediction (Lindenmayer and McCarthy 2006).

PVA models generate crude measures of species viability using a common currency of the estimated probability of the risk of extinction. Given the large number of parameters for which data are lacking or are, at best, educated guesses, estimates of extinction risk are very unlikely to be accurate. However, it is often not the accuracy of precise extinction risks that is important. From a practical perspective, the relative performance of different management options is more critical. This philosophy underpinned, for example, the extensive appraisal of management options in the extinction risk assessment of Leadbeater's Possum. Notably, the relative merit of recommended options proved robust to subsequent model testing with new data (McCarthy and Lindenmayer 1999).

KNOWLEDGE GAPS

One of the most important uses of models in wildlife conservation and resource management is the identification of key knowledge gaps (Burgman *et al.* 1993). A vitally important knowledge gap is the dispersal ability of arboreal marsupials and the extent to which animals can move between patches of suitable habitat. Past modelling work has used educated guesses at best (e.g. Lindenmayer and Possingham 1996a) and there are clear deficiencies in scientific understanding. Genetic methods can assist in resolving these issues; some work using these approaches has recently commenced for Leadbeater's Possum (Hansen *et al.* 2008; Hansen 2008) and the Mountain Brushtail Possum (Banks *et al.* 2008). Work on the Greater Glider in other forest landscapes, using a combination of genetic and demographic methods, is well underway (Taylor *et al.* 2007). Despite these advances, there is much to learn about dispersal. Such understanding is important because it can have significant impacts on fundamental population processes such as recovery following disturbance (Lindenmayer *et al.* 2005), and metapopulation persistence (Brooker and Brooker 2002; Hanski 1994).

Part of analyses using PVA involved comparing extinction risk between different species of arboreal marsupials (Lindenmayer and Lacy 1995a). This was instructive, but species do not exist in isolation from one another. PVA modelling of multiple interacting species has not been completed and is an interesting topic for further work.

FOREST PATTERN, ECOLOGICAL PROCESS AND LINKS TO OTHER CHAPTERS

As outlined in Chapter 9, the distribution of any given species is a function of an array of forest patterns and ecological processes that operate at a range of spatial scales (Mackey and Lindenmayer 2001). The persistence of individuals and populations in an area is one factor influencing distribution patterns. Persistence is scale-dependent – it can be at the tree level, site level, landscape level or region level. It can be the combined outcome of habitat suitability (and the processes which influence habitat suitability, e.g. natural and human disturbances and climatic conditions), population processes (e.g. mortality and reproduction) and metapopulation processes (e.g. interpatch dispersal). PVA provides a simulation modelling framework within which to integrate some of the many forest patterns and ecological processes which shape the patterns of animal distribution and abundance.

Links to other chapters

The quality of the information derived from PVA is highly dependent on the quality of information used in model parameterisation. Studies of the life history and habitat requirements of the three species of arboreal marsupials (see Chapters 3 and 9) were very important for generating some of the input data needed to run the models. Similarly, an understanding of disturbance regimes was fundamental for much of the modelling using ALEX (Lindenmayer and Possingham 1995a, 1996c). Much of the information on disturbance regimes discussed in Chapters 12–14 is relevant to studies using PVA.

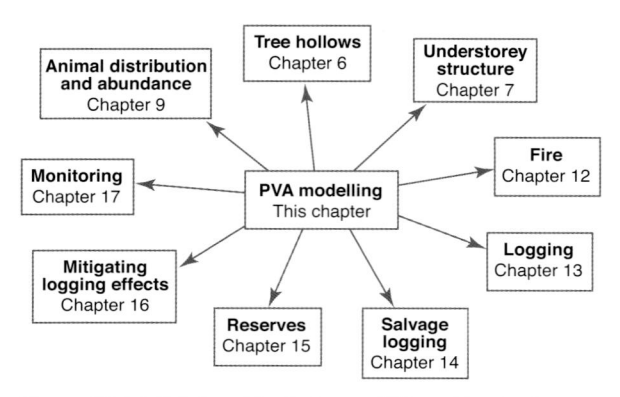

Figure 10.14: Relationships between PVA and key topics examined in other chapters.

Some PVA work has influenced the management planning for Leadbeater's Possum, particularly the setting aside of reserves for the species. Reserve design is the primary topic of Chapter 15.

Testing the accuracy of PVA predictions has entailed coupling models with data obtained from long-term population monitoring (Lindenmayer and McCarthy 2006). The framework that was developed for gathering monitoring data and the results of monitoring efforts to date are discussed in Chapter 17. Data from monitoring post-fire population ecological recovery may prove very important in future modelling of the viability of populations of arboreal marsupials.

11 Composition of animal communities

INTRODUCTION

The focus of the previous two chapters was on individual species. However, species do not exist in isolation from one another and they can be grouped together in assemblages and communities. Many interesting perspectives on forests and wildlife can be gained from looking at these levels of biological organisation. Three broad assemblages of vertebrates are the primary topics of this chapter – arboreal marsupials, small terrestrial mammals and birds. A range of aspects of assemblage research in the Central Highlands of Victoria is explored. These include:

- factors influencing the species richness of the arboreal marsupial assemblage;
- the search for an indicator species from the arboreal marsupial assemblage;
- resource partitioning among different species of arboreal marsupials;
- factors influencing the species richness of the bird assemblage;

Table 11.1: Studies directly or indirectly examining animal communities in montane ash forests

Study description	Reference
Species richness of arboreal marsupials	Lindenmayer et al. (1991c); Lindenmayer et al. (1994b)
Co-occurrence patterns of arboreal marsupials	Lindenmayer and Cunningham (1997)
Resource partitioning among different species of arboreal marsupials	Lindenmayer (1997)
Resource partitioning among different species of small terrestrial mammals	Cunningham et al. (2005)
Species richness of birds	Lindenmayer et al. (2009b)
Assembly rules for bird communities[a]	Driscoll and Lindenmayer (2009)
Resilience of bird assemblages[a]	Fischer et al. (2007)
Landscape texture and composition of bird assemblages[a]	Fischer et al. (2008)

a   Includes analyses of bird assemblages not only from montane ash forests but from elsewhere in south-eastern Australia.

- tests of ecological theory using bird assemblages, particular assembly rules, resilience and body size relationships.

BACKGROUND DATASETS

Studies of the composition of animal communities in montane ash forests have been underpinned by two major datasets. The first was the presence and abundance of arboreal marsupials at 207 sites, each measuring 3 ha, that were stagwatched (see Chapter 3) between 1983 and 1993 (Lindenmayer et al. 1994b, 1991d). The second major dataset was repeated point interval counts of birds at 81 sites (Driscoll and Lindenmayer 2009). For both datasets, counts of animal numbers were accompanied by extensive measurements of vegetation structure and composition (Chapter 3). In the case of arboreal marsupials, analyses of the ways in which different species partition resources were based on information on patterns of den tree use (Lindenmayer et al. 1991e), as well as life history attributes drawn from values published in the general ecological literature.

SPECIES RICHNESS OF THE ARBOREAL MARSUPIAL ASSEMBLAGE

Chapter 9 outlined the basis for quantifying the factors influencing the presence and abundance of individual species of arboreal marsupials. A similar kind of approach was employed to examine the factors influencing aggregate species richness of arboreal marsupials – the measured attributes which significantly influenced the total number of different species of arboreal marsupials recorded on a survey sites.

Statistical modelling indicated that the species richness increased with:

- increasing numbers of age cohorts of the dominant overstorey trees;
- an increasing number of trees with hollows on a site;
- increasing amounts of *Acacia* spp. trees in the understorey;
- increasing amounts of decorticating bark hanging from overstorey trees (Lindenmayer et al. 1991d).

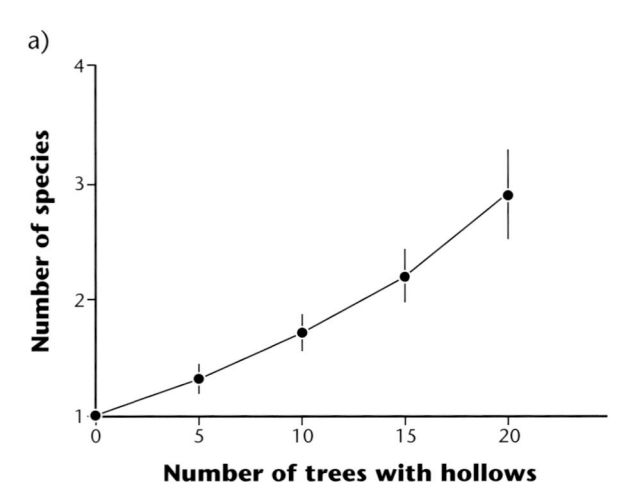

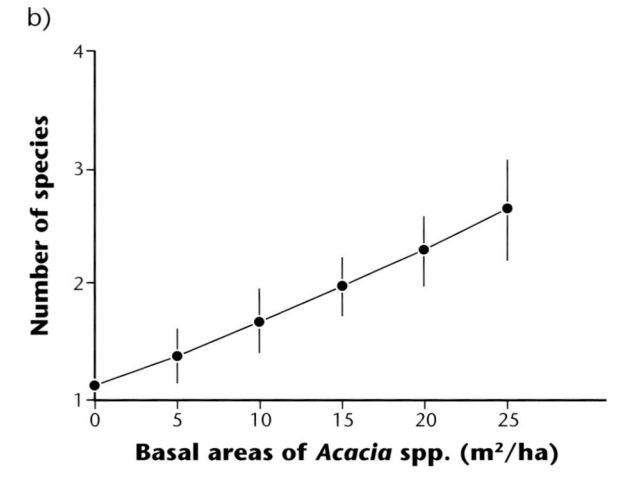

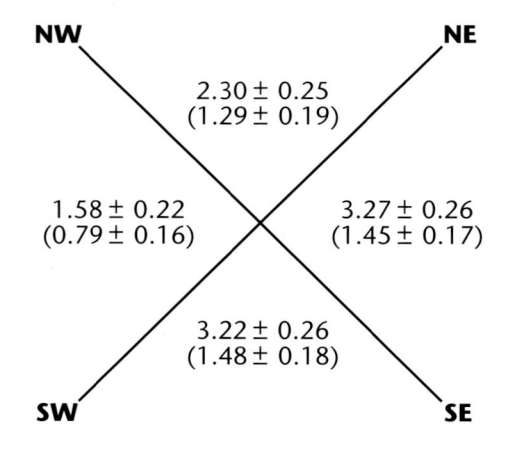

Figure 11.1: Relationships between the predicted species richness of arboreal marsupials on a site and (a) the abundance of trees with hollows, (b) the basal area of *Acacia* spp. trees in the understorey and (c) site aspect.

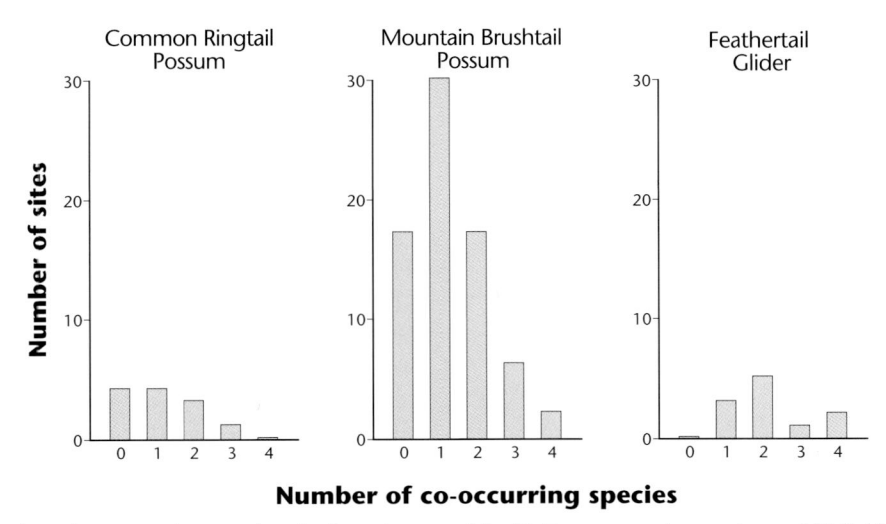

Figure 11.2: Number of co-occurring species of arboreal marsupial with three exemplar species on 151 field survey sites in the montane ash forests of the Central Highlands of Victoria.

CO-OCCURRENCE PATTERNS AND THE SEARCH FOR AN INDICATOR SPECIES

In addition, the number of species of arboreal marsupials was significantly higher on sites with an easterly and southerly aspect than it was on sites with a westerly or northerly aspect (Lindenmayer et al. 1991d).

These findings make intuitive sense as they link strongly with the known biology of many species of arboreal marsupials – the requirement of almost all species for trees with hollows as shelter and nesting sites (see Chapter 6), and the value of *Acacia* spp. trees and bark streamers as foraging substrates (see Chapters 4 and 5). The influence of site aspect is less readily explained, but may be associated with temperature effects on microclimate conditions within trees with hollows. An alternative or additional effect might be influences on site productivity and hence the availability of food resources.

CO-OCCURRENCE PATTERNS AND THE SEARCH FOR AN INDICATOR SPECIES

As outlined in Chapter 2, montane ash forests support eight species of arboreal marsupials. Most analyses of the group have focused on the habitat and other requirements of individual species, particularly the more common ones and those of conservation significance (Chapter 9). Some key structural attributes of forests, such as the abundance of large trees with hollows, were found to be important in statistical models of the occurrence of most species (see Chapters 3, 6). Given this, an obvious question was whether a particular species of arboreal marsupial might be a suitable indicator for the rest of the assemblage.

The notion of a management indicator species (Landres *et al.* 1988; Milledge *et al.* 1991) was explored in the context of identifying a species that reflected the co-occurrence of a maximum number of other members of the arboreal marsupial guild. Two questions in particular were posed:

- How many species occur together on sites surveyed for arboreal marsupials?
- Which species, if any, tended to be co-recorded?

A co-occurrence analysis was completed. One extremely uncommon species – the Eastern Pygmy Possum – was dropped from the analysis, leaving seven species to be examined (Lindenmayer and Cunningham 1997).

Figure 11.2 shows a frequency diagram for three representative species, with the number of sites on which other species also occur. These values ranged from zero to four other taxa per site for each species. For example, in the case of Leadbeater's Possum, it was the only species of arboreal marsupial on 12 sites. One other species (in addition to Leadbeater's Possum) was recorded on 25 sites. Leadbeater's Possum and three

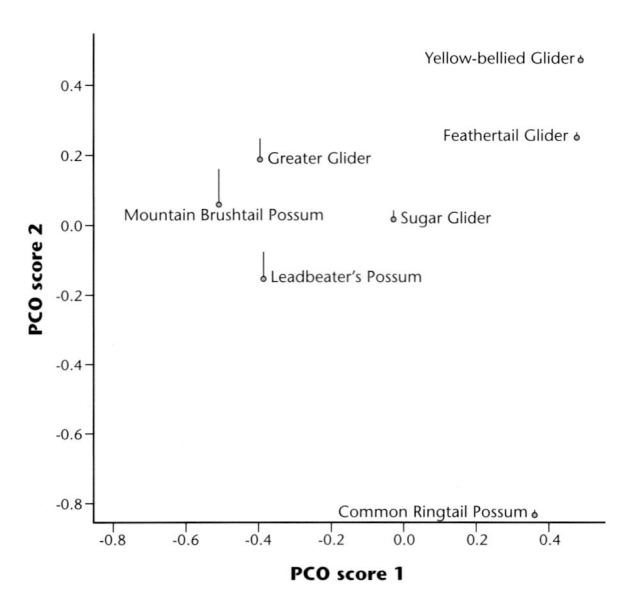

Figure 11.3: Patterns of occurrence of individual species of arboreal marsupial. (Redrawn from Lindenmayer and Cunningham 1997).

other species of arboreal marsupials were detected on only two sites. A similar pattern was observed for the Mountain Brushtail Possum, with relatively few sites supporting three and four additional species respectively. As in the case of Leadbeater's Possum, it was most commonly co-recorded with only one other species of arboreal marsupial. The most common frequency of co-occurrence category for Greater Glider, Sugar Glider, Yellow-bellied Glider and Feathertail Glider was two additional species. In summary, each site typically supported one or two species and the average number of taxa co-occurring with each species ranged from 0.7 additional species for the Common Ringtail Possum to 1.8 additional species for the Feathertail Glider (Lindenmayer and Cunningham 1997). A principal components analysis showed that there were few consistent patterns among groups of co-occurring species apart from a general tendency for Leadbeater's Possum, Mountain Brushtail Possum and Greater Glider to be found together (Figure 11.3).

A key conclusion was that none of the species could be regarded as a particularly reliable or consistent (and therefore useful) surrogate for the presence or abundance of any other members of the arboreal marsupial guild in the montane ash forests. This was, in part, because few sites support many different species of arboreal marsupials. In addition, although the taxa in these forests typically co-occurred with one or two other species, there were no strong patterns of co-occurrence (Lindenmayer and Cunningham 1997). Hence, the suitability of a given area of forest for one species does not imply that it is appropriate for another, even among this relatively closely related group. The reasons may be related to resource partitioning among the different members of the arboreal marsupial guild. There are large differences between each species in diet, body size, nest tree use, mating system and other life history characteristics. It is perhaps not surprising that attempts to identify one (or two or more) species whose presence reflects the occurrence of other taxa are unlikely to be successful. The chances that a species from this assemblage could be a useful indicator of the response of taxa in more distantly related groups (such as birds or small mammals) is even more remote, and unlikely to be based on ecologically robust relationships. This is an important outcome – it is critical for forest managers to be aware that a particular species is not a good indicator of the presence or abundance of others. Otherwise, they may believe that by conserving a so-called 'indicator species' they have effectively conserved other biota, when, in fact, they have gained a false sense of management security.

As outlined above (and discussed in more detail in Chapters 6 and 9), an important habitat component for virtually all species of arboreal marsupials inhabiting montane ash forests is the availability of hollow trees as nest and den sites. The abundance of these trees is a factor limiting animal populations in large areas of montane ash forest (Lindenmayer 2000). As a result, rather than seeking a management indicator species in montane ash forests, the prevalence of hollow trees may serve as a useful structural indicator, reflecting the response of many species to forest management actions. However, there would still be numerous non-hollow-dependent taxa that would not benefit from such an approach (see Table 9.6 in Chapter 9).

RESOURCE PARTITIONING AMONG ARBOREAL MARSUPIALS

The previous section strongly suggested that interspecific differences in the biology and ecology of arboreal marsupials were the most likely reason for the weak patterns of co-occurrence and, in turn, the lack of a robust indicator species in the assemblage. This section highlights these differences and reinforces the notion that considerable partitioning of resources occurs in montane ash forests. This enables such a diversity of species to exist in these forests.

Body size

A difference exceeding an order of magnitude occurs in the mean body mass of the species of arboreal marsupials that inhabit montane ash forests. The smaller species feed on insects, pollen, nectar and the exudates of plants and animals. These species also have the lowest field metabolic rates (Nagy 1987). The two species with the largest body masses have the highest field metabolic rates (Nagy 1987) and are characterised by a diet comprised predominantly of leaves.

Home range

The home ranges of arboreal marsupials vary from ~0.5 ha to 60 ha. The associations between diet and body mass are not mirrored by those for body mass and size of home range. For example, the Yellow-bellied Glider has the largest home range although it is only intermediate in size (Craig 1985; Goldingay and Kavanagh 1991).

Life history attributes

The larger species generally have smaller litter sizes, produce just one litter per year, suckle their young longer and have a longer lifespan. Such patterns in fecundity and other attributes of life history and associated correlations with body size are well-known among mammals (Lee and Cockburn 1985; Tyndale-Biscoe 2005).

Three types of social systems are recognised among the eight species of arboreal marsupials. Most of the smaller species are colonial and the size of groups varies from two to five for the Common Ringtail Possum (How *et al.* 1984), three to 11 for the Yellow-bellied Glider (Craig 1985), two to 12 for the Leadbeater's Possum and Sugar Glider (Smith 1984a; Suckling 1984), and two to 29 for the Feathertail Glider (Ward 1990a). Monogamy is the most common form of mating system among arboreal marsupials. However, systems may vary between populations of the same species, e.g. Yellow-bellied Glider (Goldingay and Kavanagh 1991) and within groups of animals in the same population, e.g. Greater Glider (Norton 1988). Variations in the availability of resources may influence the social structure of different populations of the same species (Lindenmayer and Meggs 1996).

Habitat and nest tree requirements

The results of extensive analysis of the habitat and nest tree requirements of arboreal marsupials are summarised in Chapter 6. They show that there are substantial differences in the attributes of forest stands and the characteristics of den trees used by different species of arboreal marsupials (Table 11.2). These differences represent additional ways in which arboreal marsupials partition montane ash forest environments.

Studies of the behaviour of arboreal marsupials have revealed a significant correlation between body size and the time after dusk animals leave the nest tree and commence foraging. Smaller species are typically the first to emerge, followed by intermediate-sized taxa. Larger animals are the last to exit the nest tree (Lindenmayer *et al.* 1991c). The reasons for these behaviour patterns are unknown, but may be related to energy use and metabolic rates. Small species feed on a high-energy diet of invertebrates, pollen and plant and animal exudates. These types of food are dispersed widely and require considerable energy and searching to find. Small species may need to emerge from the nest site and commence foraging as soon as possible after dusk to replenish sources of energy depleted during the time inside the nest (Lindenmayer *et al.* 1991c). Larger animals with greater stores of energy may be able to delay emerging from the nest to commence foraging.

Significant differences occur in the height above ground of hollow entrances used by different species; those used by volant species are higher than those used by non-volant species. Alighting higher on a trunk may enable gliding animals to both travel further and expend less energy in moving away from the nest. For non-volant taxa, emerging from hollows lower in trees may facilitate access to understorey vegetation that typically supports important sources of food.

Major differences occur in the type of nest entrance used by the different species. Large entrances such as stubs and spouts of broken hollow branches are used

Table 11.2: Stand-level attributes of forest sites statistically significant in generalised linear models of the presence and abundance of different species of arboreal marsupials

Independent variable	Leadbeater's Possum presence/abundance	Sugar Glider presence	Greater Glider presence/abundance	Mountain Brushtail Possum presence/abundance
Number of trees with hollows	X/X	X	X/–	X/X
Basal area of *Acacia* spp.	X/–			X/–
Bark	X/–	X		
Number of shrubs	X/–			
Slope	X/–			
Age of forest			X/X	

Trees with hollows measures the availability of potential nest and den sites for arboreal marsupials (see Chapter 9). Bark measures the amount of decorticating bark hanging from the lateral branches of eucalypt trees. Basal area of *Acacia* spp. trees and bark reflect suitability of foraging substrates. An X in a column signifies that the variable is significant in the model for that species. The species of arboreal marsupials are listed in ascending order of body mass.

by larger species, and this is related to the physical dimensions of cavities required by large animals. Small species typically use small nest entrances such as holes or narrow cracks, which may preclude predators or use by other larger, hollow-dependent species.

Box 11.1: Resource partitioning in other vertebrate groups

Research suggests that some level of resource partitioning occurs between different species of small terrestrial mammals and birds in montane ash forests. In small terrestrial mammals, studies of patterns of occurrence (Lindenmayer et al. 1994c), coupled with work on factors influencing the probability of capture (Cunningham et al. 2005), suggest that resource partitioning is occurring. Different plot- and site-level attributes appear in models of the occurrence of Bush Rat, Agile Antechinus and Dusky Antechinus (Cunningham et al. 2005; Lindenmayer et al. 1994c). There was evidence of gender differences in habitat requirements of the same species, such as the Bush Rat (Cunningham et al. 2005).

Birds also showed some evidence of resource partitioning. Extensive statistical modelling of the factors influencing the occurrence of over a dozen species in montane ash forests revealed a different combination of significant explanatory variables in the final model for each species (Lindenmayer et al. 2009b).

SPECIES RICHNESS OF THE BIRD ASSEMBLAGE

Analyses of species richness in montane ash forests were based on repeated point interval count surveys of 81 sites between 2004 and 2007. Species richness on an individual site in any given year varied from three to 25. Statistical analyses showed that the most important determinant of observed species richness was forest type. It was highest on sites dominated by Shining Gum and lowest in stands where Alpine Ash was the dominant tree species (Table 11.3). This was an interesting finding, as Mountain Ash and Alpine Ash forest often appear to be superficially similar in stand structure and share a large number of other plant species (Mueck 1990). However, the differences between these forest types, as well as differences with Shining Gum, were very large in terms of effect on bird responses.

Table 11.3: Estimated average bird species richness for dominant forest type and year

Dominant tree species	Species richness
Mountain Ash	14.2
Alpine Ash	11.7
Shining Gum	16.0
Year	Species richness
2004	15.1
2005	13.0
2007	13.7

Redrawn from Lindenmayer et al. (2009b).

Other significant variables influencing species richness were identified. Observed species richness on a site varied significantly between survey years – a result common for birds in highly variable Australian environments. Species richness increased significantly with increasing amounts of *Acacia* spp. and the number of Myrtle Beech stems in the understorey. The additional stand structural complexity contributed by *Acacia* spp. is a valuable foraging substrate because there are many species of invertebrates associated with it (Woinarski and Cullen 1984). It also provides places for birds to nest. The contribution to species richness of these stand attributes corroborates findings of other workers who have demonstrated that bird species richness rises with increased vertical structural heterogeneity in forests (MacArthur and MacArthur 1961; Recher 1969).

Bird species richness in montane ash forests declined significantly with increasing altitude. This kind of general result is not often reported but has been seen in some studies (reviewed by Gaston and Spicer 2004). Mountain Ash, Alpine Ash and Shining Gum occupy different bioclimatic domains (Lindenmayer *et al.* 1996b) (see Chapter 4) and these tree species typically occur in different altitudinal bands (Boland *et al.* 2006). However, the strong elevation effects held even within particular forest types. For example, for Mountain Ash sites, observed bird species richness was lowest on the highest-elevation sites. The same was true for Alpine Ash sites and those dominated by Shining Gum. This kind of response is perhaps not surprising given links between elevation, temperature regimes and key physiological processes (Green and Osborne 1994) as well as the indirect effects of temperature regimes on key forest attributes likely to influence bird species, such as the availability of insect prey, flows of nectar and pollen.

As notable as the effects of significant explanatory variables were, the lack of effects on bird species richness for covariates is regarded as important by many workers. These include stand age (Loyn 1985) and topographic position (Palmer and Bennett 2006). For example, bird species richness was not significantly higher in old-growth forests than in stands of other age cohorts, nor was species richness significantly higher in gullies than elsewhere in the montane ash forest landscapes (Lindenmayer *et al.* 2009b).

FACTORS INFLUENCING COMPOSITION OF THE BIRD ASSEMBLAGE

An investigation using correspondence analysis (Greenacre 2007) showed that the bird assemblage exhibited highly significant relationships with a number of site characteristics. These were forest type, topographic position, basal area of *Acacia* spp., elevation and number of shrubs (Figures 11.4, 11.5). Analysis showed that (Lindenmayer *et al.* 2009b):

- the bird assemblage for sites dominated by Mountain Ash was separated in multidimensional space from those where Alpine Ash and Shining Gum were the dominant tree species. Among the species which contributed to this separation were the White-naped Honeyeater, White-eared Honeyeater, Red Wattlebird and Yellow-tufted Honeyeater, each of which tended to be recorded more often in Alpine Ash sites than in those dominated by the other tree species;
- two species from the overall bird assemblage – the Leaden Flycatcher and Flame Robin – were most often recorded in gullies, and the White-naped Honeyeater, White-eared Honeyeater and Red Wattlebird were mostly recorded on midslope sites.

Analysis of the bird assemblage revealed that differences between years could be significantly linked with bird movement status. In 2004, for example, resident species were more likely to be recorded than were migratory species (Lindenmayer *et al.* 2009b).

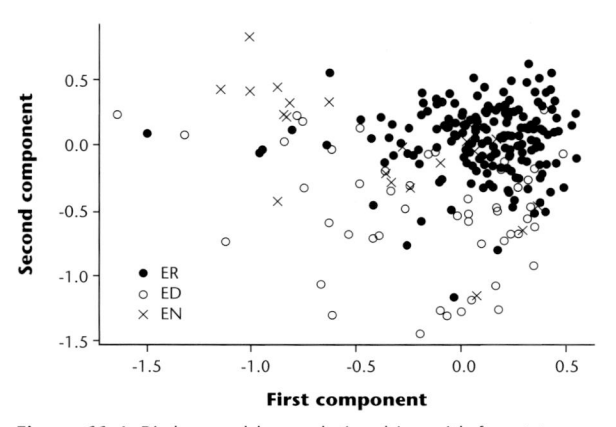

Figure 11.4: Bird assemblage relationships with forest type.

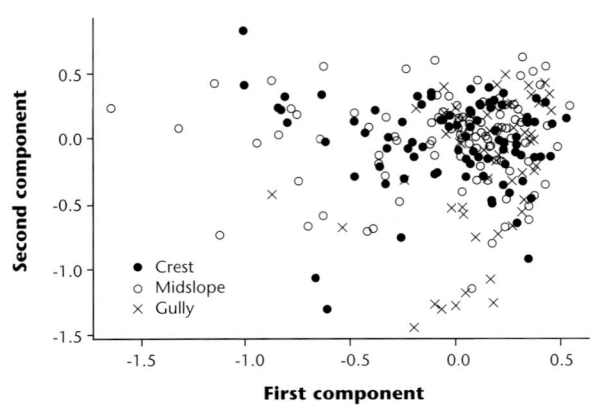

Figure 11.5: Bird assemblage relationships with topographic position.

ASSEMBLY RULES FOR BIRDS

The species that occur together at a site may reflect a range of influences. Vegetation type is an obvious one, with different species occurring in contrasting environments (Bennett *et al.* 1991). Competitive or predatory interactions also have a strong influence on the co-occurrence of species. Negative species interactions are the basis of assembly rules and stem from a study of birds by Diamond (1975) on the islands around New Guinea.

Assembly rules are general principles considered to determine or regulate which species form communities, including, for example, how new species might be added to a community as a consequence of dispersal and the order in which they colonise an area (Belyea and Lancaster 1999). Based on a study of desert rodent communities in North America, Fox and Brown (1993, p. 358) articulated a community assembly rule which stated that:

> each [new] species entering a community will tend to be drawn from a different [functional] group until each group is represented, and then the rule repeats itself.

The number of negative relationships between bird species is expected to be influenced by how similar species are, but should vary with other environmental characteristics, such as the amount of disturbance and resource availability. An interesting question is whether there are assembly rules influencing the co-occurrence of bird species in the montane ash forests of the central Highlands of Victoria.

To explore this question, birds were surveyed twice at three points in each of 57 sites in 2004 and 2005, focusing on old-growth and 1939-regrowth forests. Generalised linear models were built for each species to take into account any differences in species distributions due to environmental differences. The next step was to identify negative relationships among all the bird species and to assess the significance of the relationship, taking into account the large number of non-independent pair-wise comparisons.

Despite the potential for competition among ecologically similar species and for aggressive species to drive other birds away, there were no significant negative interactions among any pairs of bird species. In montane ash forests, particular bird species do not seem to exclude other bird species from a site. A possible explanation is that birds regularly disperse throughout the forested landscape, which prevents any one species from successfully dominating others (Driscoll and Lindenmayer 2009).

The patterns in montane ash forests contrast with results from highly modified agricultural landscapes on the western slopes of New South Wales, where aggressive honeyeaters were responsible for many negative interactions and therefore had a major impact on community composition (Driscoll and Lindenmayer 2009).

RANK ABUNDANCE DISTRIBUTION FOR BIRDS

Rank abundance distributions (RADs) are the curve that results from ranking species in order from the most to least abundant (Figure 11.6). These are usually based on survey data and often approximate a negative exponential shape. Rank abundance distributions may be useful for measuring community-level properties and for comparing communities from sites in different conditions. The extent to which RADs are more useful than simpler community metrics such as abundance and species richness has not been explored.

Using bird data from 27 field sites in montane ash forests, rank abundance distributions were described

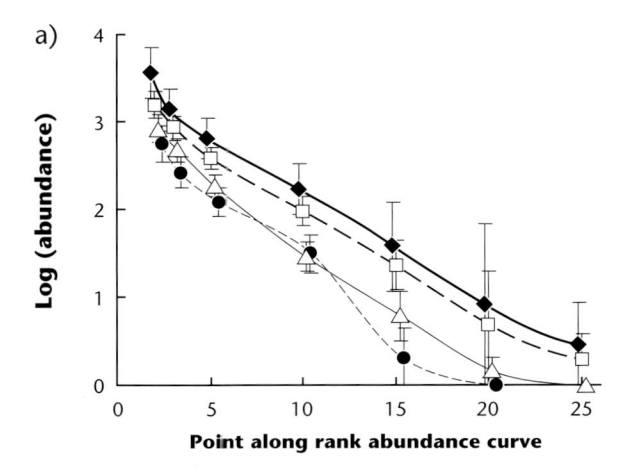

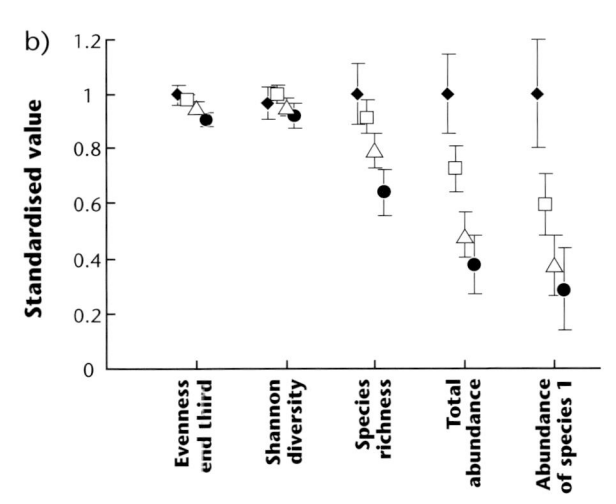

Figure 11.6: Rank abundance curve (a), standardised to the most abundant species (b), for bird survey data from one forest site.

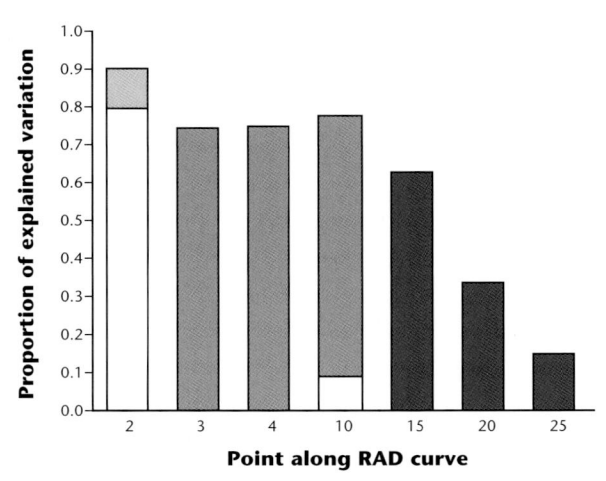

Figure 11.7: Proportion of variation in RAD among sites explained by abundance of species 1 (light grey), abundance of species 1 + species 2 (white), total abundance (dark grey) or number of species (black). Results are shown for seven points along the RAD curve.

by fitting a smoothing spline. Points along the smoothing spline were correlated with four simple community metrics: species richness, the abundance of the first and second most abundant species, and overall abundance. These metrics explained most of the variation among sites in rank abundance distributions (Figure 11.7).

Although less of the variation could be explained in the tail of the curve, there was very little ability to detect remaining differences among sites using a t-test, after the simple community metrics had been taken into account. This implied no additional information in the curve beyond that described by abundance and species richness (Driscoll and Lindenmayer 2009).

RELATIVE RESILIENCE OF THE BIRD ASSEMBLAGE

Ecological resilience is the ability of ecosystems to absorb disturbances while maintaining their characteristic functions and feedback (Folke et al. 2004). Fischer et al. (2007) completed a detailed analysis of bird assemblages using data from five large-scale studies in south-eastern Australia, including birds in the montane ash forests of the Central Highlands of Victoria.

The work revealed that functional richness and relative resilience of bird assemblages was reduced at all species-poor sites, including species-poor sites in montane ash forests (Fischer et al. 2007). Some of the most substantial reductions in relative resilience were not from montane ash forests but from agricultural areas characterised by large-scale vegetation loss and modification (Fischer et al. 2007). In these environments, small-bodied birds were most negatively affected by a gradient of least-to-most landscape modification (Fischer et al. 2007). The negative effects could be partly reversed by replanting programs, as revealed by analyses of sites in revegetated areas, highlighting the value of restoration efforts in agricultural regions (Cunningham *et al.* 2008).

LANDSCAPE TEXTURE HYPOTHESIS: BIRD BODY SIZE PATTERNS AND THE COMPOSITION OF BIRD ASSEMBLAGES

The texture of a landscape is a combination of the vertical structural heterogeneity of the vegetation and its spatial grain. The textural discontinuity hypothesis states that biophysical features in the environment are scaled in a discontinuous way, and that discontinuities in the body size distribution of animals mirror the biophysical discontinuities (Holling 1992). This proposition was tested using bird data gathered from five large-scale studies in south-eastern Australia, including the montane ash forests (Fischer *et al.* 2008). The landscapes in the five studies encompassed a gradient from high to low texture; data analyses showed that in landscapes with high texture, the number of detections of small-bodied birds was higher than expected and the number of detections of large-bodied birds was lower than expected (Fischer *et al.* 2008). The analyses were rerun with only the birds common to all five study areas and the general relationships between landscape texture and body size remained. Hence, the results were not an artifact of different species pools in the different study areas (Fischer *et al.* 2008). Thus, the results very strongly supported the textural discontinuity hypothesis.

SUMMARY

The studies of animal communities and assemblages provide a different set of insights from those gained from studies of individual taxa. For example, examining the entire assemblage of arboreal marsupials made it possible to uncover some of the ways in which resources are partitioned by the different species and partly, as a result of such partitioning, why different approaches to the management of forests are needed to conserve the suite of species of arboreal marsupials. It showed that not only is the search for an indicator species futile, but it is also misleading for informed wildlife conservation and forest management (Lindenmayer *et al.* 2000c).

A major study of the bird assemblage in montane ash forests has highlighted that the superficially similar Mountain Ash and Alpine Ash forests are characterised by different suites of birds. Other factors, such as the understorey and environmental conditions, such as elevation, have an important influence on bird species richness and the composition of the bird assemblage.

Other work on bird assemblages has been more theoretical than that on arboreal marsupials and has drawn upon large datasets collected not only from montane ash forests but elsewhere in south-eastern Australia, spanning other forest types, woodlands, Radiata Pine (*Pinus radiata*) plantations and temperate woodlands (Fischer *et al.* 2007, 2008). The work has produced important insights that would not have been gained from single-species investigations.

LESSONS LEARNED

Two important lessons have been learned from the suite of studies on the composition of animal assemblages. First, it is clear there are no valid indicator species in montane ash forests. The kinds of significant explanatory variables influencing both the bird assemblage and bird species richness were often markedly different from those in models of the species richness of arboreal marsupials. For example, the abundance of trees with hollows, stand age and the extent of multi-agedness did not appear in the final models for birds but were prominent in models of species richness of arboreal marsupials. Such differences reinforce the argument that neither group, nor particular taxon within each group, would be a robust surrogate or biodiversity indicator for other species or groups. There are no easy 'biological shortcuts' to guide management. This is an important lesson because it means that wildlife and forest managers need to be much better focused on the precise objectives of management. They need both a clearly articulated vision for landscape management and conservation, and quantifiable objectives that offer unambiguous signposts for measuring progress. As Andersen *et al.* (2005) stated, informed management must go beyond simply saying 'we need to conserve biodiversity'. It is important to be more explicit in stating which elements of the biota are appropriate targets for management actions. This is crucial because approaches taken to, for example, increase species richness may not be good ones to enhance the persistence of particular taxa of conservation concern. Indeed, species richness as an aggregate measure of the total number of species at a site might not be a good measure to guide management because it gives equal weight to exotic and native taxa.

A second important lesson from the work on assemblages is that it has a high degree of complementarity with single-species work. There has been much debate about the relative merits of single-species versus assemblage-based and ecosystem-oriented research for conservation. This debate has become increasingly important as resource managers and policy-makers in some jurisdictions focus on ecosystem-level problems. Making decisions about priorities for research, priorities for conservation management and associated allocation of scarce funds is not a trivial task. The scarcity of funding, time and expertise means it is impossible to study and manage each species, ecological process or ecological pattern separately. A combination of approaches is ideal, particularly when it is focused on closing important knowledge gaps and helps to promote synergies between single-species, multi-species and ecosystem-oriented research (Lindenmayer *et al.* 2007b).

KNOWLEDGE GAPS

The large datasets gathered for birds throughout south-eastern Australia have proved extremely valuable in examining such theoretical concepts as assembly rules, rank abundance distributions, resilience and textural discontinuity hypothesis. Further studies of this kind are planned, including studying relationships between site productivity and species richness and between vertical structural complexity and species richness. In each case, data gathered on birds from the montane ash forests will be a valuable component.

The composition of animal and plant communities can often be markedly influenced by disturbance events and disturbance regimes (Fox 1982). In some ecosystems, there can be a strongly defined series of species replacements over time associated with post-disturbance succession (Fox 1982; Friend 1993). It is not known whether such successional patterns exist in montane ash forests, particularly for animal communities. This is a key knowledge gap, but will be addressed through careful monitoring of ecological recovery in the years following the February 2009 wildfires.

FOREST PATTERN, ECOLOGICAL PROCESS AND LINKS TO OTHER CHAPTERS

Much of the work on aggregate measures such as species richness and community composition essentially describes patterns of species co-occurrence at varying levels of detail. Hence, it attempts to better quantify patterns of community composition and how it changes through space, over time or both. This is important, but extremely difficult. It is even harder to quantify the ecological processes which give rise to these emergent patterns. In many cases, work at this level results in linking a given pattern to another pattern; the underlying reasons for that link remain obscure. For example, work on the spatial pattern of landscapes (the landscape texture hypothesis) suggests that the composition (pattern) of landscapes influences the composition of bird communities (by shifting species according to their size). The reason why size should be a strong factor governing that shifting process remains unclear, and competing hypotheses giving rise to these patterns are very hard to test in a formal empirical way (Fischer *et al.* 2008). The spatial pattern of community assembly comprises the composite occurrences of many individual species, and the kinds of ecological processes and other patterns which influence individual species are often the same ones which are important in shaping communities of species. Thus, the work reported in this chapter is closely linked with that on the factors influencing the occurrence of individual species (Chapter 9). Many of the insights generated from community and assemblage-level analyses underpin strategies to best manage biodiversity in wood production forests – themes which underpin much of the discussion in Chapters 15–17. The response of species assemblages is often linked with features of the overstorey and understorey of montane ash forests, the core topic of Chapters 9–11.

PART V

DISTURBANCE REGIMES

Disturbance is a characteristic of almost all ecosystems and can have major impacts on biodiversity and key ecosystem processes (Moore *et al.* 2008). Montane ash forests are no exception. They have been substantially affected by wildfires for millions of years, and were badly affected by the 2009 Black Saturday wildfires (Colour Plate 11). They have also been substantially affected by logging for well over a century.

Part V examines three broad kinds of disturbances in montane ash forests. Natural disturbance, particularly wildfire, is the focus of Chapter 12. The subsequent two chapters summarise the effects of logging. Chapter 13 outlines the impacts of traditional forms of logging (clearfelling). The effects of a combination of human disturbance and natural disturbance – logging following wildfire (salvage logging) – are examined in Chapter 13.

12 Natural disturbance regimes: fire

INTRODUCTION

Natural disturbances are increasingly recognised by ecologists as critically important ecosystem processes that help create habitats and resources for biodiversity (Connell 1978; Moore et al. 2008; Parr and Andersen 2006; Sousa 1984). Many aspects of composition and structure in forests at the tree, stand and landscape scales are shaped by natural disturbances (Cary et al. 2003; Frelich 2005; Parminter 1998). In addition, many species have strong associations with natural disturbance regimes – some positive and some negative (Bergeron et al. 1993; Bunnell 1995; Moore et al. 2008; Spies et al. 2004).

The initial section of this chapter summarises some of the key attributes of natural disturbances. Much of the focus is on wildfire, the primary form of large-scale natural disturbance in montane ash forests (Ashton 1981b). Some important aspects of the fire regime in montane ash forests comprise the second section of this chapter.

Table 12.1: Studies directly or indirectly examining natural disturbances in the montane ash forests of the Central Highlands of Victoria

Study description	Reference
Disturbance pathways	Lindenmayer (2009); Lindenmayer and Franklin (1996)
Stand attributes of forests of different ages	Lindenmayer et al. (2000a)
Natural disturbance and environmental factors	Mackey et al. (2002)
Natural disturbance and multi-aged forest	McCarthy and Lindenmayer (1998)
Fire intervals in Mountain Ash forest	McCarthy et al. (1999)
Fire and viability of populations of arboreal marsupials	Lindenmayer and Possingham (1995a); McCarthy and Lindenmayer (2000)
Carbon storage and fire regimes	Mackey et al. (2008)

BACKGROUND DATASETS

A range of data has been used in studies of natural disturbance regimes in montane ash forests. These include:

- information on the prevalence of multi-aged stands at more than 520 3-ha sites throughout the Central Highlands of Victoria (Lindenmayer *et al.* 2000a, 1991g; McCarthy and Lindenmayer 1998);
- information on variation in the diameter of dead trees with hollows on over 300 field sites dominated by 1939-regrowth montane ash forest (Lindenmayer and McCarthy 2002);
- the occurrence of fire scars on measured living large trees with hollows on those sites (Lindenmayer and McCarthy 2002; Lindenmayer *et al.* 1991g);
- data on stand structure and plant species composition within the 520 sites based on a stratification of the age of the dominant cohort of overstorey trees (Lindenmayer *et al.* 2000a);
- data on the environmental attributes of the 520 sites based on simulation and mathematical models of climate, terrain, temperature, rainfall and incoming solar radiation;
- simulation modelling of the effects of fires of varying intensities and frequencies on the suitability of forest environments as habitat for different species of arboreal marsupials (Lindenmayer and Possingham 1995a; McCarthy and Lindenmayer 1999, 2000);
- historical information on the frequency and spatial extent of past fires in the Central Highlands of Victoria (McCarthy *et al.* 1999).

IMPORTANT FEATURES OF NATURAL DISTURBANCES

Natural disturbances vary spatially and temporally in all forests. They range from frequent low-intensity gap-forming disturbances operating at the scale of individual trees (Denslow 1987; Runkle 1982; Yamamoto 1992) to infrequent landscape-scale high-intensity events (fires, windstorms, insect attacks) that can radically alter stands and landscapes (Spies and Turner 1999; Stocks *et al.* 2002; Turner *et al.* 1997). In addition, a region may experience unprecedented disturbances, such as glaciation or the introduction of exotic herbivores or pathogens.

This chapter is primarily concerned with infrequent large-scale, high-intensity natural disturbance events. This broad category of natural disturbances can be defined as:

discrete events that do not have a substantial human origin and which alter ecosystem structure and resource availability (White and Pickett 1985).

Large-scale and/or high-intensity natural disturbances typically result in a 'rapid release or reallocation of community resources' (Shiel and Burslem 2003). They are characteristic of many terrestrial ecosystems (Bradstock *et al.* 2002; Pickett and Thompson 1978) and include:

- wildfires (Gill 1999; Lindenmayer *et al.* 2008b; Luke and McArthur 1977; Wadleigh and Jenkins 1996);
- windstorms, hurricanes and cyclones (Foster and Boose 1992; Schoener *et al.* 2004);
- volcanic eruptions (Cristafulli *et al.* 2005; Franklin and MacMahon 2000);
- floods (Calhoun and deMaynadier 2008);
- landslides (Geertsema and Pojar 2007; Veblen *et al.* 1996);
- widespread insect attacks (Holling 1992; Shore *et al.* 2003);
- prolonged droughts (Gordon *et al.* 1988).

Natural disturbance regimes are inherently variable and the local intensity of most disturbances is largely dictated by historical contingency and chance (Kulakowski and Veblen 2007). No two natural disturbances will be identical (Schoener *et al.* 2004). This variability also produces multiple (and often simultaneous) post-disturbance pathways (Lecomte *et al.* 2006; Noble and Slatyer 1980; Turner *et al.* 1998). This is illustrated in many mesic forest types such as those found in the Pacific north-west of the US (Morrison and Swanson 1990), the boreal forests of Canada (Burton *et al.* 2003) and the wet eucalypt forests of south-eastern Australia (Mackey *et al.* 2002).

Large-scale and/or high-intensity natural disturbances vary substantially in:

- timing (time of year or even time of day when they occur);
- frequency (or its inverse, return interval);
- intensity;
- size;
- heterogeneity (variation in intensity and impact within the limits of the total area affected);
- duration (fires and windstorms can be relatively short-lived, drought and insect attack can be prolonged).

These factors act together, rather than in isolation. For example, fire frequency and fire intensity usually covary. Many landscapes are characterised by recurrent low-intensity disturbances but few landscapes naturally experience frequent high-intensity wildfires. Factors such as climatic conditions and topography further influence how variables interact within particular landscapes (Lindenmayer *et al.* 1999f; Rülcker *et al.* 1994). The particular combination of cause (agent), size, frequency and severity of disturbances that prevail in a landscape characterises 'disturbance regimes' (Gill 1975; Swanson *et al.* 1994).

Variation in disturbance regimes leads to marked differences in landscape and stand conditions by significantly influencing the number, type and spatial distribution of habitat patches (e.g. age cohorts of stands) and stand-level biological legacies (Franklin *et al.* 2000; Turner *et al.* 2003). This temporal and spatial variability in natural disturbance regimes helps explain regional differences in vegetation patterns and species assemblages (Boose *et al.* 2004).

Non-uniform patterns of organism distribution and abundance partially result from spatial variation in environmental conditions, such as climate, terrain, soils and the availability of moisture and nutrients (Florence 1996; Hansen and Rotella 1999; Woodward and Williams 1987). Such spatial variation leads to natural ecosystems being heterogeneous at the landscape scale (Huggett and Cheeseman 2002; Parr and Andersen 2006). Disturbance regimes overlay and interact with the patterns created by environmental regimes to further shape heterogeneous landscapes and influence the distribution of species (Boose *et al.* 2004; Wardell-Johnson and Horowitz 1996). Factors such as topographic variability and weather changes during a disturbance event, such as a fire or windstorm, can result in undisturbed patches within the broad boundaries of a disturbed area (Eberhart and Woodard 1987; Syrjanen *et al.* 1994). Recent research has revealed considerable internal variability in the severity and residual structure of disturbance events typically mapped as a single change in land cover (Kafka *et al.* 2001; Schmiegelow *et al.* 2006).

It is a challenging task to characterise the historical disturbance regime of forest stands and landscapes, but it is essential for understanding how site, succession and disturbance interact to produce the land patterns seen today. Such work may be based on the dendrochronological dating of tree ring scars caused by fires, treefalls, avalanches and insect outbreaks (Lorimer and Frelich 1994; Veblen *et al.* 1994) or by mapping historically observed disturbance events (Morgan *et al.* 2001; Schulte and Mladenoff 2005). In most cases, there is an attempt to characterise the causes and levels of tree damage and/or mortality (and sometimes other measures of severity or ecosystem impact), their spatial extent and frequency. The many combinations of disturbance agents (and their severity, homogeneity, extent and frequency) combine with underlying differences in terrain and climate to generate distinctive stand development trajectories (Frelich 2005; Noble and Slatyer 1980). We recognise these legacies of past disturbance as distinctive forest types, varying in age, composition and structure.

WILDFIRE: THE PRIMARY FORM OF NATURAL DISTURBANCE

Several kinds of natural disturbance occur in montane ash forests. These include insect attacks and windstorms, but the most prevalent is wildfire (Ashton 1981b; Attiwill 1994b; Mackey *et al.* 2002). Many fires have occurred in the Central Highlands over the past 400 years, with the largest and most extensive in 1939 (Gill 1981) and more recently in February 2009. The intensity of wildfires is highly variable, the interval between fires varies across landscapes (McCarthy *et al.* 1999), and several kinds of disturbance pathways

Forest Pattern and Ecological Process

are known to occur (Attiwill 1994b; Lindenmayer and Franklin 1997; Mackey *et al.* 2002) (Table 12.2).

Stand-replacing wildfires

Major conflagrations can be stand-replacing events in which virtually all overstorey trees are killed, producing a stand of even-aged regrowth (Colour Plate 10). Large areas of predominately even-aged montane ash forest that regenerated after the 1939 Black Friday wildfires characterised the landscape in the Central Highlands of Victoria prior to the 2009 wildfires (Commonwealth of Australia and Dept of Natural Resources and Environment 1997). Young seedlings germinate from seed released from the crowns of burned mature eucalypts to produce a new even-aged regrowth stand (Ashton 1976). This makes it possible to readily determine the age of the dominant overstorey in a stand. For example, Lindenmayer *et al.* (2000a) identified 11 different fire-derived age classes of montane ash forest – the mid 1700s, 1824, 1851, 1895, 1905, 1908, 1926, 1932, 1939, 1948 and 1983 (see Chapter 7).

Although high-intensity fires in montane ash forests can be overstorey-replacing events, they also leave many other kinds of biological legacies (Lindenmayer *et al.* 2008a; Ough 2002). For example, a wide range of plant species survive and resprout vegetatively after wildfires (Ough 2002). Similarly, large-diameter fire-damaged living and dead standing trees remain in many stands of young regrowth montane ash forest (Lindenmayer *et al.* 1997b; Lindenmayer *et al.* 1991b) (see Chapter 6). Such trees contain hollows that provide den and nest sites for many species of arboreal marsupials (Lindenmayer *et al.* 1991e) as well

Figure 12.1: Stands of montane ash forest burned in the 1939 wildfires.

Figure 12.2: Typical stand profile for an even-aged stand of post-disturbance 1939 regrowth montane ash forest.

Table 12.2: Different disturbance pathways in montane ash forests

Disturbance pathway	Starting stand condition	Disturbance type	Post-disturbance stand condition	Dead biological legacies	Live biological legacies	Response of cavity-dependent fauna
P1	Regrowth	High-intensity fire (frequency > 25 years)	Regrowth eucalypt overstorey, understorey rejuvenation	Limited, standing life < 50 years	None	Poor. Prolonged absence of tree hollows
P2	Regrowth	Moderate-intensity fire (frequency > 25 years)	Regrowth eucalypt overstorey, understorey rejuvenation	Limited, standing life < 50 years	None	Poor. Prolonged absence of tree hollows
P3	Regrowth	Moderate- to high-intensity fire (frequency < 25 years)	Understorey rejuvenation (*Acacia* spp.), eucalypt overstorey lost	None	None	Poor. Prolonged absence of tree hollows
P4	Old growth	Low-intensity fire	Mixed-aged overstorey, understorey rejuvenation	High levels, standing life ~50+ years	Large numbers, standing life ~100+ years	Strong. Large numbers of hollow-bearing trees of varying characteristics. Live trees persisting for a prolonged period through stand development
P5	Old growth	Moderate-intensity fire	Mixed-aged overstorey, understorey rejuvenation	High levels, standing life ~50+ years	Large numbers, standing life ~100+ years	Strong. Large numbers of hollow-bearing trees of varying characteristics. Live trees persisting for a prolonged period through stand development
P6	Old growth	High-intensity fire	Regrowth eucalypt overstorey, understorey rejuvenation	High levels, standing life ~50+ years	Low levels, standing life ~50+ years	Moderate. Large numbers of dead hollow-bearing trees. Limited persistence time through stand development

as birds and bats (Gibbons and Lindenmayer 2002; Loyn 1985) (see Chapter 6).

Wildfires in montane ash forests not only produce important standing biological legacies, but significantly influence conditions on the forest floor. Trees that are killed and collapse onto the forest floor in burned stands become key habitat components for a range of vertebrates (Lindenmayer et al. 2002a). Large decaying logs in montane ash forests are important substrates for the germination of rainforest plants (Howard 1973), tree ferns (Ashton 2000) and the development of dense and luxuriant mats of bryophytes (Ashton 1986, 2000) (see Chapter 8).

Intermediate-intensity wildfires: an additional disturbance pathway

High-intensity overstorey-replacing fires represent one disturbance pathway in montane ash forests. Lower-intensity fires lead to partial stand replacement because some trees survive (Bowman and Kirkpatrick 1984; Smith and Woodgate 1985). Regeneration of young trees in these forests creates multi-aged stands comprising eucalypts of two or more distinct age cohorts (Lindenmayer et al. 1999f) (Figure 12.3). The understorey is also multi-aged in these forests: some plants regenerate from seed after each fire event and others survive many fires (Mueck et al. 1996). Recent studies have shown that up to 30% of stands of montane ash burned in the 1939 wildfires may have been multi-aged prior to that conflagration (McCarthy and Lindenmayer 1998). However, the widespread use of post-fire salvage logging following the 1939 wildfires (Lindenmayer and Ough 2006; Noble 1977) means that these stands are now largely even-aged (Lindenmayer and McCarthy 2002).

The traditional view of natural disturbance in Mountain Ash forests is that of high-intensity stand-replacing wildfires that can result in the regeneration of even-aged regrowth forests (Attiwill 1994b). There is no doubt that large areas of montane ash burned in February 2009 will result, in some cases, in stand replacement. However, some disturbance pathways are only partially stand-replacing. Prior to the 2009 wildfires:

- more than 7% of ash-type forests in the Central Highlands of Victoria were distinctly multi-aged (McCarthy and Lindenmayer 1998);
- almost all old-growth stands showed signs of past disturbance (Lindenmayer et al. 2000a);
- many individual trees were burned in earlier major wildfires such as those in 1983, yet survived (Smith and Woodgate 1985) (Figure 12.4);
- dendrochronology showed that some large Mountain Ash trees have survived many fires (Banks 1993).

Two related factors influence where partial stand-replacing disturbance pathways may manifest (Figure

Figure 12.3: Typical stand profile for a multi-aged stand of montane ash forest.

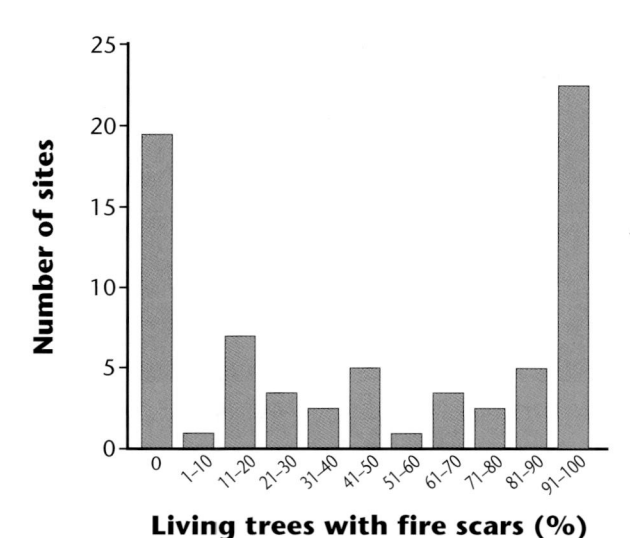

Figure 12.4: Prevalence of fire scars on large living montane ash trees. Data are from surveys of more than 2000 trees measured on 521 sites throughout the Central Highlands of Victoria. (Redrawn from Lindenmayer and Franklin 1997).

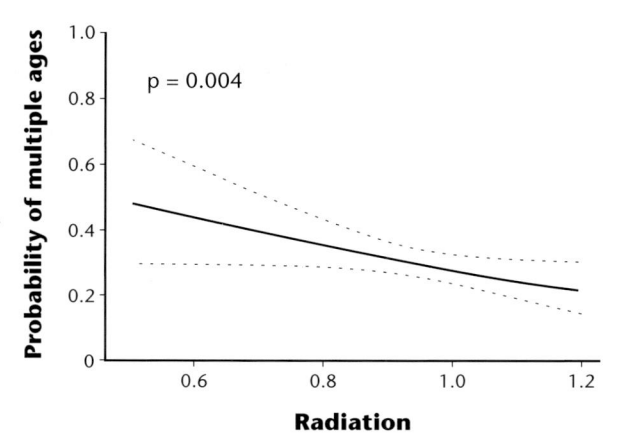

Figure 12.6: Relationships between multi-aged forest and levels of incident solar radiation. (Redrawn from Lindenmayer et al. 1999f).

12.5). First, landscape-level factors influence variation in fire intensity in montane ash forests. Integrated statistical and environmental modelling studies have shown that multi-aged stands are most likely to occur in parts of forest landscapes characterised by low levels of incident solar radiation (Lindenmayer et al. 1999f; Mackey et al. 2002) (Figure 12.6). Thus, the prevalence of different structural conditions varies between stands in response to fire intensity which is, in turn, influenced by stand location in the landscape.

Second, if a fire occurs when a stand is at an old-growth stage, it is less likely to be a complete stand-replacing event (thereby producing a multi-aged stand

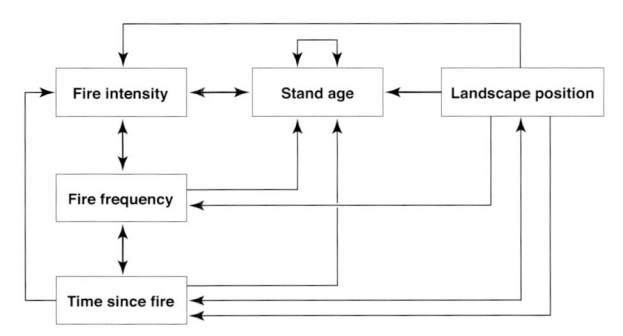

Figure 12.5: Relationships between landscape and fire effects, which influence stand age.

structure). This is partly because old trees have thicker bark and are more fire-resistant than younger stems (Ashton 1981b). Old trees are taller than younger ones and fire intensity has to be much greater to result in flame heights sufficient to kill the canopy (Mackey et al. 2002). Thus, the impacts of a fire in a regrowth forest on subsequent stand development, and the types and quantities of biological legacies that are left after a disturbance, will be very different from a similar-intensity fire in an old-growth stand (Chapter 12).

Fire intervals

The emphasis above was on fire intensity, but there are many other aspects of fire regimes. For example, the interval between stand-replacing fires can have a wide range of effects on stand structure and plant species composition in montane ash forests (Table 12.2). McCarthy et al. (1999) calculated that the mean fire interval in these forests was 75–150 years for stand-replacing wildfires but 37–75 years for all fires, because many stands have been subject to partial stand-replacing events. The emphasis here is on the mean fire interval, because other work indicates that not all parts of landscapes are likely to burn, or to burn at the same intensity (Mackey et al. 2002).

Given the influence of fire interval, another disturbance pathway in montane ash forests is the complete absence of any moderate or major natural disturbance. Under this pathway, old-growth stands may

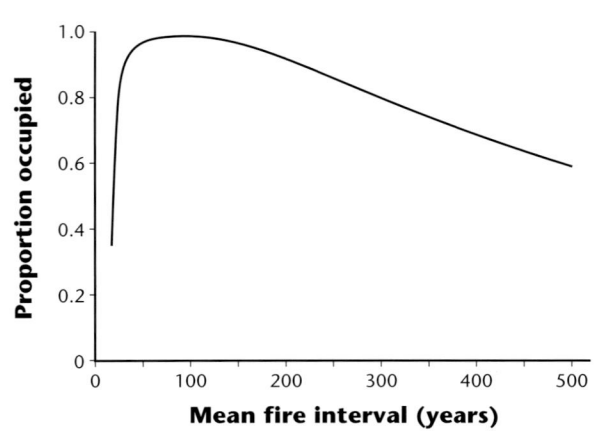

Figure 12.7: Predicted occupancy of the landscape by stands of Mountain Ash forest as a function of the interval between high-intensity wildfires. (Redrawn from McCarthy et al. 1999.)

eventually be replaced by rainforest. However, the recurrence of fires in many old-growth stands (Banks 1993), including in February 2009, indicates that rainforest replacement may not occur in all parts of montane ash landscapes. In addition, climatic and related analyses of key rainforest plants such as Myrtle Beech suggests that the development of cool temperate rainforest is unlikely to occur in all parts of Mountain Ash landscapes (Lindenmayer et al. 2000b) (see Chapter 7).

Finally, if the interval between stand-replacing fires is very short – less than 20–30 years, the period required for trees to reach sexual maturity and begin producing seed – then stands of montane ash will be replaced by other species. This pathway of frequent high-intensity disturbance has resulted in the creation of 'unstocked' stands of *Acacia* spp. understorey that lacks an overstorey of ash-type eucalypts. This is reflected in the modelling by McCarthy et al. (1999), which shows a significant reduction in the proportion of the landscape occupied by montane ash forest in response to the mean interval between stand-replacing fires.

Importance of different disturbance pathways

Different disturbance pathways lead to marked differences in stand structure and stand composition (Lindenmayer and Franklin 1997). This is important for several key reasons. First, it highlights why old-growth stands of montane ash forest are characterised by particular features, such as a range of age cohorts of trees (Lindenmayer et al. 2000a) (Chapter 4). Second, it emphasises a key ecological mechanism underpinning spatial variability in the occurrence of old-growth montane ash stands across forest landscapes. Third, it provides a basis for understanding temporal variation in the types and abundance of important biological legacies, such as large trees with hollows which are, in turn, critical habitat components for many forest-dependent vertebrates (Franklin et al. 2000) (see Chapter 6).

FIRE AND THE VIABILITY OF ARBOREAL MARSUPIAL POPULATIONS

Although wildfire is a form of natural disturbance in montane ash forests, it is a potential threat for arboreal marsupials and other elements of the biota in those forests. Modelling of wildfire has been at the heart of several major studies of extinction risk in montane ash forests (Chapter 10), including those on the Greater Glider (McCarthy and Lindenmayer 1999; Possingham et al. 1994) and Leadbeater's Possum (Lindenmayer and Possingham 1995a; McCarthy and Lindenmayer 2000). Modelling of both species indicated that wildfire significantly increases their risk of extinction (Figure 12.8), particularly local extinction,

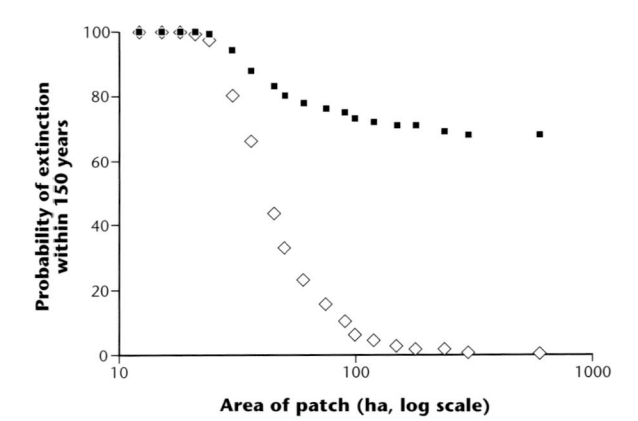

Figure 12.8: Relationships between the probability of extinction of Leadbeater's Possum and the size of isolated patches of old-growth forest of varying sizes when a fire was (open diamonds) and was not (solid squares) modelled. (Redrawn from Lindenmayer and Possingham 1995a.)

> **Box 12.1: Old-growth forest, logging and fire – a highly controversial perspective on mega-fires**
>
> Fire intensity varies across ash-type forest landscapes. Lower-intensity fires occur where levels of incoming solar radiation are reduced, such as on flat plateaux and in deep valleys (Lindenmayer et al. 1999f). Old-growth and multi-aged forests are most likely to be found in these areas if logging operations have not altered landscape cover patterns (Lindenmayer et al. 2000a). Old-growth forests are characterised by trees with open crowns, wet understoreys of rainforest, and high levels of moisture within fallen logs (Ashton 1986; Lindenmayer et al. 1999d) – features which reduce fire intensity and slow rates of fire spread. In contrast, forests on ridges and valley slopes burn more frequently and at higher intensities.
>
> Clearfelling of ash forests targets accessible areas such as those on valley bottoms. Relatively fire-resistant old-growth and multi-aged forest that would traditionally have occurred in these areas has largely been replaced by post-logging regrowth stands. Recent work in the North American coniferous forests (a close coniferous analog of Australian ash-type forests) has shown that dense self-thinning stands of regrowth trees are significantly more likely to reburn and to burn at higher intensities after logging than are other parts of landscapes (Thompson et al. 2007). Research is urgently needed to test for similar patterns in Australia. It is possible that fire-prone post-logging regrowth located in valleys has been spatially connected to fire-regrowth forests on valley sides and ridges – leading to far greater spatial contagion in fire than occurred prior to European forest modification. That is, past and current forest management may have promoted the spread of fire through landscapes. If this is true, a possible response would be to expand presently unloggable riparian buffers to develop wider, less fire-prone old-growth stands in valley floors. These may act as natural firebreaks to help limit the spatial extent of wildfires. This approach would influence the location of logged areas and their spatial juxtaposition with places for water and biodiversity.
>
> This is a speculative hypothesis about forest flammability, old-growth distribution and logging. It may have no scientific support. Alternatively, it may be important. Research is therefore warranted, particularly given the risks of large-scale mega-fires that may arise from future changes in climate. Such research is particularly important given the massive conflagration that occurred in February 2009 and the implications of spatial contagion in fire behaviour for protection of timber and pulpwood assets within production forest, the conservation of forest biodiversity, the safety of firefighters and the protection of human infrastructure. Following the 2009 wildfires and spatial mapping of fire severity classes, it should be possible to study relationships between fire severity and past land use such as logging history, reservation history and other factors.

in forest landscapes with limited areas of old-growth forest. For example, species such as Leadbeater's Possum had a high chance of persistence in the O'Shannassy Water Catchment where forest cover was dominated by extensive old-growth stands but were at a very high risk of extinction in wood production forest blocks with limited old-growth stands (Lindenmayer and Possingham 1995c). Old-growth stands are likely to be important future refugia for the species as a consequence on the ongoing losses of trees with hollows (see Chapter 6). A fundamental issue is the cumulative effects of limited areas of habitat, ongoing deterioration in habitat suitability (with eventual habitat loss), ongoing logging operations, and direct impacts of wildfire on both populations and habitat suitability (Lindenmayer and Possingham 1995c).

Wildfire effects have been an important influence on the kinds of management recommendations made for the conservation of arboreal marsupials in areas broadly designated for wood production. Lindenmayer and Possingham (1995c) recommended that intermediate-size reserves for Leadbeater's Possum be dispersed across each wood production forest block to limit the risk of them all burning in a single conflagration. Similarly, McCarthy and Lindenmayer (1999) highlighted the importance of significantly increasing the size of patches of reserved old-growth forest to promote the conservation of the Greater Glider, by ensuring that at least some parts of the expanded areas would remain unburned in a wildfire (see Chapter 10).

CARBON BUDGETS AND DISTURBANCE REGIMES

Carbon storage in forests is an increasingly important issue as a consequence of greenhouse gas-driven rapid climate change (Chazdon 2008; Jackson and Schlesinger 2004; Mackey *et al.* 2008; Strengers *et al.* 2007). The effects of wildfires on carbon stocks in forests is poorly understood, although considerable effort is being directed toward altering fire regimes in some jurisdictions in an attempt to reduce greenhouse gas emissions resulting from fires. An example is the West Arnhem Fire Management Agreement in the tropical savannas of the Northern Territory (West Arnhem Fire Management Agreement 2009).

The relationships between carbon emissions and wildfire in montane ash forests are unknown, but recent empirical analyses of standing biomass data for sites across the O'Shannassy Water Catchment have revealed some unexpected outcomes (Keith *et al.* 2009). For example, the highest values for biomass carbon came from mixed-aged forests with a history of natural disturbance. These stands comprise large living and dead trees including many fire-scarred stems that have survived wildfires (Lindenmayer *et al.* 2000a). It appears likely that past fires in these forests have killed some trees, scarred but not killed others, stimulated the regeneration of new cohorts of overstorey trees and rejuvenated the understorey. This would account for average values exceeding 1000 tonnes for living and dead biomass per ha – the highest biomass carbon forests in the world (Keith *et al.* 2009). Carbon emissions would have been generated from the burning of living plant tissue (e.g. leaves and small branches) but considerable stores of carbon remained, including in the standing and fallen dead trees. A key part of remeasuring long-term monitoring sites in montane ash forests will be recalculating values for above-ground carbon biomass and comparing them with pre-fire values to estimate carbon emissions following the February 2009 wildfires. It will also be important to make repeated measurements of the vegetation for a prolonged period to ascertain if values for above-ground biomass do indeed increase substantially as the carbon biomass of newly regenerating stands is added to the dead or fire-scarred biomass left as biological legacies from the pre-fire stand.

> **Box 12.2: Postscript – the myth of prescribed burning**
>
> A recurrent theme after all major conflagrations in Australia is a public outcry that insufficient prescribed burning was done prior to the wildfires, and that if it had been done then 'none of this would have happened'. These are unfortunate and ill-informed comments, generally made in the absence of any understanding of the ecology of montane ash forests. The reality is that montane ash forests cannot be prescribed-burned – it is an ecologically inappropriate management strategy in those ecosystems. The forests are generally very wet and burn only under extreme conditions, when the only kind of fire is a stand-replacing wildfire or partial stand-replacing perturbation. If ash-type forests are burned too frequently (more than every 20 years), stands will be replaced by unstocked forests supporting only *Acacia* spp. trees.

SUMMARY

Fire is a fundamental ecological process in montane ash forests. Key attributes of fire regimes are intensity, frequency and the number and type of biological legacies that remain after fire. These attributes of fire regimes are influenced by other factors, such as topography and the age of the forest when burned.

Fire shapes the spatial distribution of stands of different ages across landscapes and influences the structure and composition of those stands, particularly through its impacts on the types and numbers of biological legacies (Franklin *et al.* 2000). This, in turn, influences the suitability of forest stands as habitat for biodiversity and affects the probability of persistence of some species across forest landscapes.

LESSONS LEARNED

One of the most important lessons learned from the body of work conducted on wildfire in montane ash forests has been the recognition of a greater range of disturbance pathways than previously acknowledged. In particular, the existence of intermediate-intensity disturbance pathways has provided new insights on how and where stands comprising multiple age cohorts of trees are likely to develop (Lindenmayer et al. 2000a). The work suggested that multi-aged forests were far more prevalent in the past than they are currently (McCarthy and Lindenmayer 1998) and may possibly be after the February 2009 wildfires. It also demonstrated that the impacts of salvage logging operations, especially those following the 1939 wildfires, may have been far more pronounced than previously recognised (Lindenmayer and Ough 2006) (see Chapter 14).

KNOWLEDGE GAPS

As outlined in Box 12.1, a controversial perspective on montane ash forests is the possibility that the spatial bias in past logging operations may have made forest landscapes more susceptible to spatial contagion in wildfires. It is important to reiterate that there are limited data to support this proposition, but the impacts of so-called mega-fires on carbon budgets, human property and life, economies, biodiversity and the timber industry indicate that it requires urgent but careful examination, particularly given the effects of the 2009 wildfires in the Central Highlands of Victoria. The importance of such a study is magnified by the prospects of altered fire regimes as a consequence of climate change (Pittock 2005), in which wildfires become more frequent (Cary 2002), widespread (Flannigan et al. 2005) and intense, or all of these (Franklin et al. 1991; Lenihan et al. 2003; Pittock 2005). There is statistical evidence from western North America of an increase in the extent and intensity of wildfires as a result of climate change (Westerling et al. 2006). Adaptation strategies to climate change and altered fire regimes are especially pertinent in ash-type forests as repeated high-intensity fires in those environments risk the complete elimination of the valuable forests (McCarthy et al. 1999).

FOREST PATTERN, ECOLOGICAL PROCESS AND LINKS TO OTHER CHAPTERS

Natural disturbance is a fundamentally important ecological process in montane ash forests and the key driver underlying an array of other ecological processes. Natural disturbances can alter carbon budgets and nutrient cycling (Adams and Attiwill 1984; Keith et al. 2009) as well as hydrological regimes (Vertessey and Watson 2001). A given natural disturbance may exert a strong influence on subsequent natural disturbances, for example through changes in fuel loads or the spatial juxtaposition of different age classes of forest (Whelan 1995). Natural disturbances also affect human disturbances such as (post-fire) salvage logging as well as traditional logging, through their influence on forest age class distribution and the time when particular stands are suitable for harvesting.

Many spatial patterns in montane ash forests are strongly influenced directly or indirectly by the ecological process of natural disturbance. These include the spatial cover of ash-type forest and rainforest, the structure and composition of stands, and the distribution and abundance of animal and plant communities and individual species within those communities.

As outlined in earlier chapters, the ecological process of natural disturbance is itself influenced by other ecological processes such as climate regimes and topographic influences and by forest patterns such as the spatial arrangement of age classes (e.g. old-growth forest) and the occurrence of rainforest. Past human disturbances like traditional logging and salvage logging might influence the intensity and spatial extent of subsequent natural disturbances, as has been found in parts of western North America (Thompson et al. 2007).

Links to other chapters

The pivotal influences of natural disturbance regimes outlined above mean that this chapter is very strongly linked to the content of many other chapters in this book. Chapter 10 outlines how wildfire was modelled in studies of the viability of several species of arboreal marsupial. Stand-replacing and partial stand-replacing fires influence the spatial distribution of forest age classes in montane ash forests, as outlined in Chapter 4.

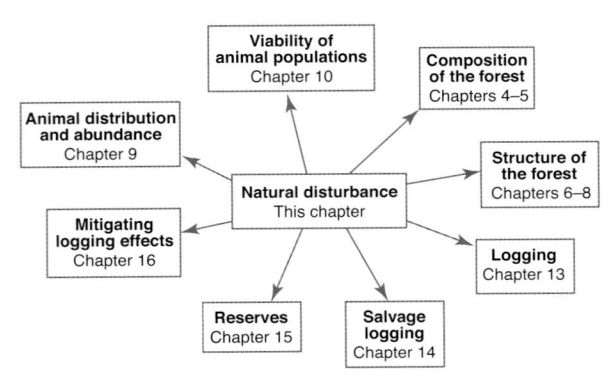

Figure 12.9: Relationships between natural disturbance regimes and key topics examined in other chapters.

Wildfire is integral to the creation as well as the loss of trees with hollows, as discussed in Chapter 6 (Gibbons and Lindenmayer 2002). It also influences the structure and composition of the understorey in montane ash forests (Chapter 7). These factors, in turn, can be important drivers of the distribution and abundance of some species (Chapter 9).

A general principle for conservation management is that human disturbance regimes should be as consistent as possible with natural disturbance regimes (Lindenmayer *et al.* 2007a). This concept is discussed while examining the effects of logging and salvage logging (Chapters 13 and 14). It is touched on in Chapter 16, which explores ways to mitigate the impacts of logging.

Natural disturbances influence the effectiveness of reserves and other management actions – these interactions are discussed in Chapters 15 and 16.

13 Human disturbance: logging

INTRODUCTION

Logging is the predominant form of human disturbance in montane ash forests. There are approximately 121 000 ha of Mountain Ash forest in the Central Highlands of Victoria, the vast majority in public ownership (Land Conservation Council 1994). Approximately 20–25% of the area of montane ash forest occurs within the Yarra Ranges National Park and is exempt from logging, including post-fire salvage logging (Land Conservation Council 1994; Lindenmayer and Ough 2006). The remaining ~75–80% of the resource is in places broadly designated for wood production. Of this, it has been estimated that ~35% is actually available for timber harvesting; some areas are unharvestable because they are in streamside zones or on steep and rocky terrain, or occur in special protection zones for biodiversity protection (Commonwealth of Australia and Dept of Natural Resources and Environment 1997).

The Central Highlands region is estimated to produce 132 400 m^3 net of Mountain Ash sawlogs per annum (information provided by VicForests and Dept of Primary Industries, unpublished data, June 2006). These logs make a significant contribution to the operation of 15 mills and the towns in which they are located, by directly employing approximately 600 people in sawmills alone. It is estimated that the Mountain Ash forests of the Central Highlands region employ up to 1000 people in harvesting, haulage, sawmilling, secondary processing and pulp and paper manufacture. The Victorian Government receives approximately $11 million pa in revenue from the sale of Central Highlands Mountain Ash sawlogs and approximately $4 million from the sale of residual logs. After processing and value-adding, the net value of the Central Highlands Mountain Ash resource makes an estimated contribution of ~$485 million to the Victorian economy (information provided by VicForests and Dept of Primary Industries, unpublished data, August 2006).

Over the past three decades, forestry and forest management have been among the most socially divisive resource management issues in Australia (Ajani 2007; Dargarvel 1995; Lindenmayer and Franklin 2003; Lunney 2004; Routley and Routley 1975). The montane ash forests of the Central Highlands of Victoria are no exception. There have been two broad areas of conflict:

- land allocation (setting aside forest reserves versus maintaining forest in production areas);

Table 13.1: Studies directly or indirectly examining impacts of traditional clearfell logging in the montane ash forests of the Central Highlands of Victoria

Study description	Reference
Impacts of spacing of trees with hollows	Lindenmayer et al. (1990s)
Impacts of abundance of trees with hollows	Lindenmayer et al. (1991b)
Impacts on collapse of trees with hollows within retained linear strips	Lindenmayer et al. (1997b)
Impacts on prevalence of multi-aged stands	McCarthy and Lindenmayer (1998); Lindenmayer and McCarthy (2002)
Impacts on understorey plants	Lindenmayer et al. (2000a)
Impacts on abundance of animals in retained linear strips	Lindenmayer et al. (1993a); Lindenmayer et al. 1994d); Lindenmayer et al. (1994c)

- the impacts of logging on other values, particularly biodiversity conservation and water production.

Clearfelling is the conventional way of logging montane ash forests (Campbell 1997; Flint and Fagg 2007; Squire et al. 1991). This chapter briefly outlines the impacts of traditional (green tree) clearfelling practices on biodiversity in the montane ash forests. These effects have been the subject of detailed studies over the past 25 years (Lindenmayer 2000; Lindenmayer and Franklin 2002; Ough 2002). Impacts are discussed at two scales – the stand level and the landscape level. There may be interactions between impacts at different scales, and the potential for these is explored. Discussion of a special kind of clearfelling that takes place directly after natural disturbances such as wildfire and windstorms – salvage logging – has been reserved for Chapter 14. This is because salvage logging may have effects that are different from, or additional to, traditional clearfell logging (Lindenmayer et al. 2008b). The material presented in this chapter is a prelude to Chapter 16, which discusses ways to mitigate logging impacts in montane ash forests, including using harvesting methods other than clearfelling.

BACKGROUND DATASETS

The impacts of conventional clearfell logging practices in montane ash forests have mostly been inferred from analyses of large datasets gathered over the past 25 years. These datasets include:

- information on the overall abundance and spacing of trees with hollows on a large number of sites subject to different disturbance regimes, including those where clearfelling has or has not taken place (Lindenmayer et al. 1991b, 1990a);
- data on the collapse rate of trees with hollows on 49 linear strips of forest retained between recently clearfelled stands (Lindenmayer et al. 1997b);
- information on the prevalence of multi-aged stands in areas that had or had not been subject to past clearfelling operations (Lindenmayer and McCarthy 2002);
- data on plant species composition and stand structure in 520 forest sites assigned to different dominant overstorey age cohorts (Lindenmayer et al. 2000a);
- the abundance of small terrestrial mammals and arboreal marsupials within 49 retained linear strips surrounded by recently clearfelled stands (Lindenmayer et al. 1993a, 1994c).

CONVENTIONAL CLEARFELLING OPERATIONS

Many steps are involved in the planning and execution of logging operations in montane ash forests. The extent of forest resources is estimated in several ways including aerial surveys, inventory plots and stand simulation models. The location of potential cutblocks or coupes is mapped with respect to logging exclusion areas such as special protection zones, permanent monitoring sites and other considerations

Figure 13.1: The principal stages in clearfelling operations. (a) Removal of all merchantable trees on site with living and dead trees left standing. (b) High-intensity regeneration burning of the logging slash on site. (c) Reseeding of the broadcast burn to promote regeneration after logging. (Photos by David Lindenmayer).

which limit where timber harvesting can take place. On-ground pre-logging assessments result in some areas planned for logging being withdrawn from harvesting (e.g. because they contain suitable habitat for Leadbeater's Possum). The location of coupe boundaries is marked to guide logging contractors where trees can and cannot be cut.

Clearfelling is the conventional harvesting method in montane ash forests (Campbell 1997; Flint and Fagg 2007). Of the areas of montane ash forest subject to logging, more than 95% is clearfelled (Lutze *et al.* 1999). Virtually all standing trees are removed over 15–40 ha in a single operation (Lutze *et al.* 1999; Squire *et al.* 1991). The average size of coupes in Central Highlands of Victoria in the 1990s was 16 ha because of topographic constraints on where logging could take place. Logging coupes can be aggregated up to 120 ha and harvested over a period of five years (Dept of Natural Resources and Environment 1996). Logging is followed by a high-intensity fire to burn logging debris (bark, tree crowns and branches), creating a nutrient-rich ash seedbed to promote the regeneration of a new stand of eucalypts (Figure 13.1). The rotation time between clearfelling operations is nominally 80 years (Dept of Natural Resources and Environment 2002; Govt of Victoria 1986), although the Timber Industry Strategy permits logging below the nominal rotation age to regulate age classes and provide for smooth timber flows (Dept of Natural Resources and Environment 2002).

STUDIES ON THE IMPACTS OF CLEARFELLING

The impacts of clearfelling in montane ash forests have largely been inferred from observational studies and detailed empirical measurements of animal occurrence, plant species composition and vegetation structure on large numbers of sites throughout the Central Highlands of Victoria (reviewed in Lindenmayer 1994c, 2000). The majority of studies have been comparative retrospective ones in which statistical modelling was used to contrast animal occurrence and stand structural complexity between areas with different characteristics, including the presence or absence of previous logging disturbance (Lindenmayer 1994d).

Box 13.1: The 'natural advantage' of studying logging effects in montane ash forests

A vast number of investigations worldwide have tried to quantify logging impacts on biodiversity and key ecological processes by comparing logged and unlogged sites. This seems a straightforward way to compare sites with different histories of human impact. However, there can be serious environmental confounding with human disturbance, which thwarts a valid interpretation of logging effects in many studies. That is, some areas have remained unlogged for a good reason – they are steep, rocky and characterised by low forest productivity. Indeed, prescriptions for forest management frequently preclude these kinds of areas from timber harvesting. Populations of many species of plants and animals can be naturally lower in these areas (even in the absence of human disturbance). By contrast, areas of high productivity and on flat terrain are where populations of some species can be naturally higher (Braithwaite 1984), but are also where logging operations are likely to take place. Thus, there is confounding between site productivity and logging history. This can seriously influence inferences about the effects of logging. For example, a comparison of animal presence and abundance may find no significant difference between the two kinds of sites, and infer there has been no effect from logging. An alternative explanation may be that the initially higher (pre-logging) densities of animals on subsequently logged sites have been reduced to match the initially lower densities of unlogged lower-productivity sites.

Importantly, the water catchments within the Yarra Ranges National Park in the Central Highlands of Victoria support extensive areas of high-productivity forest on flat terrain, with no history of human disturbance (logging). Sites in these areas are a valuable baseline for comparison with logged sites outside the national park. As part of the design of initial field surveys in the late 1980s and the subsequent establishment of the long-term monitoring program in the Central Highlands, there was careful matching of key environmental conditions, such as slope and topographic position. Thus, it has been possible to avoid confounding between environmental conditions and disturbance history.

Other kinds of investigations have been used to infer clearfelling impacts. These include studies involving intensive radio-tracking of animals on sites (Lindenmayer *et al.* 1996a) (see Chapter 6) and simulation modelling of predicted population persistence in logged and unlogged landscapes (Lindenmayer and Possingham 1995c) (see Chapter 10).

STAND-LEVEL IMPACTS OF CLEARFELLING

The stand-level impacts of conventional clearfelling operations on biodiversity can be broadly summarised as effects on:

- overstorey hollow tree diversity, spacing and abundance;
- the prevalence of multi-aged stands;
- understorey trees and other plants;
- ground cover and coarse woody debris.

Impacts on overstorey trees

Overstorey trees with hollows are significantly reduced in abundance by clearfelling (Lindenmayer *et al.* 1991b) and can be significantly lower than levels characteristic of unlogged stands (Figure 13.3). The rotation time between logging events means that trees are removed before they become old enough to begin developing hollows suitable for use by many species of vertebrates (120–190+ years old). Current tree retention rates in cutover areas are typically five or fewer trees per ha. However, altered microclimatic conditions (e.g. increased wind speeds) coupled with scorching from regeneration burns result in high rates of mortality and attrition among trees retained on logged sites (Ball *et al.* 1999) and contribute to accelerated rates of tree loss in forests immediately adjacent to harvest units (Lindenmayer *et al.* 1997b) (see Chapter 6). This leads to low numbers of retained trees over time both on logged sites and in uncut surrounding areas.

Figure 13.2: The accelerated attrition of trees with hollows on logged sites and on areas immediately adjacent to logged coupes. (Photo by David Lindenmayer).

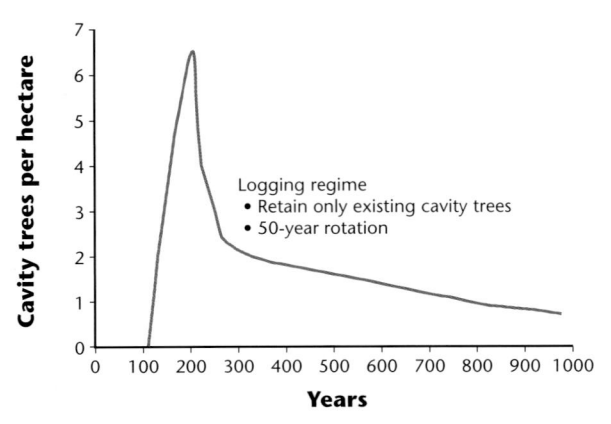

Figure 13.3: Long-term predicted decline in the abundance of trees with hollows in a logged area. (Redrawn from Ball *et al.* 1999).

Large trees with hollows are nesting and denning sites for arboreal marsupials. Thus, areas of forest are essentially rendered unsuitable for cavity-dependent animals in areas which are clearfelled on a 50–80 year rotation. These areas are highly unlikely to become suitable for the entire suite of cavity-dependent fauna, including Leadbeater's Possum.

The range of types of trees with hollows in stands (i.e. those with different external morphological characteristics) is substantially reduced in forests subject to

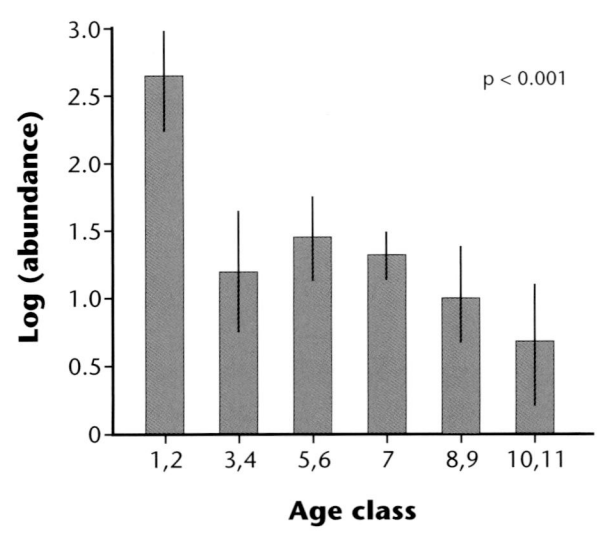

Figure 13.4: The abundance of trees with hollows in montane ash stands of different ages. Age classes 8+9 and 10+11 correspond to areas of post-logging regrowth. (Redrawn from Lindenmayer *et al.* 2000a).

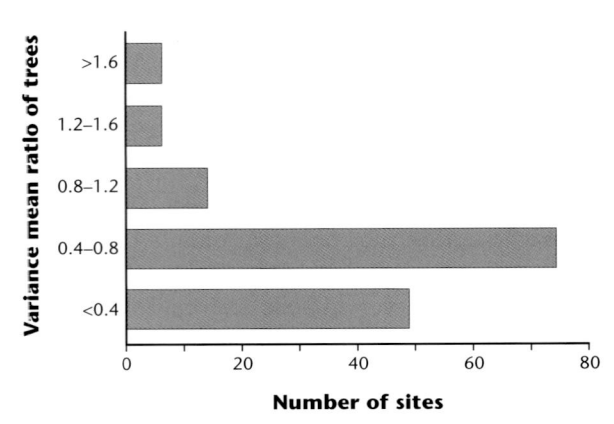

Figure 13.5: The distribution of trees with hollows in montane ash forests. Values of less than 1 for the variance:mean ratio of trees with hollows means their distribution is regular. A clustered distribution corresponds to values of 1 or greater than 1; these sites predominantly correspond with clearfelled sites. (Redrawn from Lindenmayer *et al.* 1990a).

clearfelling. This can deplete the diversity of arboreal marsupials, as each species requires trees with hollows in different stages of decay and condition (Lindenmayer 1997) (Chapter 9).

Clearfelling operations typically alter the spacing patterns of trees with hollows. For example, the spacing of these trees is typically regular in unlogged mature or old-growth forest, whereas the spacing pattern is clustered in clearfelled and regenerated areas (Figure 13.5). This is often because of attempts to better protect trees with hollows on logged sites by retaining them within clumps in particular parts of coupes. Tree spacing patterns can be important because the social organisation of some species of animals, coupled with their propensity to swap frequently between dens (Lindenmayer and Meggs 1996; Lindenmayer *et al.* 1996c) (Chapter 9) means that rates of tree occupancy are reduced when trees are clustered (Lindenmayer *et al.* 1990a).

Impacts on the prevalence of multi-aged stands

Clearfelling operations in montane ash forests produce areas with a uniform, even-aged stand structure with the young regenerating trees belonging to a single cohort (Squire *et al.* 1991). This contrasts with the more complex multi-aged stand structure sometimes produced by wildfires in parts of Mountain Ash landscapes (Lindenmayer et al. 1999f; McCarthy and Lindenmayer 1998) (Chapter 12). Multi-aged montane ash stands are particularly important for biodiversity conservation and support the highest densities of native mammals (Macfarlane 1988). Large living trees that survive the effects of fire and form a second age cohort within multi-aged forests are not only nest sites for vertebrates, but also provide large numbers of flowers (Ashton 1975a) which can be a critical food resource for pollen and nectar-feeding vertebrates. Microchiropteran bats spend more time foraging in multi-aged stands where there is considerable variation in vertical vegetation structure (Brown et al. 1997). The highest diversity of species of arboreal marsupials is found in multi-aged montane ash forests (Lindenmayer et al. 1991d). This is possibly because different species of arboreal marsupials require hollow-bearing trees with markedly different external morphological features (e.g. diameter and stage of senescence) (Lindenmayer et al. 1991e) (Chapter 6) and multi-aged stands may be more likely to support a range of tree types (Chapter 6).

Impacts on understorey vegetation

Clearfelling can markedly alter understorey vegetation (Ough 2002; Ough and Murphy 2004; Ough and Ross 1992; but see Harris 2004). Weed and sedge species are more common on clearfelled sites and populations of resprouting shrubs, tree ferns and many species of ground ferns are depleted (Ough 2002).

Mechanical disturbance from logging machinery, together with the effects of high-intensity slash fires following harvesting, significantly lowers the abundance of tree ferns (Lindenmayer and Ough 2006; Ough and Murphy 1996, 2004) (Chapter 11). A loss of tree ferns substantially reduces the availability of nursery sites for other plants (Ough and Murphy 1996) such as epiphytic plant species (Ough and Murphy 2004) and eliminates the sheltered moist microhabitats that support fungi and other food resources for forest animals (Lindenmayer et al. 1994d).

Thickets of long-lived fire-resistant understorey plants can be significantly depleted or lost as a result of clearfelling (Mueck et al. 1996; Ough and Murphy 1998). Intact thickets of understorey vegetation are important foraging sites for some species of forest-dependent vertebrates, such as the Mountain Brush-tail Possum (Lindenmayer et al. 1994d). Stands of understorey rainforest plants such as Myrtle Beech may be negatively influenced by clearfelling operations. Their occurrence is significantly reduced in young clearfelled stands, relative to burned old-growth ones (Lindenmayer et al. 2000a). Myrtle Beech may be slow to re-establish on intensively clearfelled sites.

Impacts on ground cover and coarse woody debris

Young logged stands support significantly smaller-diameter logs than old-growth forests (Lindenmayer et al. 1999d) (Figure 13.6). The development of moss mats on these logs is also lower than in old-growth forests (Lindenmayer et al. 1999d) (Figure 13.7). Changes in forest floor conditions in logged forests may influence the suitability of substrates for reptiles (Brown and Nelson 1993), small mammals (Cunningham et al. 2005; Lindenmayer et al. 2002a), the germination of rainforest plants (Howard 1973) and the development of mats of moss cover (Ashton 1986).

Many regrowth stands of montane forest recovering after the 1939 wildfires are characterised by high volumes of large-diameter logs, often exceeding 1000 m^3/ha (Lindenmayer et al. 1999d). The size and accumulated volume of these logs is greater than the

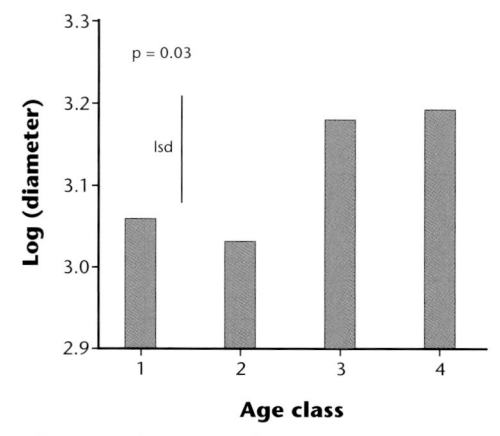

Figure 13.6: Log diameter and age class. Category 1 corresponds to 20–25 post-logging regrowth forest and categories 3 and 4 are mature and old-growth stands. (Redrawn from Lindenmayer et al. 1999d).

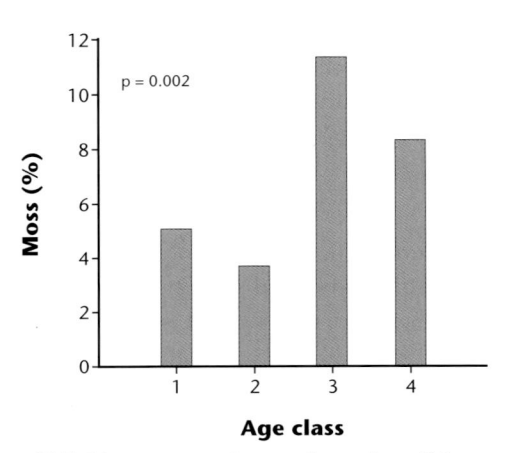

Figure 13.7: Moss cover on logs and age class. Category 1 corresponds to 20–25 post-logging regrowth forest and categories 3 and 4 are mature and old-growth stands. (Redrawn from Lindenmayer et al. 1999d).

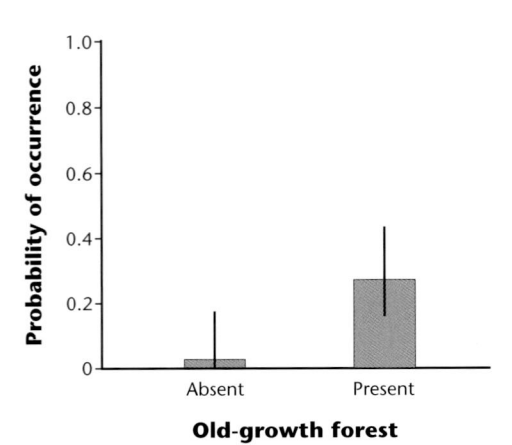

Figure 13.8: Relationships between occurrence of the Yellow-bellied Glider and amount of old-growth forest in 80-ha areas of montane ash forest. (Redrawn from Lindenmayer et al. 1999a).

size and volume of the standing living trees, reflecting a major cohort of biological legacies on the floor of these post-fire regrowth stands (see Chapter 8).

LANDSCAPE-LEVEL IMPACTS OF CLEARFELLING

The impacts of clearfelling may extend beyond effects at the stand level. However, it can be much harder for humans to understand the potential for landscape-level impacts of clearfelling because landscapes are difficult entities for many people to perceive and comprehend how they function. For example, a landscape may have a complete cover of native forest yet no available habitat for some species because suitable old-growth stands are absent (Lindenmayer et al. 1999a).

Clearfelling may alter landscape composition, with remaining areas of old-growth forest (now exempt from logging) becoming isolated among extensive stands of young forest recovering after harvesting. This is likely to be the reverse of landscape composition patterns prior to white settlement. That is, montane ash landscapes are likely to have formerly been dominated by large areas of old-growth forests with comparatively limited areas of regrowth. The converse was true except in places like the O'Shannassy Water Catchment, because it had been closed to almost all human access for over a century and prior to the February 2009 Black Saturday wildfires it sup-

ported extensive stands of old-growth forest (Lindenmayer and Possingham 1995c). For example, prior to Black Saturday, the single largest patch of old growth in the 80% of the montane ash forest estate broadly designated for wood production was only 57 ha. Similarly, some wood production forest blocks in the Central Highlands of Victoria (3000–7000-ha planning areas) supported <2–5% cover of old growth (Lindenmayer and Possingham 1995c).

Such differences in landscape composition can have negative effects on some wide-ranging vertebrates, such as the Sooty Owl and Yellow-bellied Glider, which are strongly associated with large areas of old-growth forest (Incoll et al. 2000; Lindenmayer et al. 1999a; Milledge et al. 1991) (Figure 13.8). Old-growth forests are also important habitat refugia for the Mountain Brushtail Possum and the Greater Glider. Populations of these species in late-successional stands that are fragmented by widespread clearfelling may not be viable in the medium to long term (Chapter 5) (Lindenmayer and Lacy 1995a, 1995b; McCarthy and Lindenmayer 1999; Possingham et al. 1994).

Clearfelling may create patches of retained forest that are unsuitable for occupancy by some species. For example, narrow linear strips are left as buffer strips between logging coupes in wood production forest blocks. These strips of forest are suitable habitat for species such as the Greater Glider and the Moun-

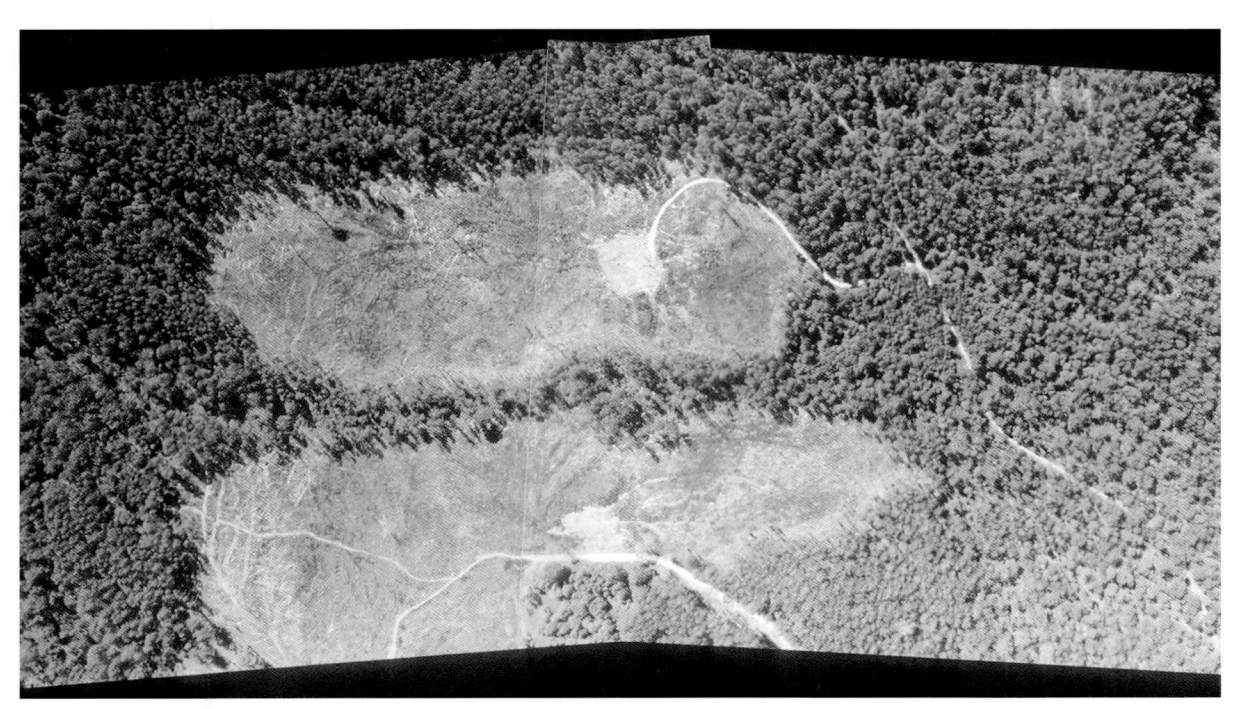

Figure 13.9: Narrow retained linear strip between two recently clearfelled areas. (Photo by Ross Meggs).

tain Brushtail Possum, but appear to be rarely occupied by Leadbeater's Possum (Lindenmayer and Nix 1993). It is thought that the complex colonial social system of Leadbeater's Possum, coupled with its reliance on widely dispersed food resources, make narrow linear strips unsuitable for the species (Lindenmayer et al. 1993a).

Clearfelling may have other kinds of landscape-level effects that are broadly associated with spatial variability in environmental conditions. These kinds of effects are related to the fact that not all parts of landscapes are created equal in terms of their suitability for biodiversity. First, clearfelling operations are typically targeted at sites on flat or undulating terrain and are exempt from steep and rocky areas (Dept of Conservation 1989). However, the abundance of some species is negatively associated with these kinds of areas. For example, the abundance of Leadbeater's Possum is significantly lower on steep slopes than on flat terrain (Lindenmayer et al. 1991d). Second, bioclimatic analyses indicate that the climatic envelope occupied by Leadbeater's Possum encompasses a warm and wet subset of environmental conditions within montane ash forests (Lindenmayer et al. 1991f). Therefore, if clearfelling is targeted at the parts of forest landscapes which are likely to be most suitable for certain species (e.g. Leadbeater's Possum) it will have a disproportionately negative effect on those species.

Policies of dispersing logging coupes in space and time may mean that there are cumulative effects on biodiversity resulting from ongoing logging of the more productive parts of landscapes. Thus, stand simplification (which impairs habitat suitability for many taxa) can occur in many stands and eventually lead to homogenisation of large areas of the landscape. Localised extinction–recolonisation metapopulation dynamics (Beissinger and McCullough 2002; Hanski 1994) may be impaired and potential population source areas may be removed. Their role in helping to recover populations in disturbed areas elsewhere in the landscape may therefore be reduced. It is extremely difficult to demonstrate empirically these kinds of cumulative and interacting stand and landscape-scale effects at the spatial scales relevant to forest wildfire conservation. However, a raft of modelling studies, including many undertaken in the Central Highlands of Victoria, suggest that landscape-level processes may threaten the medium- and long-term viability of some taxa in montane ash forest (McCarthy and Lindenmayer 1999) (Chapter 10).

SUMMARY

The impacts of clearfelling in montane ash forests are substantial and multi-scaled. Clearfelling alters the prevalence of individual large trees, changes the composition of the ground, tree fern and understorey layers of stands, reduces the prevalence of multi-aged stands and alters landscape composition through its influence on the spatial pattern of stands (e.g. the prevalence and spatial juxtaposition of old-growth and regrowth stands) (Lindenmayer and Franklin 1997).

These changes can have a range of negative impacts, best documented at the stand level, on cavity-dependent arboreal marsupials (Lindenmayer 1994b) and on plants (Ough 2002; Ough and Murphy 2004). Landscape-level impacts are also likely to be important but are much harder to quantify with empirical data (but see Lindenmayer et al. 1999a; Milledge et al. 1991).

LESSONS LEARNED

The impacts of clearfelling operations appear to manifest at a greater range of spatial and temporal scales than has often been recognised. The immediate impacts of forestry practices on stands are obvious to any observer. But the changes in stand structure and plant species composition, particularly several decades after trees have begun to regenerate, require careful scientific measurement and analysis, as do the responses of the biota to those changes. It has become clear over the past 25 years that the negative impacts of clearfelling operations can be more substantial and more prolonged than formerly appreciated. This is particularly true for structures such as large trees with hollows that take a long time to develop, live for a long time and can stand for many years after their death but are readily lost during logging operations (Chapter 6). Similarly, the time required to develop complex stands comprising trees from multiple age cohorts may exceed 200 years (RAC 1991, 1992).

A more recent insight has been the potential for cross-scale effects of clearfelling, such as the cumulative impact of aggregating many logged stands across landscapes, with subsequent losses of landscape heterogeneity and consequent broad-acre stand simplification.

KNOWLEDGE GAPS

Almost all the work summarised in this chapter has inferred logging impacts from observational studies and from computer simulation modelling. While these are undoubtedly important, they are less powerful than experiments for inferring impacts and designing strategies to mitigate those impacts (Margules et al. 1994; Schmiegelow et al. 1997). Moreover, inferences for mobile species like birds can be problematic in these kinds of studies. A major experiment on logging and biodiversity responses commenced in 2003 (Lindenmayer 2007) (Chapters 16 and 17) but will take time to generate results.

An interesting knowledge gap is the fate of animals on sites that are logged. No studies examining this issue have been conducted in montane ash forests. It is possible that a general public increasingly concerned with animal welfare issues may begin to demand this kind of work. Studies on the direct and immediate impacts of logging elsewhere in Australia suggest that the individuals of many species probably die *in situ* during harvesting or immediately after it (How 1972; Tyndale-Biscoe and Smith 1969). This may or may not influence the longer-term persistence of populations of particular taxa on logged sites; for example, some species may recolonise regenerating areas if suitable habitat develops (Loyn 1985).

The work on logging effects in montane ash forests has focused on data on the occurrence of animals in logged sites and compared it with data from unlogged areas. However, the social organisation of populations can have marked impacts on population growth and dynamics (Banks et al. 2008). Information on the disturbance recovery of the social structure of populations is currently lacking.

Finally, all the work discussed in this chapter has concerned vertebrates and plants. Almost nothing is known about logging effects on invertebrates, by far the most species-rich animal group in montane ash forests. Work elsewhere in the world, as well as research in equivalent wet forests in Tasmania, indicates that some invertebrate taxa can be sensitive to logging (Meggs 1997; Niemela *et al.* 1996, 1993). Studies of invertebrates, such as carabid beetles, and their response to disturbance are well overdue. The potential for logging effects on other kinds of organisms has received limited attention. Further work is required to examine, for example, the magnitude and persistence of logging effects on moss mats and bryophyte communities in montane ash forests (reviewed by Lindenmayer *et al.* 2002a).

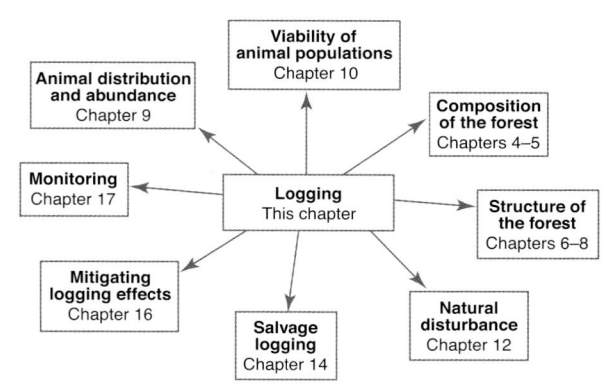

Figure 13.10: Relationships between logging effects and key topics examined in other chapters.

FOREST PATTERN, ECOLOGICAL PROCESS AND LINKS TO OTHER CHAPTERS

As in the case of natural disturbance, human disturbance (logging) regimes exert a major influence on many other ecological processes and an array of forest patterns, as shown in the links between this chapter and others in this book (Figure 13.10). Human disturbances affect ecological processes such as carbon storage, nutrient cycling and natural disturbance regimes. They also affect patterns such as those of forest cover, stand structure and composition, animal and plant distribution and abundance. Some of these patterns also influence logging operations, for example where areas can and can't be harvested (old-growth exclusion zones).

Links to other chapters
Developing an understanding of logging impacts has been at the heart of much of the work in the montane ash forests over the past 25 years and therefore the content of this chapter is intimately linked with that of many other chapters.

Because logging shapes stand structure and the composition of landscapes, it is a key process influencing the distribution and abundance of forest biota – the core topic of Chapter 9. Logging threatens the viability of some forest-dependent species in montane ash forests, discussed as part of the work on population viability analysis in Chapter 10. Relationships between logging and plant species composition and vegetation structure are a common thread through the five chapters which comprise Parts II and III.

It is not possible to ignore the many links between natural disturbances and human disturbances in montane ash forests. Information in the previous chapter on wildfire highlights the extent of contrasts between human disturbance and natural disturbance – a topic which features prominently in Chapter 16, about ways to mitigate logging impacts.

Two ways to better quantify (and mitigate) the effects of logging are to conduct experiments and rigorous monitoring studies. These topics are addressed in Chapters 16 and 17, respectively.

14 Salvage logging effects

INTRODUCTION

As outlined in Chapter 12, large-scale natural disturbances occur at varying intervals in most ecosystems worldwide. In many cases, major efforts are mounted to clean up after natural disturbances (Beschta *et al.* 2004; Lindenmayer *et al.* 2008a; Robinson and Zappieri 1999). This is particularly true in forest landscapes, where salvage logging of disturbed stands is widely practised for such reasons as recouping economic losses before serious deterioration of trees occurs (Shore *et al.* 2003; Ulbricht *et al.* 1999).

Salvage logging is defined by the Society of American Foresters (Helms 1998) as 'the removal of dead trees or trees damaged or dying because of injurious agents … to recover economic value that would otherwise be lost'. In practice, salvage logging often results in the removal of undamaged live trees along with the dead or damaged ones (Bunnell *et al.* 2004; Foster and Orwig 2006).

This chapter has two primary themes. The first concerns general differences in the impacts of salvage logging and those of conventional (green tree) logging. The second summarises the range of salvage logging effects in montane ash forests.

POTENTIAL DIFFERENCES BETWEEN CONVENTIONAL AND SALVAGE LOGGING

Chapter 13 outlined the range of impacts that can arise from conventional clearfell logging. Salvage logging can differ from conventional logging of green (unburned) stands in four ways that are important to the maintenance of ecosystem processes and biodiversity. These are described below (after Lindenmayer *et al.* 2008a).

Conditions preceding logging operations

Major disturbances may be associated with unusual environmental conditions. For example, extensive soil-wetting occurs before the high winds associated with hurricanes and cyclones (Elliott *et al.* 2002). Prolonged droughts and high temperatures are typical before wildfires in some forest types (Wallace 2004) and can exert strong influences on many organisms (Rubsamen *et al.* 1984). Consequently, plants and animals are often under stress at the time of disturbance and may not have recovered (or have the potential to recover) from the dual impacts of environmental stress and the disturbance, before salvage logging operations begin.

Conditions under which logging takes place

Salvage logging is conducted in disturbed ecosystems where the organic component of soils may have been burned (Beschta et al. 2004) or mineral soil exposed (James and Norton 2002). This may make soils more vulnerable to impacts associated with salvage logging, such as compaction and erosion (McIver and Starr 2000, 2001; Shakesby et al. 1996). Salvage logging may take place around piles of fallen trees which may make timber removal more difficult (Holtam 1971), thereby requiring more skidding and increasing the soil disturbance.

What is logged

Salvage logging involves the removal of particular trees or stand components that are often uncommon in the landscape, such as charred standing stems, insect-killed or dying trees, recently windblown trees or the largest trees that remain – because of their economic value (Morissette et al. 2002). Conditions following stand-replacing disturbances in many regions are among the most biologically diverse and most uncommon of all forest conditions (Franklin and Agee 2003). Recent studies have shown that such ecosystems support distinctive biotic assemblages that are clearly different from those characteristic of other kinds of stands (Saint-Germain et al. 2004).

Prescriptions for logging practices

Salvage logging sometimes occurs in ways that are more intensive at the stand level or extensive at the landscape level than green tree logging is (Dept of Sustainability and Environment 2007; McIver and Starr 2001; Schmiegelow et al. 2006). Salvage logging may be allowed in areas where green tree logging would normally be prohibited (Foster and Orwig 2006). For example:

- cutover sizes can be larger (Govt of British Columbia 2008);
- forests may be cut at much younger ages than normal (Radeloff et al. 2000);
- larger or older trees may be removed when it is not otherwise allowed (Thrower 2005);
- larger quantities of slash may be left behind and sometimes burned due to greater tree breakage and defect, or smaller tree size (Lindenmayer et al. 2008a);
- areas previously designated as roadless may be roaded, providing access for future logging (Karr et al. 2004);
- road networks may be more extensive and more intensively utilised, but may not be constructed to high engineering standards;
- particular kinds of trees, stands or areas normally reserved from logging (e.g. old-growth reserves) may be logged (Eggler 1948; Foster and Orwig 2006).

SALVAGE LOGGING EFFECTS IN MONTANE ASH FORESTS

While there have been instances of salvage logging following windstorms and insect (psyllid) outbreaks in montane ash forests, the vast majority of salvage operations have taken place after wildfires (Lindenmayer et al. 2008a) in 1926, 1932, 1939, 1983 and 2009. The largest operation was in forests damaged in the 1939 Black Friday wildfires. The forests were intensively salvaged until the late 1950s (Lindenmayer et al. 2008a). Noble (1977) noted that 'salvage of fire-killed timber went on until the deterioration in logs and the damage being done to regenerating forests called a halt'.

Salvage logging in montane ash forests is conducted to offset some of the economic losses created through the conversion of high-quality merchantable timber to fire-damaged timber (McHugh 1991). The sooner the burned trees are removed (preferably within two years), the higher their value as timber (Dept of Sustainability and Environment 2003). Extensive salvage logging will take place after the 2009 Black Saturday wildfires in montane ash forests.

Salvage logging resembles clearfelling, but the disturbance order is reversed. Stands are initially burned by an unplanned wildfire. Fire-damaged stands are then clearfelled with all merchantable timber removed (as described above for conventional clearfelling, see also Chapter 13). The intensity of logging and size and pattern of logged areas varies according to accessibility and fire intensity. In some cases, eucalypt regeneration is inadequate and regeneration burns or mechanical site preparation methods are used to re-establish eucalypt stands (Figure 14.1).

The purpose of salvage logging has not changed since the 1930s, but harvesting methods have. Teams of bullocks and horses, cable systems and tramways

Figure 14.1: Burned and salvaged logged stand in the Powelltown State Forest following the 1983 Ash Wednesday fires. Regeneration failure at this site required subsequent mechanical site preparation prior to a second attempt at stand establishment. (Photos by David Lindenmayer).

have been replaced with bulldozers and skidders. Mechanical harvesters often replace hand-felling. Different logging methods produce different degrees of physical disturbance – a fact to be mindful of when comparing outcomes from different decades.

POTENTIAL IMPACTS OF SALVAGE LOGGING

Given that salvage logging typically involves clearfelling, the impacts associated with conventional clearfelling discussed in the previous chapter will characterise salvage logging operations in montane ash forests. A key issue is whether salvage logging has impacts additional to those documented for conventional clearfelling (Lindenmayer and Noss 2006). This issue remains largely unresolved because there have been no specific studies targeted at quantifying the impacts of salvage logging in Central Highlands montane ash forests (one current study in eastern Victoria is investigating the response of the vegetation of Alpine Ash forests to salvage logging after the 2003 wildfires). A study planned to quantify salvage logging effects after the February 2009 wildfires has yet to be implemented. Therefore, much of the following discussion is a synthesis of effects that are *hypothesised* to occur on animal species, plant species and the structure of individual stands.

Impacts on stand structure

Salvage logging may have a range of impacts on stand structure. It can result in the depletion or loss of key forms of biological legacies such as large living and dead fire-scarred trees with hollows, large burned logs and thickets of semi-fire-resistant vegetation. It can also result in the conversion of multi-aged stands to even-aged ones, with negative implications for biota associated with complex stand conditions.

Loss of biological legacies

Franklin *et al.* (2000) discuss the array of biological legacies in forests, and the key ecological roles they play. The biological legacies found in naturally disturbed montane ash forests of the Central Highlands of Victoria are no exception. Wildfires can promote the development of cavities in large trees in these forests (Lindenmayer *et al.* 1993b), providing sheltering, nesting and foraging sites for more than 40 species of vertebrates (Gibbons and Lindenmayer 2002) (Chapter 9). Wildfires and windstorms may generate pulses of large fallen logs (Lindenmayer *et al.* 1999d) which have an array of roles (Chapter 8).

Conventional clearfelling operations can result in the depletion of important biological legacies such as trees with hollows, old understorey shrubs, tree ferns and large fallen logs (Chapter 13). The reduced abundance of such legacies can have flow-on negative effects on dependent biota (Lindenmayer and Franklin 1997; Ough and Murphy 2004). Salvage logging can have similar and potentially additional impacts. A study of the vegetation structure and composition of montane ash stands of different age cohorts was completed by Lindenmayer *et al.* (2000a) (Chapter 4). From that dataset, it was possible to more closely examine stands that regenerated following the 1939 wildfires. Based on historical information on logging activities and known boundaries of several water catchments (where logging has never occurred), it was possible to determine where post-fire salvage logging took place and where it did not. For 1939-aged sites, where it was possible to unequivocally determine logging history, there was a clear and highly significant difference in the abundance of trees with hollows between salvaged and unsalvaged stands (Figure 14.2).

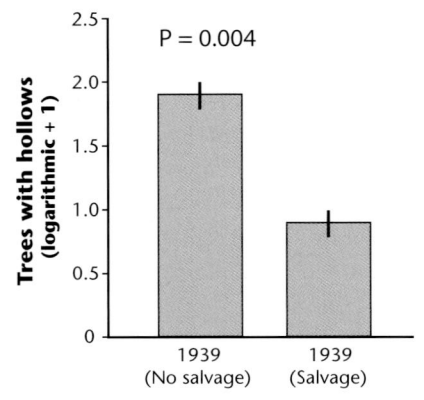

Figure 14.2: Difference in the abundance of trees with hollows between burned and naturally regenerated 1939 montane ash forest (Mountain Ash and Alpine Ash combined) and burned and salvaged stands dating from the 1939 fires. (Redrawn from Lindenmayer and Ough 2006).

Another study of mammal response to stand conditions (van der Rhee and Loyn 2002) drew similar conclusions about the differences in hollow tree abundance between areas of montane ash forest that have and have not been salvage logged.

Differences in the abundance of hollow trees in logged and unlogged areas is critical for many elements of vertebrate biodiversity in montane ash forests. As discussed at length in Chapter 6, highly significant relationships have been established between the abundance of trees with hollows in montane ash forests and the presence and abundance of a range of species of arboreal marsupials, including the endangered Leadbeater's Possum (Lindenmayer et al. 1991d).

Large living and dead trees with hollows in montane ash forests are subject to natural rates of collapse estimated to be between 2.2% and 4% of the population annually (Lindenmayer et al. 1997a; Lindenmayer and Wood 2009). It is likely that collapse rates are higher for hollow trees within burned stands of forest that are subject to salvage logging. Burned hollow trees in these stands are exempt from cutting unless they pose a safety hazard to timber workers, but have increased exposure to altered microclimatic conditions (Parry 1997) when the surrounding stand is cut down. Such exposure also occurs with conventional clearfelling operations (Lindenmayer et al. 1997a).

Conversion of multi-aged stands

The development of stand structure in montane ash forests is heavily influenced by both natural and human disturbance regimes. In the case of wildfire, the range of intensities and frequencies give rise to a wide range of stand conditions (Ough 2002) (Chapter 12). High-intensity wildfires that occur, on average, every 100–120 years or more frequently (McCarthy et al. 1999) typically result in even-aged montane ash forests (Ashton 1976). Low to moderate fire intensities (Smith and Woodgate 1985) over moderate to long frequencies (>150 years) are more likely to give rise to a multi-aged overstorey because not all dominant trees are killed (McCarthy and Lindenmayer 1998) (Chapter 12).

Up to 30% of stands of montane ash burned in the 1939 wildfires may have been multi-aged (Mackey et al. 2002; McCarthy and Lindenmayer 1998). However, the widespread use of post-fire salvage logging following those fires means that many stands have been converted to largely even-aged forest (Lindenmayer et al. 2008a). Even if salvage logging were conducted on

> **Box 14.1: Simulations of salvage logging effects on wildlife persistence**
>
> The impacts of salvage logging on stand structure, multi-agedness and the recruitment of trees with hollows, related to the persistence of populations of Leadbeater's Possum, was simulated using the computer program ALEX for population viability analysis (Lindenmayer and Possingham 1996c). The work modelled the species' risk of extinction in forest blocks with varying amounts of old-growth forest and varying amounts of burned forest (Chapter 10). It showed that salvage logging significantly elevated extinction risk for the species (Table 14.1) (Lindenmayer and Possingham 1996c).

Table 14.1: Estimated probability of extinction (%) of populations of Leadbeater's Possum in the Steavenson Forest Block.

Proportion of patches burnt in a wildfire	Extinction probability at various times					Mean[a]
	P150	P300	P450	P600	P750	
Without salvage logging						
75%	42	79	91	97	99	60
50%	17	49	72	82	91	42
With salvage logging						
75%	58	86	96	99	100	69
50%	34	71	89	95	99	58

The annual probability of a fire was set to 1% and the number of patches burned in a conflagration was 50% or 75%.

a random basis (which it was not), such activities would have reduced the number of trees and the overall range of stem diameters on a site. After the 1926, 1932 and 1939 wildfires, salvage logging was biased toward cutting larger fire-damaged stems, reducing stem diameter variation and the prevalence of multi-aged stands. Only 7% of montane ash forests in the Central Highlands of Victoria was multi-aged prior to the 2009 wildfires (Lindenmayer et al. 1991b).

Impacts on plants
Several studies (Ough 2002; Ough and Murphy 2004) have established that plant species composition and the developmental trajectory of burned (but not salvaged) areas of montane ash forest are different from those of clearfelled stands. There are no studies which have specifically examined the impacts of salvage logging on plant communities in montane ash forests of the Central Highlands. The response of many species of plants to salvage logging will vary, depending on:

- intensity of cutting;
- intensity of wildfire;
- logging machinery used;
- composition and proximity of the surrounding vegetation (e.g. as a source of propagules);
- interval between wildfire and salvage logging;
- season/s in which wildfire and logging occur.

The remainder of this section broadly discusses the responses that might be generally expected.

Salvage logging within the first few years of a wildfire is likely to reduce the survival of many plant species. Salvage logging is a double disturbance in which mechanical disturbance takes place after fires have stimulated many plant species to regenerate, either by germinating from seeds or by resprouting. Mechanical disturbance, such as by heavy logging machinery and log movement, will kill many recently germinated or recovering plants.

Several ecologically important taxa, such as tree ferns, rely on resprouting to regenerate after wildfire (Ough and Murphy 2004). For resprouting species, salvage logging is likely to have effects similar to conventional clearfelling (Ough 2002). The physical disturbance associated with logging will uproot and damage vegetative propagules, particularly aboveground structures and shallow rhizomes, resulting in reduced survival. Careful mechanical logging during the salvage operation and the reduced need to clear already burned understorey for visibility and safety purposes may reduce these impacts.

The main difference between plant community composition after conventional clearfelling and salvage logging is the probable additional impact on seed regenerators. Seed regenerators, which typically regenerate well after fire or conventional clearfelling, would be expected to decline after a combination of fire and salvage logging because the stimulation for germination (fire) takes place prior to the mechanical disturbance (logging). Most species will not have reached sexual maturity in the short time between the disturbance events. Recovery of the understorey in areas disturbed by salvage operations will rely on the seed that remains in the soil seedbank, dispersal in from outside the logged area and survival of deeply buried rhizomes. Unless further seed can be supplied from the soil-stored seedbank or is dispersed into the area, the representation of these species in newly established stands is likely to decline (Whelan 1995).

Species which have a reduced soil seed store after fire (all or most of the seed has been stimulated to germinate or has been killed by the fire), rely on canopy-stored seed or have limited dispersal from surrounding areas are likely to decline as a result of salvage logging. Conversely, some short-lived herbs may increase due to decreased competition and increased light if maturation time is short and seed has been set before logging (e.g. *Senecio* spp. and weed species like *Cirsium vulgare* and *Hypochoeris radicata*). These taxa are likely to decline again as the canopy closes (Appleby 1998; Harris 2004). Vehicle-borne weed flora can be substantial (Wace 1977); fire-damaged areas subject to salvage logging may provide many germination opportunities for pest plant species. Ough (2002) showed that areas subject to conventional clearfelling practices were characterised by more weedy plants than burned areas (not subject to salvage logging). Future field studies need to examine whether additional weed problems arise in salvage-logged areas.

OTHER POTENTIAL IMPACTS

Neumann (1991, 1992) examined the arthropod fauna of burned and salvaged Mountain Ash forests in the Central Highlands of Victoria and compared it with that of sites subject to clearfell logging and follow-up slash regeneration burns. The work demonstrated large differences in the composition of communities of ants and litter-dwelling arthropods before and after disturbance. Although community composition was altered, Neumann speculated that such changes were likely to be relatively short-lived. A key problem with this work was the absence of burned but unsalvaged sites for comparison, which limited the extent of inferences about salvage logging impacts.

There may be other impacts on fauna associated with salvage logging. For example, extensive road systems are constructed to facilitate salvage logging. These may provide a conduit for movement of feral animals (May and Norton 1996) which may have major impacts on native fauna, particularly when levels of vegetation cover were depleted by the preceding wildfire event.

SUMMARY

Quantitative field data show that salvage logging significantly alters the abundance of some types of biological legacies (e.g. large living and dead trees with hollows) and can reduce the prevalence of multi-aged stands. Therefore, salvage logging can simplify otherwise complex stand structures that develop after wildfires. There appear to be impacts on particular species of animals and plants that may be additional to, and possibly more detrimental than, the impacts of conventional clearfelling methods.

Plant communities in salvage logged areas are likely to be dominated by a smaller suite of species, particularly those that are wind-dispersed, have viable soil-stored seed remaining after salvage logging or have deep rhizomes. There will be fewer individuals and taxa that use regeneration strategies such as fire-stimulated germination of soil- or canopy-stored seed, or resprouting from above ground or shallow vegetative propagules. The extent of the impact on plant communities depends on the extent and intensity of mechanical disturbance to the site.

LESSONS LEARNED

The impacts of salvage logging in montane ash forests, particularly those from the prolonged operations after the 1939 wildfires (Noble 1977), may have been far more substantial than appreciated by many forest managers and conservation biologists (Lindenmayer and Ough 2006). The removal of considerable numbers of large fire-damaged trees and the loss of many areas of what might have otherwise been multi-aged forest (Lindenmayer and McCarthy 2002) have undoubtedly contributed to the shortage of trees with hollows in substantial areas of montane ash forest in the region – a shortage that is likely to become increasingly problematic in the future (Lindenmayer *et al.* 1997a; Lindenmayer and Wood 2009).

Because cavity development in montane ash trees takes more than 120 years (Ambrose 1982) (Chapter 6),

> **Box 14.2: Postscript – salvage logging and the 2009 wildfires**
>
> Fire in montane ash forests is inevitable, particularly if forecasts of altered fire regimes as a consequence of climate change prove accurate. This was realised in very tragic circumstances in February 2009. The extent of the Black Saturday wildfire in montane ash forests means that extensive salvage logging operations are inevitable. It will be critical to reduce both the immediate and long-term impacts of such activities – Chapter 16 outlines ways in which this might be done in the next 10 years. It will be crucial to study the effects of salvage logging operations and use the knowledge gained to further reduce its impacts.
>
> A design for a salvage logging experiment had been developed prior to the 2009 fires (Lindenmayer *et al.* 2008a) and was critical for informing a salvage logging experiment planned for establishment in mid 2009. The experimental design was important because many studies of salvage logging have been problematic due to a lack of disturbed but unsalvaged sites (Elliott *et al.* 2002; Greenberg *et al.* 1995) and/or the absence of pre-disturbance data (reviewed by McIver and Starr 2000).

Table 14.2: Hypothesised impacts of salvage logging on plants in the montane ash forests of the Central Highlands of Victoria

Species	Life-form	Most common regeneration strategy after fire	Propagule or organ of regeneration	Likely (initial) population response after wildfire[a]	Likely (initial) population response after salvage logging[b]	Likely (initial) population response after clearfelling[c]
Cirsium vulgare[d]	Herb	Germinate	Seed	Increase	Further increase (short maturation time and wind-dispersed seed)	Increase
Hypochoeris radicata[d]	Herb	Germinate	Seed	Increase	Further increase (short maturation time and wind-dispersed seed)	Increase
Dianella tasmanica	Herb	Resprout	Rhizome	Stable	Decline	Decline
Sambucus gaudichaudiana	Herb	Resprout and germinate	Deep rhizome, seed	Stable	Stable	Stable
Rubus fruticosus spp. Agg.[d]	Scrambler	Resprout and germinate	Rhizome, stolon and seed	Increase	Further increase (animal-dispersed seed, growth from small pieces of detached rhizome or stolon)	Increase
Blechnum cartilagineum	Ground fern	Resprout	Erect shallow rhizome	Stable, sometimes decline	Decline	Decline
Polystichum proliferum	Ground fern	Resprout	Erect shallow rhizome	Stable, sometimes decline	Decline	Decline
Pteridium esculentum	Ground fern	Resprout	Long, creeping deep rhizome	Stable	Stable or increase (deep rhizome)	Stable or increase
Cyathea australis	Tree fern	Resprout	Terminal bud on trunk	Stable	Decline	Decline
Dicksonia Antarctica	Tree fern	Resprout	Terminal bud on trunk	Stable	Decline	Decline
Cassinia aculeate	Shrub	Germinate	Seed	Increase	Increase if mature populations nearby (wind-dispersed seed)	Increase
Correa lawrenciana	Shrub	Resprout and germinate	Aerial buds, seed	Stable	Decline	Increase (in germinants) but decline in older population (resprouters)

Species	Life-form	Most common regeneration strategy after fire	Propagule or organ of regeneration	Likely (initial) population response after wildfire[a]	Likely (initial) population response after salvage logging[b]	Likely (initial) population response after clearfelling[c]
Goodenia ovata	Shrub	Germinate	Seed	Increase	Increase, but less than wildfire and clearfelling	Increase, but much less than wildfire
Olearia argophylla	Shrub	Resprout	Lignotuber	Stable	Decline	Decline
Olearia phlogopappa	Shrub	Germinate	Seed	Increase	Increase if mature populations nearby (wind-dispersed seed)	Increase
Zieria arborescens	Shrub	Germinate	Seed	Increase	Decline (or stable if seed store remains)	Increase, but less than wildfire
Lomatia fraseri	Shrub/small tree	Resprout	Lignotuber	Stable	Decline	Decline
Bedfordia arborescens	Tree	Resprout and germinate	Epicormic buds and seed	Stable, sometimes increase	Decline	Decline
Acacia dealbata	Tree	Germinate	Seed	Increase	Decline (or stable if seed store remains)	Increase
Persoonia arborea	Tree	Germinate (some resprout)	Seed, stem buds	Stable or decline	Decline	Stable or increase (if mature plants present pre-logging)
Pomaderris aspera	Tree	Germinate	Seed	Increase	Decline (or stable if seed store remains)	Increase

a Response will depend on fire intensity; response given is based on moderate intensity.
b Assumes logging-free period of at least two decades prior to wildfire and salvage logging; assumes moderate intensity wildfire.
c Assumes wildfire-free period of at least two decades prior to logging; assumes logging followed by regeneration burn.
d Introduced species.
Based on Lindenmayer and Ough (2006).

the creation of simplified structural stand conditions by past salvage logging operations, combined with current widespread clearfelling, may take hundreds of years to redress (Lindenmayer and Ough 2006) and may therefore significantly threaten the population viability of some species. The persistence of many species of hollow-dependent vertebrates in montane ash forests during this period will remain a significant conservation issue in the Central Highlands of Victoria (Lindenmayer and Wood 2009).

KNOWLEDGE GAPS

The impacts of salvage logging on stand structure and particular species of plants and animals have been inferred from other kinds of studies in montane ash forests (Lindenmayer and Ough 2006). Although extensive research has been undertaken on the impacts of many types of disturbances, remarkably little work has been conducted on the salvage operations that often follow them. Experiments are needed to better quantify the nature and magnitude of salvage logging effects.

FOREST PATTERN, ECOLOGICAL PROCESS AND LINKS TO OTHER CHAPTERS

Salvage logging and its relationships with ecological processes and forest pattern shares many themes common to the discussions on natural disturbance and human disturbance (Chapters 12 and 13). This is because salvage logging is a combination of a natural disturbance event quickly followed by a human disturbance event. Salvage logging therefore influences many of the same ecological processes and forest patterns elucidated in the chapters on natural and human disturbance. Examples include carbon storage, nutrient cycling, forest age class distribution, stand structure and composition, and animal distribution and abundance (Lindenmayer *et al.* 2008a).

Salvage logging may interact with the processes of natural and human disturbance in complex ways.

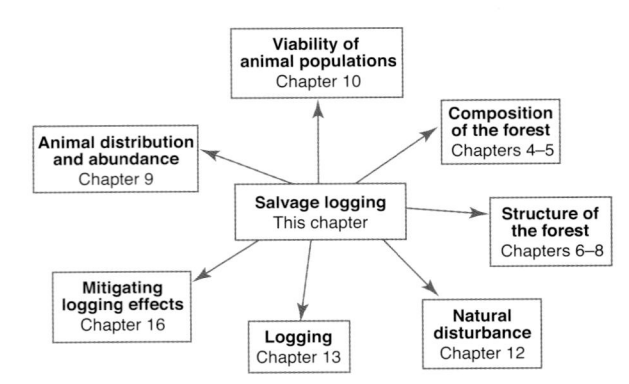

Figure 14.3: Relationships between salvage logging effects and key topics examined in other chapters.

Examples exist in North America, where stand conditions and landscape cover patterns modified by traditional logging may make forests more susceptible to high-intensity fires (Noss *et al.* 2006); once these areas are salvage logged, the further-modified stands and landscapes may be even more susceptible to subsequent wildfires (Thompson *et al.* 2007).

Links to other chapters

The impacts of salvage logging may extend beyond those described for conventional (green tree) logging in Chapter 13. Both kinds of human disturbance can alter the structure and composition of stands and landscapes (Chapters 6–8) and influence the distribution and abundance of organisms in montane ash forests (Chapter 9). Salvage logging may threaten the viability of populations of some species – a topic tackled in detail in Chapter 10.

Salvage logging results in a combination of natural disturbance (e.g. wildfire) and human disturbance (logging), important processes discussed in Chapters 12 and 13 respectively.

If salvage logging operations are well planned and carefully managed both spatially and temporally, impacts on stand structure, plant species composition and biota may be reduced. Approaches to mitigate logging impacts are discussed in detail in Chapter 16.

PART VI

FOREST MANAGEMENT AND BIODIVERSITY CONSERVATION

While some of the work conducted over the past 25 years has had a theoretical focus, most attention has been focused on improved biodiversity conservation and its relationships with forest management. Forest management is the key theme of Part VI, which contains chapters covering three broad but strongly interrelated areas of work – reserves (Chapter 15), the mitigation of logging effects in off-reserve areas (Chapter 16) and monitoring and adaptive management (Chapter 17).

A broad principle underpinning forest biodiversity conservation is that management is required at multiple scales (Lindenmayer and Franklin 2002) – a common thread through all three chapters. There are several important reasons why forest biodiversity conservation needs to be multi-scaled. First, different species have different spatial and other requirements (Allen and Starr 1988). Suitable habitat may vary from extensive intact stands for area-sensitive organisms such as some wide-ranging carnivores (Milledge et al. 1991) to the moisture and decay conditions provided by individual logs for invertebrates (Lindenmayer et al. 1999a). Second, any individual species responds to factors at multiple spatial scales (Lindenmayer 2000). Third, there are multiple ecological scales for different ecological processes (Urban et al. 1987). Fourth, a management strategy implemented at one scale may generate positive benefits for another strategy implemented at a different spatial scale. Conversely, if a strategy at one scale is found to be ineffective, others at a different scale might better protect sensitive elements of forest biodiversity. This reduces overreliance on a strategy which may be of limited value in meeting specific conservation objectives. Finally, multi-scale conservation strategies may produce a heterogeneous landscape containing the spatially dispersed array of resources needed by some species (Bennett et al. 2006; Law and Dickman 1998). Collectively, these five reasons demonstrate that there is no single 'right' or 'sufficient' scale for forest and conservation management. A single conservation strategy adopted at a single spatial scale will meet a limited number of stand and landscape management goals (Christensen et al. 1996; Tang and Gustafson 1997) and will, for example, provide suitable habitat for a limited number of different taxa.

The catchcry in modern natural resource management is for evidence-based management. The extensive body of research over the past 25 years indicates there is more than enough information for the management of montane ash forests to be truly evidence-based. Much of the discussion in Part VI describes how evidence has influenced management, or how it should influence management.

15 Reserves

INTRODUCTION

The creation and maintenance of reserves is a critical conservation strategy for almost all ecosystems worldwide, including forests. It is perhaps not surprising that reserves have been the primary focus of conservation biologists since the discipline began (Fazey *et al.* 2005). In this chapter, a hierarchically scaled approach is used in discussing the strengths and limitations of the reserve system in the Central Highlands of Victoria. This is in keeping with the general philosophy which recognises that successful biodiversity conservation will be underpinned by the implementation of a range of management strategies at multiple spatial scales (Lindenmayer and Franklin 2002).

The first part of the chapter focuses on large ecological reserves and presents some general reasons why they are important. The Yarra Ranges National Park is the primary large ecological reserve in the ash-type forests of the Central Highlands of Victoria; the second part of the chapter outlines some reasons why it is so important for the conservation of the region's biodiversity.

There are several kinds of places outside the Yarra Ranges National Park which are set aside for conservation or managed primarily for conservation within areas broadly designated for wood and paper production. These are meso-scale reserves and the reasons why they are important are explored in the third part

Table 15.1: Studies directly or indirectly examining reserves in the montane ash forests of the Central Highlands of Victoria

Study description	Reference
Overview of conservation values of large ecological reserves	Lindenmayer and Franklin (2002)
Overview of conservation values of meso-scale reserves	Lindenmayer and Franklin (2002)
Importance of retained linear strips for arboreal marsupials	Lindenmayer *et al.* (1993a)
Importance of retained linear strips for terrestrial mammals	Lindenmayer *et al.* (1994c)
Zoning systems as a meso-scale reserve system for Leadbeater's Possum	Lindenmayer and Cunningham (1996a)
Reserves and viability of populations of arboreal marsupials	Lindenmayer and Possingham (1996c); McCarthy and Lindenmayer (1999); Lindenmayer and McCarthy (2006)

of this chapter. Such meso-scale reserves include wildlife corridors, unloggable areas and special protection zones, including specifically designated reserves set aside for Leadbeater's Possum.

The finest spatial scales in a hierarchy of reservation are conservation strategies at the stand and tree levels. These do not feature in this chapter but are discussed in the next one on mitigating logging effects.

WHY LARGE ECOLOGICAL RESERVES ARE IMPORTANT

Large ecological reserves are an essential part of all comprehensive and credible biodiversity conservation plans. They are critical for at least four key reasons (Lindenmayer and Franklin 2002):

- reserves support some of the best examples of ecosystems, landscapes, stands, habitat and biota;
- some species find optimum conditions only within large ecological reserves which become strongholds for them. The effects of human disturbance on biodiversity are poorly known and some impacts may be irreversible. Others, such as synergistic and cumulative effects, can be extremely difficult to quantify or predict. These factors make large ecological reserves a valuable safety net, relatively free from human disturbance, especially if mistakes are made in off-reserve management;
- some species are intolerant of human intrusions, making it imperative to retain areas which are largely exempt from human activity;
- large ecological reserves provide control areas against which the impacts of human activities in managed forests can be compared. Many examples highlight the contribution of large ecological reserves to increased scientific understanding and enhanced natural resource management (Crisafulli *et al.* 2005; Turner *et al.* 2003).

LARGE MONTANE ASH RESERVES: YARRA RANGES NATIONAL PARK

The Yarra Ranges National Park is the major ecological reserve in the Central Highlands region. It is an

Table 15.2: Number of patches of old-growth montane ash forest in various size classes in four forest management areas, before 2009 wildfires

Patch size[a] (ha)	No. of patches
Murrindindi Block	
0–1	38
1–3	6
Total area	24
Ada Forest Block	
1–3	14
4–6	4
7–10	3
11–14	3
>14	3
Total area	162
Steavenson Forest Block	
0–3	141
3–6	10
6–9	3
9–12	2
12–24	2
>25	2
Total area (ha)	264
O'Shannassy Catchment	
1–5	6
6–10	1
11–15	2
16–20	5
21–50	3
51–100	5
101–200	2
201–500	1
500–1500	2
>1501[b]	1
Total area (ha)	6420

a Different patch size intervals are given due to differences in the amount of old-growth forest in the study areas.
b The size of this largest patch is approximately 3534 ha.

iconic park in the Victorian and national reserve systems and was gazetted in 1995. It encompasses three large water catchments with some areas, such as

Figure 15.1: View across the Yarra Ranges National Park. (Photo by David Lindenmayer).

the O'Shannassy Water Catchment, having been closed to logging and general public access for over 100 years (Land Conservation Council 1994; Parks Victoria 2002). In 2000, the Yarra Ranges National Park was Victoria's ninth-largest National Park, encompassing 76 000 ha. The Yarra Ranges National Park supports populations of 10 species of nationally threatened plants and animals and is one of the least weed-infested areas in Victoria (Parks Victoria 2000).

The Yarra Ranges National Park supports about 20% of the existing total area of 170 000 ha of ash-type forest in the Central Highlands of Victoria (Macfarlane et al. 1998) and prior to the February 2009 wildfires contained the most extensive areas of old-growth, mature and multi-aged montane ash forest in mainland Australia. For example, the largest areas of old-growth ash forest in wood production zones are approximately 50 ha, whereas contiguous areas of old growth covering 500 ha or larger characterised parts of the Yarra Ranges National Park before the 2009 fires.

A number of studies over the past 25 years have highlighted the value of the Yarra Ranges National Park for biodiversity conservation. First, extensive survey work and subsequent data analyses demonstrated that the Yarra Ranges National Park is a stronghold for two wide-ranging species strongly associated with old-growth montane ash forest – the Yellow-bellied Glider (Lindenmayer et al. 1999a) (Chapter 4) and the Sooty Owl (Milledge et al. 1991).

Second, the extensive stands of old-growth forest will remain important for biodiversity, even though many of them were burned in the 2009 fires. These kinds of burned areas are those with the highest probability of becoming multi-aged stands (Lindenmayer et al. 2000a). Even if fires in these areas are stand-replacing, the large numbers of biological legacies mean they will rapidly become important habitat for species such as Leadbeater's Possum (see Table 12.2 in Chapter 12) (Lindenmayer 2009). This will not occur in large parts of the wood production forest estate because wildfires will burn extensive areas of young regrowth forest, and fire-killed or damaged stems will be unlikely to develop into large hollow-bearing trees (Lindenmayer 2009).

Third, extensive computer simulation modelling (Chapter 10) has demonstrated that the probability of persistence of Leadbeater's Possum, the Greater Glider and Mountain Brushtail Possum was significantly

Table 15.3: Estimated probability of extinction of populations of Leadbeater's Possum in response to wildfire in four forest management areas, before 2009 wildfires

Area	Probability of extinction after 150 years	Probability of extinction after 300 years
Murrindindi Forest Block		
No fire	100	100
With fire	100	100
Ada Forest Block		
No fire	34	48
With fire	78	98
Steavenson Forest Block		
No fire	2	5
With fire	48	84
O'Shannassy Water Catchment		
No fire	0	0
With fire	4	15

For those scenarios where the impacts of wildfires were modelled, there was no added impact of post-fire salvage logging operations incorporated in the analysis. The annual probability of a fire in the modelling was 1% and 75% of patches were burnt in a given fire event.

greater in large contiguous stands of old growth such as those which characterise the O'Shannassy Water Catchment (Lindenmayer and Possingham 1995a) (Table 15.3). Similarly, McCarthy and Lindenmayer (1999) showed that large patches of old-growth forest were required to maintain a high probability of persistence of the Greater Glider within stands of montane ash forests in the coming centuries. Prior to the February 2009 wildfires, only the O'Shannassy and Watts Creek Water Catchments supported the old-growth patch sizes considered appropriate by McCarthy and Lindenmayer (1999).

A fourth reason why the Yarra Ranges National Park is important for biodiversity is that human disturbance has been excluded from substantial parts of the area, facilitating the completion of valuable scientific studies. For example, the O'Shannassy and Watts Creek Water Catchments have had a prolonged history of exclusion of human access, particularly by European settlers. It appears that the O'Shannassy Water Catchment was disturbed very little by indigenous people. This has allowed valuable comparisons between stands and landscapes which have and have not been disturbed by humans. This has led to new insights about the range of disturbance pathways in montane ash forests (Lindenmayer and Franklin 1997) (Chapter 12), the prevalence of multi-aged stand conditions prior to European settlement (McCarthy and Lindenmayer 1998) and relationships between the occurrence of particular species and landscape-scale patterns of forest cover (Lindenmayer et al. 1999a; Milledge et al. 1991).

Yarra Ranges National Park: biodiversity conservation in a rapidly changing climate

There are other reasons why the Yarra Ranges National Park is important for biodiversity conservation. First, rapid climate change will lead to substantial distributional shifts of many species, with their ranges exhibiting latitudinal or elevational movements (Parmesan 2006; Parmesan and Yohe 2003). The Yarra Ranges National Park contains areas with substantial topographic gradients, which may assist species to move their ranges upslope in response to anticipated increases in temperature – something which has been recognised for more than two decades (Lindenmayer 1989). Second, medium-term climatic refugia for some species have been identified within the Yarra Ranges National Park (Mackey et al. 2002). Third, large ecological reserves, such as the park, are places where the risks of cumulative threats to biodiversity might be limited. It is well-known that the risk of decline and/or extinction of many species are great-

> **Box 15.1: Large ecological reserves, wilderness and biodiversity conservation**
>
> The word 'wilderness' derives from the old English 'wild deer-ness', which referred to uninhabited and uncultivated tracts of land occupied only by wild animals (Shea *et al.* 1997). Wilderness is a human construct that relates to areas remote from human influence and infrastructure (Mackey *et al.* 1998; Mittermeier *et al.* 2003). The concept is usually linked to the spiritual, aesthetic and recreation needs of Western peoples in natural landscapes. The role of wilderness or remote areas in conserving biodiversity is complex. Some authors assert that large wild areas are essential for conservation, even though such areas may not be species-rich (Mittermeier *et al.* 2003). The type and intensity of human disturbance will always be an important factor influencing biodiversity but, in some cases, measures of the integrity of remote areas may be outweighed by management practices in production landscapes. Lindenmayer *et al.* (2002b) found no correlation between the occurrence of any species of arboreal marsupial in montane forests and measures of the intensity of human development, taken from the National Wilderness Inventory (Lesslie and Maslen 1995), such as the distance of field survey sites from roads. Better predictors of species occurrence were factors such as the extent of matrix management practices (e.g. levels of tree retention on harvest units). Therefore, although remote areas clearly have spiritual, aesthetic and recreation values, they are not always automatically of high conservation value (Brown and Hickey 1990; Lindenmayer and Franklin 2002). Although large relatively undisturbed areas have considerable conservation value, small reserves that are not remote also have much to contribute to the conservation of biodiversity.

est when there are multiple threatening processes (Caughley and Gunn 1996). For example, several authors have warned of the potentially negative interacting effects of landscape change and climate change (Opdam and Wascher 2004; Travis 2003; Ward 2004). Finally, the conservation of biodiversity will depend on mitigating the effects of climate change. Carbon storage and sequestration are fundamental strategies in achieving this. The Yarra Ranges National Park will be critical because it supports areas with the highest biomass carbon in the world (Keith *et al.* 2009; Mackey *et al.* 2008) (Chapter 18).

LIMITATIONS OF LARGE ECOLOGICAL RESERVES

Large ecological reserve systems are rarely comprehensive, representative and adequate for all elements of biodiversity (Margules and Pressey 2000; Scott *et al.* 2001). Hence, credible plans for forest biodiversity conservation must incorporate off-reserve approaches that complement reserve-based approaches, that is, conservation strategies at the landscape and stand levels in areas outside large ecological reserves (Lindenmayer and Franklin 2002).

In the case of montane ash forests, a reserve-only approach to conservation based solely in the Yarra Ranges National Park would be highly problematic because:

- populations of some species in the area are at risk of extinction from high-intensity wildfires which could affect the entire park (Lindenmayer and Possingham 1995c) as appears to have happened in the Black Saturday 2009 wildfires;
- a large proportion (>80%) of the distribution of endangered species such as Leadbeater's Possum exists outside the Yarra Ranges National Park;
- although the national park spans a wide range of elevational gradients, climatic conditions in existing wood production forests are unique and could be vital for some species in the future as a result of rapid climate change (Mackey *et al.* 2002).

MESO-SCALE RESERVES

A key part of conservation in areas outside large ecological reserves is the protection of meso-scale reserves or smaller areas *within* landscapes broadly designated for wood production (Gustaffson *et al.* 1999). Such systems of scattered small meso-reserves provide (after Lindenmayer *et al.* 2006):

- increased protection of habitats, vegetation types and organisms poorly represented or absent in large ecological reserves;
- protection for aquatic and semi-aquatic ecosystems;
- refugia for forest organisms that subsequently provide propagules and offspring for recolonising surrounding forest areas as they recover from logging;
- 'stepping-stones' to facilitate the movement of biota across wood production landscapes.

Several kinds of meso-scale reserves characterise the wood production montane ash forests in the Central Highlands of Victoria. These are wildlife corridors, unloggable areas and special protection zones, including specifically designated reserves set aside for Leadbeater's Possum and the Baw Baw Frog.

Unloggable areas and wildlife corridors

Substantial areas within wood production forests are unavailable for logging (Colour Plate 12). Wildlife corridors, streamside reserves and forests on steep and rocky terrain are not logged. These can be valuable areas for the conservation of many (although not all) species.

Stands of old-growth forest are exempt from logging, as are other areas of high conservation value that are typically captured with the special management zoning system within wood production forests.

Special protection zones and the zoning system for Leadbeater's Possum

A management zoning system for Leadbeater's Possum has been incorporated within a more general forest zoning approach for montane ash forests. High-conservation value sites are specified as Special Management Zones; the remainder of the forest where intensive forestry can be practised is called the General Management Zone. Management zoning is a mid-spatial-scale conservation strategy and partitions wood production forests into three types of areas:

- **Zone 1:** conservation of Leadbeater's Possum is a priority;
- **Zone 2:** wood production is a priority;
- **Zone 3:** joint land use is a priority.

Identification of Zone 1 is based on a broad understanding of the habitat requirements of Leadbeater's Possum (Macfarlane *et al.* 1998), particularly the abundance of cavity trees (Lindenmayer and Cunningham 1996b) (Figure 15.2).

Even in areas where wood production is a priority (Zone 2), potential logging coupes are subject to pre-harvesting surveys. Coupes found to support suitable habitat for Leadbeater's Possum are typically excluded from harvesting, or the boundaries of these areas might be altered to maintain the subsection of the proposed coupe which contains suitable habitat.

In late 2008, parts of the zoning system for Leadbeater's Possum and other areas of special protection were amalgamated to create a special set of meso-scale reserves for Leadbeater's Possum within each forest block in the wood production areas of the Central Highlands of Victoria (Dept of Sustainability and Environment 2008) (Colour Plate 13). Other areas of Zone 1 forest not included in the permanent reserve system for Leadbeater's Possum are still managed for the species by on-ground prescriptions (Dept of Sustainability and Environment 2008). These areas were taken out of wood production, with the aims of

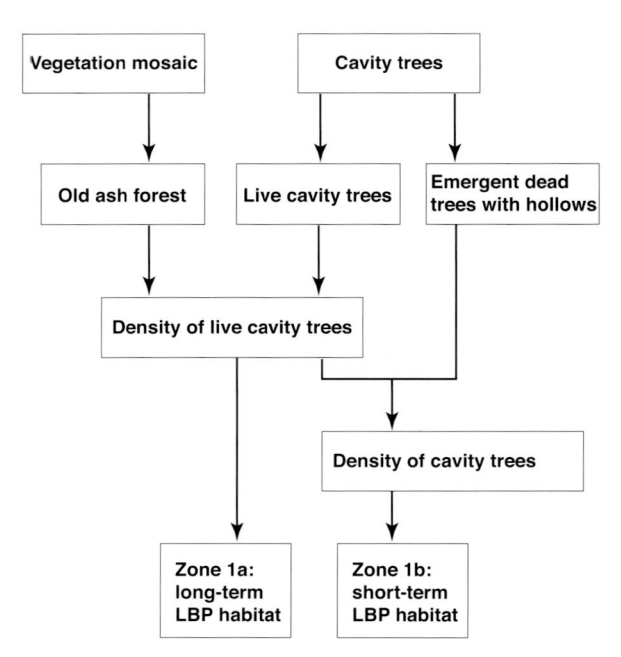

Figure 15.2: A model for the selection of meso-scale reserves set aside within wood production areas for the conservation of Leadbeater's Possum. (Modified from Macfarlane *et al.* 1998).

growing the forest through to an old-growth stage (Lindenmayer and Possingham 1995c) and expanding the amount of suitable habitat in forests broadly designated for wood production (Dept of Sustainability and Environment 2008). The basis for this set of reserves was the extinction risk assessment of Leadbeater's Possum conducted by Lindenmayer and Possingham (1995c) which called for the protection of 600–1000 ha or ~10% of each wood production forest block (see Box 15.2).

Special protection for the Baw Baw Frog

Part of the distribution of the Baw Baw Frog occurs in areas of montane ash forest broadly designated for wood production under the Central Highlands Regional Forest Agreement (Commonwealth of Australia and Dept of Natural Resources and Environment 1997). A zoning system and special logging provisions were established to modify timber harvesting in areas likely to be inhabited by the species (Dept of Sustainability and Environment 2004; Hollis 2003).

In 2008, a 5000-ha area was set aside for the conservation of the Baw Baw Frog on the southern face of the Baw Baw Plateau (Colour Plate 13). The reserve is likely to contain some areas of habitat for Leadbeater's Possum and hence should benefit the conservation of both endangered species (Dept of Sustainability and Environment 2008).

LIMITATIONS OF MESO-SCALE RESERVES

Meso-scale reserves make a major positive contribution to the conservation of biodiversity both *per se* and within the wider wood production forest area. However, they also have limitations.

- Excluding logging from up to 600–1000 ha of each management unit may still be inadequate to support long-term viable populations of species such as Leadbeater's Possum, the Baw Baw Frog and the Greater Glider, particularly in response to recurrent wildfires (Hollis 2004; Lindenmayer and Possingham 1995a; McCarthy and Lindenmayer 1999).
- Some species of animals are uncommon or rare within narrow strips or wildlife corridors located between cutover areas. For example, Leadbeater's

---

Box 15.2: Design of meso-scale reserves for Leadbeater's Possum and the Greater Glider

Extensive simulations using models for population viability analysis have explored relationships between options for the size and number of meso-scale reserves and the viability of populations of Leadbeater's Possum and the Greater Glider (Chapter 10). In the case of Leadbeater's Possum, there were important trade-offs between the size of reserves, the risk of those reserves being burned in a wildfire and the viability of populations within reserves (Lindenmayer and Possingham 1996c). A single large reserve risked being destroyed in a single fire, but many small reserves may separately be unable to support viable populations of the species. The best option appeared to be a trade-off between reserve size and number of reserves. For example, for a nominal area of 300 ha of reserves, three or four reserves each of 75–100 ha gave the best outcome in terms of probability of persistence (Lindenmayer and Possingham 1996c). This resulted in a recommendation to establish multiple meso-scale reserves within each wood production forest block in the Central Highlands of Victoria (Lindenmayer and Possingham 1995c). It was further recommended that the reserves should be in locations least likely to be burned in a wildfire and close to existing areas of old-growth and multi-aged forest, where populations of Leadbeater's Possum were most likely to occur currently.

Parallel studies on the Greater Glider also suggested a need to increase the area of suitable habitat to reduce the probability of extinction (McCarthy and Lindenmayer 1999; Possingham *et al.* 1994). Again, it was recommended that areas adjacent to existing old-growth patches be the target for reservation, to eventually increase the overall area of suitable habitat for populations of Greater Glider (McCarthy and Lindenmayer 1999).

Importantly, the accuracy of models used to explore reserve design options were checked and rerun with monitoring data gathered for a number of years after the initial modelling (Lindenmayer and McCarthy 2006). For both species, recommendations about reserve size and location based on initial assessments about reserve size remained robust to new information (Lindenmayer and McCarthy 2006) (Chapter 10).

Possum rarely inhabits these areas (Lindenmayer and Nix 1993). Species of small terrestrial mammals inhabit these areas, but levels of abundance are significantly lower than in contiguous areas of montane ash forest (Lindenmayer et al. 1994c).
- Some species, such as Leadbeater's Possum, typically avoid steep terrain (Chapter 4) (Lindenmayer et al. 1991b) where reserves are often created because of its unsuitability for logging. Similarly, riparian areas are often dominated by cool temperate rainforest (Lindenmayer et al. 2000c) (Chapter 5) that provides generally unsuitable foraging and nesting habitat for this and other species. Some species of birds have significantly reduced abundance in montane ash forests located on steep slopes (Lindenmayer et al. unpublished data).
- Very limited patches of old growth exist in the extensive areas of montane ash forests broadly designated for wood production. Many of these blocks presently contain less than 5–10% old-growth cover and are dominated by montane ash stands fewer than 20–60 years old (Commonwealth of Australia and Dept of Natural Resources and Environment 1997). Such limited areas of old-growth forest threatens the local persistence of species such as the Greater Glider, Yellow-bellied Glider and Sooty Owl (Chapter 9), especially as large trees with hollows are rapidly being lost from younger-aged stands (Lindenmayer and Wood 2009) (Chapter 6).
- While the zoning system for Leadbeater's Possum has some advantages, it also has problems. On-ground mapping errors have resulted in accidental logging of Zone 1 habitat. The system may also result in two broad categories of forest in wood production areas – cutovers aged 0–80 years (the rotation time) and old-growth stands greater than 300 years in age. This not only limits the range of age classes in the forest but could eventually create discontinuities in the ongoing recruitment of new areas of old growth.

Some of the deficiencies in meso-scale reserves can be rectified by improving approaches to design. For example, ensuring that some retained strips are not confined to streamside reserves but connect across forests in different topographic positions (e.g. from gullies to midslopes and ridges) can improve their effectiveness (Claridge and Lindenmayer 1994). Similarly, linking retained (and unlogged) areas of regrowth forest to old-growth stands will lead to the creation of larger areas of old-growth forest and benefit some species of arboreal marsupial (McCarthy and Lindenmayer 1999).

Despite these kinds of design improvements, some meso-scale reserves will remain suboptimal for wildlife conservation. The limitations of the meso-scale reserve system emphasise the importance of maintaining a range of management strategies at different spatial scales for conserving forest biodiversity. They also underscore the complementarity between large ecological reserves and meso-scale reserves (Lindenmayer and Franklin 2002).

SUMMARY

Biodiversity is a multi-scaled entity and management at multiple spatial scales is required to conserve it. In forests, including those in the Central Highlands of Victoria, the conservation of biodiversity embodies a continuum of conservation approaches from the establishment of large ecological reserves through to an array of off-reserve conservation measures, including the protection of meso-scale reserves and the maintenance of individual forest structures at the smallest spatial scale. This chapter has focused on the strengths and limitations of large ecological reserves and meso-scale reserves within areas broadly designated for timber and pulp production. Chapter 16 is dedicated to stand-level strategies for the conservation of forest biodiversity.

The Yarra Ranges National Park is the key large ecological reserve in the Central Highlands of Victoria. It is an important area for such species as the Yellow-bellied Glider, which is strongly associated with old-growth Mountain Ash and appears intolerant of the cumulative (landscape-level) effects of human disturbances like logging.

> **Box 15.3: Limitations of wildlife corridors for some species of arboreal marsupial**
>
> Wildlife corridors, or strips of retained habitat, are used to mitigate the impacts of timber harvesting on forest fauna in wood production forests in many Australian states. Typically, the corridors are unlogged strips in riparian areas (streamside reserves) that maintain water quality as well as play a role in the conservation of wildlife that may be sensitive to forestry operations.
>
> Arboreal marsupials were surveyed in 49 retained linear strips 40–250 m wide (Lindenmayer *et al.* 1993a). Each strip was surrounded by young recently cut forest that was unsuitable for animals. The aim was to determine if the strips were being used as habitat by possums and gliders. Two questions were examined:
>
> 1. Is the number of corridors occupied by different species of arboreal marsupials similar to that expected, based on the suitability of the forest in the strips for these animals?
> 2. Are there any features of the strips (e.g. width or length) that make them more likely to be used by arboreal marsupials?
>
> The habitat requirements of several species of arboreal marsupial within intact forest were known from earlier studies (Lindenmayer *et al.* 1991d) (Chapter 9), providing the basis for estimating the habitat suitability for arboreal marsupials in each of the retained strips. Several species (Mountain Brushtail Possum, Greater Glider and Sugar Glider) were observed with a frequency that did not differ substantially from expectations, based on the quality of the habitat. In the case of Leadbeater's Possum, however, 17 sites supported suitable habitat but the species was recorded in just one. Therefore, although suitable forest occurred in many sites, other factors appeared to preclude the use of the retained linear strips by Leadbeater's Possum.
>
> Leadbeater's Possum, which has a complex colonial social system and consumes widely dispersed food, was relatively rare in the retained strips. It seems likely that the narrow linear habitat strips made it difficult for the species to harvest food efficiently and undertake some aspects of group social behaviour. The species most commonly observed in wildlife corridors was the Greater Glider which has a diet comprising readily available food (e.g. leaves) and a relatively simple social system; that is, they are solitary or live in pairs. The findings indicate that although setting aside networks of wildlife corridors may be a valuable strategy for some arboreal marsupials, they may not be effective for the conservation of all species.
>
> Some features of the strips were related to their use by arboreal marsupials. Significantly more animals occurred in strips that supported numerous trees with hollows, probably because the trees provide nest sites. The position of the strips in a forest landscape also influenced their use. More animals were found in corridors that spanned forests on different parts of the topographic sequence (e.g. linked gullies to ridges) than in sites confined to a gully or a midslope. The reasons may be related to animals' needs to move through different parts of the forest to harvest a range of food (Claridge and Lindenmayer 1994). The results highlight the importance of a good understanding of the biology and ecology of the species that we hope to manage by establishing networks of wildlife corridors.

LESSONS LEARNED

The primary focus of much conservation biology over the past decades has been on large ecological reserves. This is understandable, given the importance of these areas for many elements of the biota. However, an important lesson from the body of work in montane ash forests is that conservation within areas broadly designated for timber and pulpwood production is at least as important as conservation in large ecological reserves. It may be even more so for species such as Leadbeater's Possum, which has a distribution predominantly in forests outside the large ecological reserve in the Yarra Ranges National Park. Indeed, the extensive wildfires throughout the Central Highlands of Victoria in February 2009 have underscored the crucial importance of conservation efforts that encompass reserves and off-reserve strategies.

Irrespective of the relative importance of large ecological reserves and off-reserve areas, a significant lesson is that there is critical complementarity between the two. That is, large ecological reserves and off-reserve areas are part of a continuum of approaches needed to conserve forest biodiversity in the montane ash forests of the Central Highlands of Victoria. The different scales of management are strongly interdependent. What happens at the landscape level cannot be divorced from what takes place at the stand level. and vice versa. A stand of old growth surrounded by other old-growth stands will behave quite differently (and support different species assemblages) from an old-growth stand embedded within an extensive area dominated by clearfelled forest (Lindenmayer et al. 1999a). Similarly, forest landscapes comprising an array of stands, and the structural composition of these stands can influence species occurrence at the landscape level. A lack of suitable habitat within many different stands may combine to preclude a species from entire landscapes (Milledge et al. 1991).

The lessons about multiple scales for biodiversity conservation and for forest management are a strong counter-argument to those who believe in a 'land apartheid system', that is, a tenure system in which production forests and conservation forests are quarantined with entirely separate and independent sets of values and management regimes. They are also an antidote to the belief that conservation management problems can be readily solved solely by taking a landscape-scale approach.

KNOWLEDGE GAPS

Despite the intuitive ecological appeal and logic in favour of different kinds of reserves at the meso-scale, their effectiveness remains largely unknown, both individually and collectively. That is, there is remarkably little available empirical data demonstrating the efficacy of, for example, wildlife corridors within wood production forests. Similarly, the collective contribution of unlogged areas, wildfire corridors, special protection zones and other places exempt from logging has yet to be convincingly quantified – for any forest anywhere (Lindenmayer and Franklin 2002). This remains a significant challenge for researchers and forest managers. A range of studies and approaches are urgently required to gather the kinds of data needed for better-informed judgments about the effectiveness (or otherwise) of conservation strategies such as the protection of meso-scale reserves within wood production landscapes. Some of these approaches include integrated active adaptive management experiments underpinned by rigorous long-term monitoring – topics that feature in the following two chapters.

FOREST PATTERN, ECOLOGICAL PROCESS AND LINKS TO OTHER CHAPTERS

Reserves are a human construct – more specifically, a largely Western human construct. They are a spatial pattern in a landscape or region, partly designed to protect some kinds of spatial patterns such as the occurrence of old-growth or multi-aged forest, or the occurrence of particular species or communities of conservation concern. They are also set aside to maintain particular key ecological processes such as the maintenance of population and meta-population processes that underpin species persistence and longer-term evolutionary processes (Govt of Victoria 1992).

The meso-scale reserve system within areas of montane ash forests broadly designated for wood production is strongly influenced by the ecological processes of natural disturbance and human disturbance regimes. This is because special protection zones (including areas suitable for Leadbeater's Possum) typically occur where there has been limited or no logging or salvage logging, or where there have been no recurrent high-intensity wildfires over the past 200+ years.

Links to other chapters

The primary topics in this chapter are a hierarchical checklist of management strategies for forest biodiversity. Large-scale ecological reserves and meso-scale protected areas are strategies for mitigating the broader effects of logging through focusing on places that are exempt from timber harvesting. Nevertheless, they are intimately intertwined with stand-level conservation strategies to mitigate the effects of logging – the focus of Chapter 16.

The contribution of reserves, both large ecological reserves and meso-scale reserves, featured heavily in computer simulation modelling of extinction risks (Chapter 10).

Important lessons were learned from studies conducted within the Yarra Ranges National Park, particularly about the composition and structure of old-growth stands (Chapters 4, 6 and 8) and natural disturbance regimes (Chapter 12). These lessons may become more important as post-fire stand recovery is quantified at long-term monitoring sites (Chapter 17), particularly burned ones within the O'Shannassy Water Catchment in 2009. This work will enable cross-tenure comparisons of populations of various species targeted for monitoring (Chapter 17).

16 Mitigating logging impacts

INTRODUCTION

The previous chapter, on large ecological reserves and meso-scale reserves within areas broadly designated for wood production, examined the importance of these areas for conserving forest biodiversity. However, both ecologically sustainable forest management and forest biodiversity conservation are multi-scaled propositions (Lindenmayer and Franklin 2002; Puettmann *et al.* 2009). Therefore, mitigating the impacts of logging is also a multi-scaled proposition. This chapter is focused on ways to mitigate logging impacts at the coupe or stand level, that is, on the sites where logging actually takes place. Stand-level management is important for many reasons. First, if stand structure is simplified in many areas, the problems this creates at the local level can quickly accumulate across many stands and become a landscape-wide or region-wide problem (Colour Plate 14). Second, the effectiveness of many conservation strategies is poorly known. Large ecological reserves and meso-scale reserves may fail to conserve some kinds of species, such as those that have very fine-scale patterns of distributions or that require specialised habitats. Management at the stand level acts as a kind of risk-spreading strategy – if larger-scaled approaches fail, others will be in place and may be effective. Third, it is increasingly recognised that rapid climate change and the changes this will elicit in species distribution patterns (Parmesan 2006) will severely test the effectiveness of reserve systems (Dunlop and Brown 2008). Therefore, off-reserve management (including management at the stand level) will be critical for the conservation of forest biodiversity and the development and/or maintenance of ecologically sustainable forestry practices (Franklin *et al.* 1997).

A general principle underpinning approaches to improve natural resource management is using natural disturbance regimes to guide logging practices that are more ecologically sustainable (Hunter 2007). A key part of attempts to mitigate logging impacts is increasing the levels of congruence between the effects of human disturbance and natural disturbance. This chapter opens with a discussion of ways to increase congruence between the effects of wildfires and logging regimes at the stand and landscape levels. The second part of the chapter briefly revisits the key points of Chapters 13 and 14 about the stand-level effects of clearfelling, and lists the key attributes of stand structure needed to increase levels of congruence with natural disturbance and thus enhance biodiversity conservation in montane ash forests. The recommended strategies include greater retention of trees with hollows, greater proliferation of multi-aged

Table 16.1: Studies focusing on ways to mitigate logging impacts in the montane ash forests of the Central Highlands of Victoria

Study description	Reference
General overview of multi-scaled approaches to mitigate logging impacts	Lindenmayer and Franklin (2002)
Differences between clearfelling and wildfire at stand and landscape levels	Lindenmayer and McCarthy (2002)
Increasing congruence between clearfelling and wildfire at stand and landscape levels	Lindenmayer and McCarthy (2002)
Importance of biological legacies for limiting logging impacts	Franklin et al. (2000)
Variable retention harvesting system as an alternative to clearfelling	Lindenmayer (2007)
Improved tree retention strategies in logged coupes	Gibbons and Lindenmayer (1997a, b); Ball et al. (1999)
Approaches to reduce the impacts of salvage logging	Lindenmayer and Ough (2006); Lindenmayer et al. (2008a)

stands, and increased protection of intact areas of understorey, including tree ferns. The third part of the chapter outlines a harvesting experiment, based on the retention of islands, that aims to collectively improve stand structural retention within logged forests. The final part of the chapter focuses on ways to mitigate the impacts of salvage logging in montane ash forests. Many of the ways are identical to the strategies appropriate for conventional logging, but additional ones are needed in forests that have been burned.

BACKGROUND DATASETS

Many kinds of datasets have been applied in assessments of ways to mitigate the effects of logging in the montane ash forests of the Central Highlands of Victoria. Data highlighting the habitat requirements of particular animals have been used to determine where such species are most likely to occur (Chapter 9) and, in turn, infer how changes in the structure and composition of the forest from logging operations may affect those species. Such insights have guided recommendations about how logging operations might be altered to better maintain the key attributes of stand structure and landscape composition required by elements of the biota that are of conservation and management concern. In essence, almost all the large array of datasets that have been gathered during the past 25 years inform discussions about ways to reduce the impacts of conventional clearfelling operations and better integrate conservation and wood production in montane ash environments.

The approach of inferring impacts from observational and correlation-based studies is valuable and has many advantages. It also has some limitations, such as problems associated with substituting space for time and ignoring the influence of site history on animal occurrence (Pickett 1989). However, the general approach in montane ash forests over the past 25 years has been considerably strengthened by intertwining more scientifically robust methods, such as long-term repeated measures (e.g. a sound monitoring program, Lindenmayer et al. 2003a) and rigorously designed experiments. In 2003, the first outcomes of the long-term monitoring program were published (Lindenmayer et al. 2003a) (Chapter 17) and a replicated experiment, examining the response of various biotic groups to different ways of cutting Mountain Ash forest, was instigated (Lindenmayer 2007).

NEED FOR GREATER CONGRUENCE BETWEEN HUMAN AND NATURAL DISTURBANCES

Strategies for biodiversity conservation are most likely to be successful in cases where human disturbance regimes (such as logging) are similar in their effects to natural disturbance (Hunter 1993; Korpilahti and

Table 16.2: Differences in impacts of clearfelling and wildfire on stand and landscape structure in montane ash forests

Variable	Forest response	
	After wildfire	After clearfelling
Forest floor conditions	Large-diameter logs often occur	Average number and size of logs significantly reduced
Spacing of trees with hollows	Regular or random	Clustered
Standing life of trees with hollows	Up to, or more than, 50 years	All trees removed during logging or destroyed by regeneration fire
Range of forms of living and dead trees with hollows	Often two or more morphological forms present	All trees removed during logging
Survival of trees with hollows	Variable depending on tree age and fire intensity	All stems removed or severely burnt
Vegetation structure	Multi-aged stands may occur	Even-aged stands
Plant species composition	Variable depending on fire intensity and frequency	Shrubs and ground plants typical of wet environments
Old-growth forest patchiness	Large patches of intact old-growth forest	Small patches remain, old growth limited as average forest age is reduced

Kuuluvainen 2002). Examples include the kinds and numbers of biological legacies (Franklin et al. 2000) and the spatial patterns of environmental conditions (e.g. patch types) remaining after disturbance in montane ash forests (DeLong and Kessler 2000). Organisms are likely to be best adapted to the disturbance regimes in which they evolved (Bergeron et al. 1999; Hobson and Schieck 1999) but are potentially susceptible to novel forms of disturbance (or combinations of disturbances), such as those that are more or less frequent and/or more or less intensive than normal (Lindenmayer and McCarthy 2002). Natural disturbance regimes may therefore be appropriate baselines and ranges of variability against which human disturbance regimes can be compared (Bergeron et al. 1999; Hunter 1993, 2007).

Different effects of clearfelling and wildfire on stands and landscape

The traditional view of disturbance in montane ash forests is that of high-intensity stand-replacing wildfires which produce even-aged regrowth forests. This has been used to justify the widespread application of clearfelling, because it and wildfire are considered to be 'ecologically equivalent' (National Association 1989). However, extensive empirical studies in montane ash forests have revealed major differences between clearfelled forests and those burned by wildfires (Table 16.2), particularly with respect to vegetation structure, plant species composition and landscape patchiness (Lindenmayer et al. 1991g).

Clearfelling in montane ash forests simplifies stand structure and creates a single cohort of regrowth trees. In contrast, natural disturbance by wildfire varies in intensity and effects – in some stands, all trees are killed and large dead trees remain. In others, a multi-aged stand develops – a mixture of fire-damaged living and dead trees. Many of the biological legacies typical of stands damaged by wildfires are those which are lost from, or severely depleted in, young clearfelled forest. Hence, the widespread application of clearfelling operations is not consistent with the effects of natural disturbance regimes.

Increasing congruence at the stand level

Creating greater congruence between human disturbance (logging) and natural disturbance (fire) at the stand level requires maintenance of structural features and plant species composition in logged montane ash forests that are typical of forests recovering from wildfires. Key structural attributes of montane ash forests are large living and dead trees, especially those that contain cavities. These are used as den and nest sites by many vertebrate and invertebrate taxa, and they need to be retained as part of modified silvicultural systems in montane ash forests (Gibbons and Lindenmayer

1997b, 2002). Greater levels of tree retention than the typical level of five trees per ha are required for suitable nesting habitat for cavity-dependent species, particularly given accelerated rates of mortality and collapse of retained trees after logging (Ball et al. 1999). In addition, planning silvicultural practices requires that thought be given not only to what stems are retained during the present cutting events, but what will be set aside over several successive rotations (Ball et al. 1999). This is important because old-growth trees in montane ash forests may survive several wildfires (Banks 1993) and are known to be important for wildlife – indicating that the persistence of biological legacies through multiple disturbance events could be an important part of animal species response.

Clearfelling operations on the present 50–80 year rotation in montane ash forests (Squire et al. 1991) limit the recruitment of large-diameter logs to the forest floor. This is because the frequency and intensity of repeated harvesting events prevent large-diameter standing trees from developing. In addition, large logs may be damaged by mechanical disturbance during cutting or burned as a result of high-intensity prescribed fires used to remove logging debris and promote the regeneration of a new stand of regrowth trees. The retention of trees in logged forest may partially alleviate these changes. However, at current rates of tree retention, these trees will contribute only a small fraction of the stems that would be recruited to the forest floor in an unharvested forest. The retention of significantly more stems, and marking of existing large logs for enhanced protection, will be required to ensure that conditions on the forest floor more closely resemble those of uncut stands. It is appropriate for codes of forest practice in montane ash forests (Dept of Natural Resources and Environment 1996) to better recognise the importance of log resources as part of ecosystem function (reviewed by Harmon et al. 1986) and set appropriate prescriptions for the perpetuation of this key element of stand structure.

EXISTING MANAGEMENT AT THE STAND LEVEL

Stand-level logging operations in montane ash forests are governed by a Code of Forest Practices (Dept of Natural Resources and Environment 1996) that

---

Box 16.1: Increasing congruence at the landscape level

Although the focus of this chapter is on stand-level measures to mitigate logging effects, creating greater congruence between human disturbance (logging) and natural disturbance (fire) is also important at the landscape level. Appropriate strategies include using the patch sizes and shapes created by natural disturbance regimes as a template for guiding the spatial size and location of harvested sites. For example, there may be a need to aggregate harvested areas to provide large continuous areas of suitable habitat (Franklin and Forman 1987).

In the case of wood production montane ash forests, the size and spatial extent of old-growth patches is significantly reduced relative to that characteristic of burned (but unlogged) water catchments. Some wood production forest blocks in the Central Highlands of Victoria (3000–7000-ha planning areas) support <2–5% cover of old growth (Lindenmayer and Possingham 1995c). Areas 10–100 times larger than this were known from protected water catchments prior to the February 2009 wildfires (Lindenmayer and Possingham 1995c). These factors are the likely explanation for the rarity of animals like the Yellow-bellied Glider in montane ash forests subject to wood production (Lindenmayer et al. 1999a) (Chapter 9). This indicates a need to undertake landscape reconstruction efforts in wood production forest blocks and to exempt significant areas of regrowth forest from harvesting and grow it through to an old-growth stage (Lindenmayer 2009; Lindenmayer and Possingham 1995c; McCarthy and Lindenmayer 1999).

---

encompasses a wide range of environmental considerations, including the retention of trees on logged sites. While these multi-scaled approaches are useful, there are problems at the stand level because:

- retained trees are often destroyed or badly damaged by high-intensity slash fires, and trees that do remain standing often have poor survival rates (Lindenmayer et al. 1990b);
- tree retention strategies, even if increased by 100% over those presently recommended, will still leave

Figure 16.1: Conventionally clearfelled area. (Photo by David Lindenmayer).

significantly fewer cavity trees in logged areas than occur in unmanaged stands (Ball *et al.* 1999);
- numbers of retained trees may be insufficient to meet the habitat requirements of Leadbeater's Possum and a wide range of other cavity-dependent taxa (Gibbons and Lindenmayer 1997b, 2002);
- numbers of retained trees may be insufficient to ensure the recruitment of sufficient amounts and sizes of coarse woody debris to the forest floor.

Current clearfelling operations significantly reduce levels of stand structural complexity and widespread application of clearfelling operations appears to be inconsistent with the variable effects of natural disturbance regimes. This suggests a need to change cutting and regeneration methods to more closely resemble natural disturbance regimes and promote structural complexity in stands of harvested forest, to enhance their value for wildlife.

> **Box 16.2: Are nest boxes an effective source of cavities for hollow-dependent fauna on logged sites?**
>
> The abundance of trees with hollows is a significant factor influencing the presence and abundance of cavity-dependent vertebrates in montane ash forests (Chapter 6). There is a rapid rate of decline in the abundance of this resource, which will be followed by a prolonged period of a shortage of trees with hollows within extensive areas of regrowth ash forest. The recruitment of trees with hollows will be further delayed in the extensive stands of trees that are 70 years old (dating from the 1939 wildfires) or younger, that were badly damaged in the 2009 wildfires.
>
> A frequent question is whether nest boxes can help overcome the shortage of trees with hollows and recover and maintain populations of cavity-dependent fauna until new trees become available. This was examined in the long-term nest box studies discussed in some detail in Chapter 6. Occupancy rates are low among target species such as Leadbeater's Possum. Infestation rates by pest invertebrates are relatively high and attrition rates are substantial after nest boxes have been in place for six or more years (Lindenmayer *et al.* 2009c). Hence, there is a short period (~5 years) during which nest boxes are suitable for occupancy. This means, for example, in stands of 1939 regrowth where it will be a further 50–100 years before cavities develop that are suitable for occupancy by arboreal marsupials (Ambrose 1982), 10–20 nest box replacements will be required to maintain a perpetual supply of artificial cavities in a given area. McKenney and Lindenmayer (1994) argued that a program requiring nest boxes be replaced very frequently was unlikely to be maintained by a government agency. The costs of such a program may considerably exceed those of a tree retention strategy in logged areas.

ALTERED SILVICULTURAL SYSTEMS: SILVICULTURAL SYSTEMS PROJECT

The Silvicultural Systems Project was established in Victoria in the 1980s in response to concerns about the environmental impacts of clearfelling in Victorian forests. A wide range of projects associated with improved silvicultural practices and reduced impacts on the environment were completed (reviewed by Campbell 1997). The broad objectives of the Silvicultural Systems Project were:

To identify and develop silvicultural systems with clear potential as alternatives to the clearfelling (clearcutting) system and model those systems against clearfelling in terms of the long-term balance between socio-economic and environmental considerations (Squire 1990).

A range of forest logging treatments was investigated, including clearfelling, shelterwood, small gap selection, large gap selection, seed trees and strip-felling. These methods of logging differ in the size of the area cut and the amount of forest that remains after the initial entry into a stand, and the time elapsed before another harvesting event (Flint and Fagg 2007). Squire (1987, 1990) and Flint and Fagg (2007) give further details of the treatments.

Although the Silvicultural Systems Project was laudable in exploring alternatives to clearfelling, the study was limited by its focus on a restricted set of traditional silvicultural methods (Lindenmayer 1992). Each treatment resulted in the removal of all stems in a given area on a 50–80 year rotation. For example, one of the shelterwood treatments removed retained trees only three years after the regeneration felling (Saveneh and Dignan 1998). Given removal of all stems, coupled with the requirement for nest sites in large old cavity trees by virtually all species of arboreal marsupial (including Leadbeater's Possum), all silvicultural practices tested had detrimental long-term on-site impacts (Lindenmayer 1992). It has been established that adequate regeneration can be obtained under cutting regimes other than clearfelling (Campbell 1997) and high-intensity slash fires (Squire 1993). However, these alternative methods have not been widely embraced in montane ash forests (Lutze *et al.* 1999).

ALTERED SILVICULTURAL SYSTEMS: VARIABLE RETENTION HARVEST SYSTEM

The deficiencies in the range of logging systems examined in the Silvicultural Systems Project meant there was an imperative to experiment with other silvicultural methods (Lindenmayer 2007). An approach with considerable promise for application in montane ash forests is the Variable Retention Harvest System (VRHS). Alternative forms of logging, such as the VRHS, that retain key structural elements of native forests can promote the conservation of structure-dependent biota in wood production forests (Franklin *et al.* 1997) (Figure 16.2). Implicit in the VRHS is acceptance of the idea that some of the productive capacity and economic value of a stand will be devoted to the maintenance of biodiversity, ecosystem processes and other values, rather than maximising the regeneration and growth of commercial tree species. This is entirely consistent with the underpinning philosophy of a modern application of ecologically sustainable forest management (Lindenmayer and Franklin 2002).

VRHS originated in North America and typically includes:

- retention of sufficient structural features for the practice to be socially credible and ecologically effective;
- retention of particular stand structural attributes (e.g. dominant living trees and large dead trees with hollows);
- reasonable spatial distribution of retained structures (i.e. retention cannot be concentrated only along drainage lines or along the edges of a harvest unit);
- retention of structures for at least one rotation (i.e. structures that are retained only temporarily, such

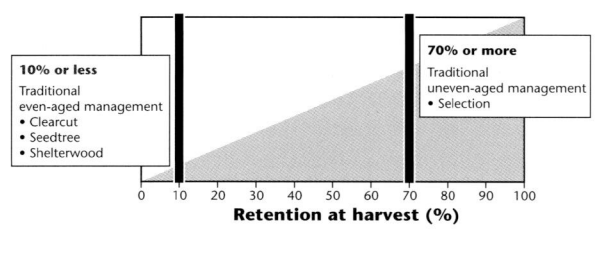

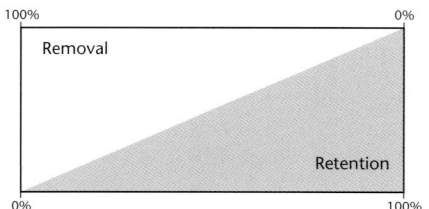

Figure 16.2: The variable retention harvest concept. (Modified from Franklin *et al.* 1997).

Table 16.3: International examples of application of VRHS

Location	Reference
Coastal forests, British Columbia, Canada	Dunsworth and Beese (2000); Bunnell et al. (2003); Outerbridge and Trofymow (2004); Swift (2006)
Interior and boreal forests of Canada	Hollstedt and Vyse (1997); Sullivan and Sullivan (2001); Coxson and Stevenson (2005); Deans et al. (2005)
Washington and Oregon, US	Halpern and Raphael (1999); Lazaruk et al. (2005); Nelson and Halpern (2005); Schowalter et al. (2005)

as a shelterwood overstorey, do not meet the goal of structural retention).

Beyond this general consensus, VRHS encompasses a broad continuum of silvicultural prescriptions. It is flexible in terms of levels of stand retention and the array of structural conditions that can be created (even-aged, multi-aged or all-aged). This flexibility allows adaptation in applying the best approach for each stand throughout a forest estate.

The VRHS is widely applied in western and eastern North America, South America and many parts of northern Europe (Table 16.3). There are examples of VRHS in native forests in south-eastern Australia, such as Tasmania (Hickey et al. 1999).

THE CUTTING EXPERIMENT

A major experiment examining the effectiveness of the VRHS in the Mountain Ash forests of the Central Highlands of Victoria commenced in 2003. The experiment had three broad aims:

- to promote the retention of key structural attributes on logged stands, particularly large trees and understorey vegetation;
- to grow these retained structures through several rotations so that they become old-growth structures;
- to assess the logistical and operational feasibility of retaining islands of original forest within otherwise clearfelled coupes.

The experiment also had general objectives, including the creation of more areas of multi-aged forest and reducing the amount of time it takes for regenerating forest to become suitable habitat for species such as Leadbeater's Possum.

The VRHS experiment was implemented in the Toolangi, Marysville and Powelltown districts and was constrained to one age class of forest – 1939 regrowth Mountain Ash – where the bulk of harvesting activity presently takes place. In addition, because different types of ash forest support different faunal compositions (Chapters 4, 9 and 11), it was important to constrain the project to a single forest type to avoid treatment/forest type confounding.

The experiment comprised four treatments:

- existing clearfelling practice;
- a logging coupe where one 1.5-ha habitat island is retained;
- a logging coupe where three 0.5-ha habitat islands are retained;
- a 'pseudo-coupe' or a location in 1939-regrowth forest in which no harvesting takes place.

There were six replicate blocks in the experiment. A block consisted of a broadly homogenous area dominated by 1939-regrowth Mountain Ash forest where each of the three treatments was established, plus the control. The six blocks, each with three treatments, gave a total of 18 coupes plus the 6 control pseudo-coupes, giving a total of 24 coupes. The general aim was to establish two or three complete blocks in each year over the five years since the experiment commenced.

Habitat islands and surveys plots within treatment classes

Staff from the Australian National University, Victorian Department of Sustainability and Environment and VicForests, as well as logging contractors, helped identify the best locations on coupes for habitat islands. Habitat islands were marked out before logging commenced to ensure that harvesting contractors were

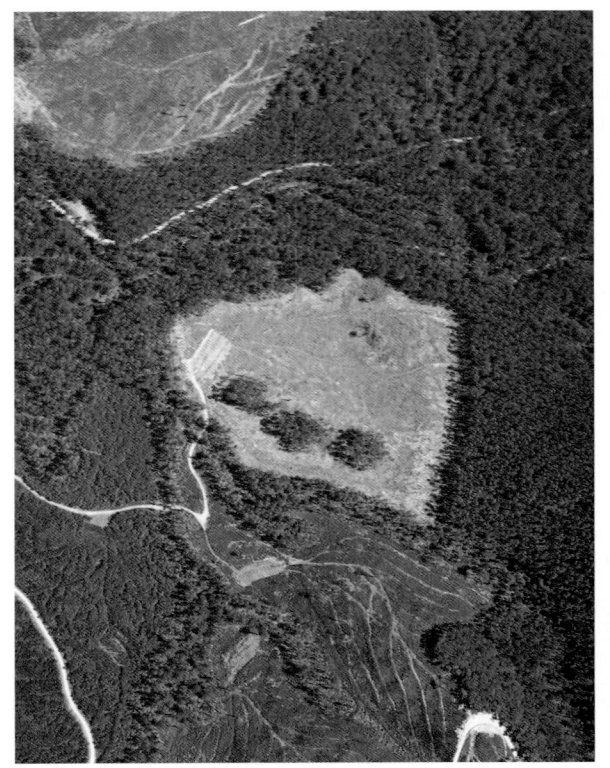

Figure 16.3: Harvested coupes with three retention islands at South Spur near Marysville. (Photo by Wally Notman).

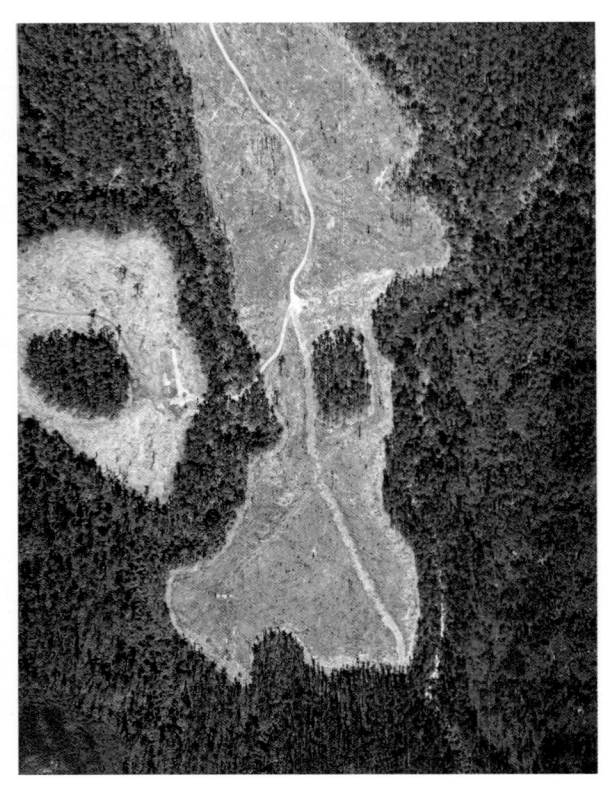

Figure 16.4: Harvested coupes with retention islands at Middle Paddock near Marysville. (Photo by Wally Notman).

well aware of the locations. The selection of islands was based on:

- habitat values for wildlife, for example places that currently support large trees with hollows;
- locations where retained vegetation was likely to be best protected from regeneration fires;
- ensuring that habitat islands were at least 50 m from coupe boundaries. The experiment focused on coupes that were 15 ha or larger so that retained forest patches would be some distance from the neighbouring unharvested forest;
- safety considerations for timber workers and field staff.

There were five wildlife survey points per coupe:

- one within each of the three 0.5-ha islands or three within the single 1.5-ha island;
- one within a plot located in the clearfelled part of a coupe;
- one within a plot at the margins of a coupe.

This approach to plot establishment made it possible to contrast the value of retention within coupes (the islands) with retention at the margins of coupes and to compare plots within islands and within coupe.

Like the coupes with retention islands, there were five sample plots per coupe in the traditional clearfelling coupes and pseudo-coupes. This ensured that sampling effort was the same for all treatments. In total the experiment comprised 24 coupes with five plots per coupe – 120 permanent survey plots in the entire study.

Target response groups

The response of four broad groups of vertebrates was quantified in the Cutting Experiment – birds, reptiles, terrestrial mammals and arboreal marsupials. Each group was counted:

- before harvesting commenced;

> **Box 16.3: Origins of the Cutting Experiment**
>
> The Cutting Experiment began with a Forestry Roundtable meeting at Marysville in the Central Highlands of Victoria in September 2002. The meeting brought together forest researchers from around the world, as well as forest managers and park managers from the Victorian Government. Representatives of the timber industry and conservation groups also attended (Lindenmayer *et al.* 2004). A key aim was to highlight ways in which alternatives to clearfelling had been implemented in wet forest environments in many parts of the world, including Tasmania, western North America and Scandinavia, as part of transitions to ecological sustainability. A book based on participants' collective experiences was published after the meeting (Lindenmayer and Franklin 2003).
>
> There was broad recognition of a need for the Cutting Experiment by some senior officials from the Victorian Government: the experiment was championed by Ian Miles, then head of the Forest Service within the Department of Sustainability and Environment. Following an extensive period of planning the appropriate statistical design, the project commenced in late 2003 with the support of many funding organisations, including the Australian Government Department of Agriculture Fisheries and Forestry, the Forest and Wood Products Research and Development Corporation (now Forest and Wood Products Australia), the Victorian Department of Sustainability and Environment and The Australian National University. The project met some resistance from field staff, who considered that changes to the way forests are logged was a slight on their professionalism as forest managers. There was not-inconsiderable resistance from some conservation groups which held strong beliefs that there should be no native forest logging in Australia. This led to some setbacks in progress, but by January 2009 the experiment had been fully implemented. Unfortunately, the February 2009 wildfires burned a subset of the experimental blocks in the Cutting Experiment.

- after harvesting was completed but before a regeneration burn was applied;
- after a regeneration burn;
- repeatedly after regeneration of the cut stand.

This gave a before-and-after component to the experiment, to increase its ability to quantify 'treatment' effects. The intention was to maintain monitoring at yearly intervals in perpetuity after the complete establishment of the experiment. However, damage to the study from the 2009 wildfires has made the future of the work unclear.

In addition to intensive data-gathering for vertebrates, the experiment is recording the structure, composition and condition of the vegetation at all 120 survey plots. This allows not only the response of the vegetation to the experimental treatments to be quantified, but also the relationships between vegetation cover and animal response.

Preliminary findings

Preliminary findings after six years of the Cutting Experiment suggested that the terrestrial small mammals – Agile Antechinus, Dusky Antechinus and Bush Rat – could persist within the retained islands after the harvesting process although densities were lower than in unharvested forest. The composition and number of bird species changed substantially post-harvest but a number of taxa continued to use the islands following logging of the surrounding forest. Reptiles respond poorly to the harvesting process and were generally not detected within the cutover area or within the retained islands post-harvest.

The persistence of some species, such as small mammals, within the retention islands following logging in the experiment was surprising. These residual animals promote population recovery following disturbance. Alterations in the presence and abundance of other species in the islands following harvesting was expected. However, a key issue was that the retained islands in the VRHS were not intended to retain viable populations of particular species. Rather, the aim was to promote the future structural complexity of logged and regenerated forests, for example, 10–40 years following harvesting, and thus improve the habitat quality of regrowth forests for otherwise logging-sensitive species such as Leadbeater's Possum. Indeed, data on the habitat requirements of Leadbeater's Possum and other vertebrate species in Mountain Ash forests suggested that the VRHS had the potential to actually create suitable habitat. Hence, as indicated by Smith *et al.* (1985) and Lindenmayer (1994a), the conservation of Leadbeater's Possum is a rare example where altered silvicultural systems (logging methods not based on traditional clearfelling) could significantly benefit the species.

FURTHER WORK AND THE IMPLEMENTATION OF VRHS

Although the experimental implementation of the VRHS in Mountain Ash forests was reasonably advanced prior to the Black Saturday 2009 wildfires (Flint and Fagg 2007a), the effectiveness of altered silvicultural systems for biodiversity conservation had yet to be demonstrated. Considerable ongoing monitoring over a prolonged period (5–10+ years) was planned to assess progress and feed key findings back to forest managers and policy-makers.

There were clear issues with logistics, worker safety and habitat island protection associated with the application of the VRHS in montane ash forests. For example, it was inappropriate to apply the system on steep terrain where there were safety issues for timber workers and high risks of retained islands being badly damaged by regeneration burning. Thus, VRHS will be applicable in places where it is safe and logistically feasible and hence where the retention of habitat islands within logging coupes can be most successfully achieved. Coupes on flat terrain, on south-facing slopes and where there are low levels of incoming radiation are good candidate areas for the application of the VRHS. Importantly, such areas are also where multi-aged stands are most likely to develop (Mackey *et al.* 2002). The ratio of coupes subject to clearfelling versus VRHS should be guided by data on natural disturbance regimes. This indicates that ~30% of coupes might be appropriate for the application of the VRHS.

Logistical issues

Experiments are one of the most powerful ways to generate understanding about ecological systems and how

> **Box 16.4: New-generation thinning operations and the Cutting Experiment**
>
> Thinning is an increasingly common form of logging in montane ash forests. It involves removing approximately every third 'row' of trees from the forest, using a mechanical harvester. Remaining trees are released from competition from neighbouring stems, which promotes increased diameter and more rapid increases in tree height. Thinning changes light conditions to the understorey and ground layers and can change some aspects of soil conditions. The effects of thinning on biodiversity are unknown. However, studies conducted in North America indicate that innovative approaches to thinning can have positive benefits for stand structure and biodiversity – if they are conducted in an appropriate way (Carey *et al.* 1999). Research is urgently required to determine ecologically appropriate methods to conduct thinning operations in montane ash forests.
>
> Thinning operations are a prelude to final harvesting of stands at the end of a rotation. An aim of VRHS in general, and the Cutting Experiment in particular, is to ensure the long-term retention of islands within otherwise logged coupes. There is an important opportunity to identify where such islands should be located, not only before cutting at the end of a rotation but some decades prior to that, when thinning operations are first proposed. This is tantamount to whole-of-rotation on-site coupe planning. If this is done appropriately and underpinned by good science, such planning could have major benefits for the multiple values that need to be maintained in wood production ash forests. For example, the locations of islands could be strategically anchored around attributes of stand structure that are to be maintained and perpetuated in the long term, for example large trees with hollows, recruit trees, large logs and long-lived understorey thickets.

to best manage them. However, large-scale logging experiments are rare. They can be expensive and labour-intensive. They can be jeopardised by unforeseen events, such as the February 2009 wildfires. The long-term nature of forestry experiments makes them difficult to establish and to maintain ongoing financial and logistical support. Indeed, many forestry experiments have been abandoned for these reasons. The Cutting Experiment attempted to avoid some of these problems by overlaying the experimental design and implementation on existing logging plans and operations. This was done in an attempt to limit the costs of establishment, conduct research that is relevant to management and create a research program that is economically sustainable in the medium to long term. However, as outlined above, the future of the Cutting Experiment remains unclear given the extent of forest damage resulting from the 2009 wildfires.

Feedback on the Cutting Experiment was provided by regional staff from VicForests and the Department of Sustainability and Environment as well as by logging contractors. Management of regeneration burns to limit damage to retained islands was a key issue. It was possible to protect islands, but required additional staff and resources to do so effectively, particularly on coupes with three 0.5-ha islands. The single 1.5-ha island on a coupe appeared to be logistically more straightforward for forest managers to work around than coupes with multiple smaller patches. Logging slash removal away from island boundaries was important in protecting the islands from regeneration burning. The strategic location of islands was critical to their protection during burning. Islands on flatter areas appear easier to protect from fire than those on steeper areas. This issue highlighted the value of communication between logging contractors and forest management staff in best locating islands as part of the application of the VRHS in Mountain Ash forests. Finally, consolidated larger islands were less likely to suppress the extent and vigour of regenerating stems in the remainder of the logged coupe, thereby limiting losses of timber that can be harvested in the subsequent rotation (Van der Meer and Dignan 2007).

VRHS as an adaptive management experiment and a model experimental system

Adaptive management involves the integration of research, monitoring and management to improve the management of resource management prescriptions (Holling 1978). Briefly, active adaptive management works as follows. Results from experience or formal

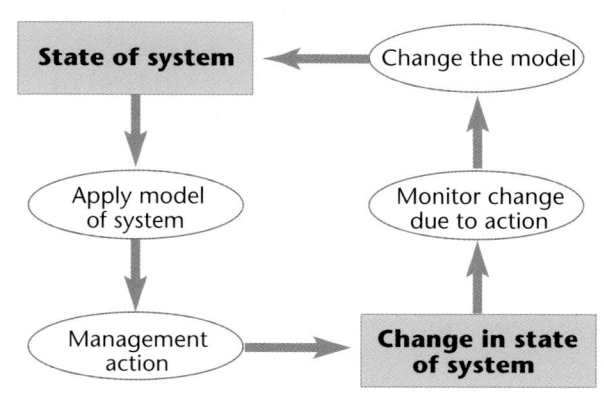

Figure 16.5: Feedback between policy, management and research in an adaptive management approach.

research inform improvements in forest management policies which, in turn, lead to improved management prescriptions. New prescriptions are rigorously assessed in new experiments or operational trials, with the results fed back to policy-makers and managers – thereby closing the cycle of knowledge-gathering, policy development and improved management application (Figure 16.5).

Although few organisations have truly embraced active adaptive management (Stankey *et al.* 2003; Parr and Andersen 2006), the approach represents a logical and potentially powerful way to manage forests and other natural resources. Adaptive management is often discussed by researchers and resource managers, but true adaptive management studies are extremely rare in practice (Bunnell *et al.* 2003).

The VRHS experiment in the Mountain Ash forests has involved experimenting with logging practices, monitoring responses to modified practices, and close communication between researchers and forest managers on the positive and negative aspects of the experimental harvesting. The experiment had many of the ingredients of a true adaptive management study (Walters 1986). This factor alone made the VRHS experiment a model that should not only be embraced in the long term, but be far more widely embraced elsewhere in Victoria, Australia and overseas. Indeed, Lindenmayer *et al.* (2000c) argue that a commitment to adaptive management should be a key requirement of ecologically sustainable forest management.

Future re-entries into coupes where VRHS has been applied will be a continuing challenge. Specific

> **Box 16.5: An important caveat – not all logging problems can be solved by logging experiments**
>
> The VRHS experiment was, in many respects, successful in examining alternative ways of harvesting montane ash forests. However, logging experiments are not a panacea. For example, a logging experiment was proposed as part of an adaptive management approach to identify the effects of logging on the highly endangered Baw Baw Frog (Hollis 2003). However, a risk assessment indicated that population declines of the species were so severe that experimenting with the forest to quantify impacts was an unnecessary risk. The experiment was abandoned (Greg Hollis, *pers. comm.*). More traditional conservation approaches, based on meso-scale reserves and zoning systems, were implemented (Chapter 15).

treatments will depend on management objectives. Computer visualisations may assist managers by providing images of likely future stand conditions under various management alternatives (Ball *et al.* 1999). In addition, management of stand information, such as precise location and extent of retained areas, will be greatly aided by technologies such as Global Positioning Systems and Geographic Information Systems. However, empirical data from VRHS applications, monitoring and experiments are likely to provide the most useful information. This is why long-term experimental monitoring and adequate record-keeping on applications of VRHS, such as why particular treatments were applied within a harvest unit and what additional operations were conducted, is so important. This information is essential to guide future generations of forest managers and wildlife managers in evaluating the success of various prescriptions, including responses of species to stand conditions.

STRATEGIES TO MITIGATE THE IMPACTS OF SALVAGE LOGGING

As outlined in Chapter 14, the impacts of salvage logging in montane ash forests, particularly those from the prolonged operations after the 1939 wildfires (Noble 1977), may have been far more substantial than previously appreciated by many forest managers and conservation biologists (Chapter 14). The removal of considerable numbers of large fire-damaged trees and the loss of many areas of what might have otherwise been multi-aged forest has contributed to the shortage of trees with hollows in substantial areas of montane ash forest in the region, a shortage which is likely to become increasingly problematic in the decades to come (Lindenmayer *et al.* 2008a). Because cavity development in montane ash trees takes more than 120 years, the creation of simplified stand structural conditions by past salvage logging operations, combined with current widespread clearfelling, may take hundreds of years to redress (Lindenmayer and Ough 2006).

The impacts of conventional clearfell logging methods and the potential effects of salvage logging were not well understood in the past. However, our significantly improved understanding of the biota and dynamics of montane ash forests should be used to better inform more ecologically sustainable logging practices, including those applied in salvage logging operations such as those after the February 2009 wildfires. Better policies and on-ground strategies for improved management of salvage logging operations after fires should provide for:

- varied salvage logging intensity, including within large areas of intensely burned forest;
- increased retention of biological legacies such as large living and dead trees, large logs and thickets of unburned or partially burned understorey vegetation;
- minimised mechanical disturbance;
- minimised seedbed preparation;
- consideration of timing and season of salvage logging.

Varied salvage intensity: exclusion and low-intensity areas

Given the limited understanding of the impacts of salvage logging in montane ash forests (Chapter 14), it is important there are areas where such practices do not occur and others where the extent of removal of post-fire biological legacies is reduced. This should be the case even where a wildfire has been intense over a large area. The Land Conservation Council (1994) recognised this for the Yarra Ranges National Park, where there are policies to exclude salvage logging. This is an important step. There should also be better-developed salvage logging policies in the 70% of montane ash forest outside the reserve system in the Central Highlands of Victoria. Some areas should be afforded increased protection from logging and others be subject to only low-intensity logging. This is particularly important for stands dominated by large old trees, which are uncommon in wood production montane ash forests. They are presently exempt from conventional (clearfell) logging and are often located within Special Protection Zones that have high conservation value within broadly designated wood production areas (Macfarlane *et al.* 1998), such as for Leadbeater's Possum. Salvage logging prescriptions developed for Alpine Ash forest burned in a recent fire states that salvage logging will not be undertaken in Special Protection Zones (Dept of Sustainability and Environment 2003). This should be extended to all stands dominated by large old trees in the whole Central Highlands of Victoria after the 2009 wildfires, for several reasons.

- Large trees in burned old stands have a higher chance of surviving a fire than smaller regrowth trees because of their girth and thicker bark. Hence, burned old-growth forests have a high chance of developing into multi-aged stands after a wildfire if salvage logging does not take place. Such multi-aged stands would be valuable habitat for many species, including Leadbeater's Possum.
- The large-diameter trees in burned old-growth stands have a high probability of containing hollows or developing them after a wildfire (Lindenmayer *et al.* 1993b). Conversely, the loss of such hollow trees through salvage logging will further contribute to the increasing and prolonged shortage of cavity trees in montane ash forests (Lindenmayer and Wood 2009).
- Large-diameter living and dead trees in montane ash forests have a significantly higher probability of remaining standing for longer periods of time than smaller-diameter trees (Lindenmayer and Wood 2009) (Chapter 6). This promotes long-term continuity in the availability of key resources such as tree hollows.
- Even if large fire-damaged trees collapse soon after a natural disturbance, they play many important ecological roles as large fallen logs in montane ash forests (Ashton 1986) (Chapter 8).
- The understoreys of old-growth forests are often dominated by tree ferns and other resprouting species. These taxa support large epiphyte populations, characteristic of old-growth montane forests. Logging will result in a decline in the abundance of these taxa (Ough and Murphy 2004) (Chapter 7).

If reconstruction analyses by Lindenmayer and McCarthy (2002) are broadly correct, then predicted past levels of multi-aged forest (30%) are much lower than they were immediately prior to the 2009 wildfires (7%). A key additional aim of excluding salvage logging from old-growth stands and creating other salvage exemption zones must be to expand the amount of multi-aged forest from the current low levels. The objective of this landscape construction work would be to create more suitable habitat, not only for Leadbeater's Possum but for numerous other species which require old-growth forest or elements of old-growth stands, particularly large-diameter trees.

Legacy retention in salvage logged areas

An important part of biodiversity conservation in wood production forests is the retention of biological legacies such as large trees with hollows. Current prescriptions allow for the retention of five such trees per ha of cutover forest (Dept of Natural Resources and Environment 1996). However, given that salvage logging may lead to the accelerated loss of retained burned trees, efforts to ensure the maintenance of five trees per ha in perpetuity demands that tree retention rates at the time of cutting should be much higher than specified under current prescriptions (Lindenmayer *et al.* 2008a). However, such retention rates may create issues for worker safety in cutover areas; retained trees may need to be arranged in clusters or aggregates to avoid the potential problems (Lindenmayer and Franklin 2002).

Mechanical disturbance

Salvage logging methods that minimise soil disturbance and the clearing of vegetation will reduce the impacts on understorey composition and legacy retention. Physical disturbance damages understorey vegetative propagules, kills seedlings and compromises or minimises legacy retention. Given this, moving and heaping of slash into windrows, for example, is not recommended.

Post-harvest seedbed preparation

Post-harvest regeneration burns to reduce logging slash and enhance eucalypt regeneration are generally not required after salvage logging (Dept of Sustainability and Environment 2003), but have sometimes been utilised. Mechanical seedbed preparation has been employed, particularly where eucalypt regeneration has been inadequate in the years following salvage logging. Regeneration burns and mechanical seedbed preparation generate a third consecutive disturbance, with regeneration generally occurring between each. The detrimental impact of another disturbance on stand structure and floristic composition is likely to be cumulative.

The application of a high-intensity regeneration burn as part of site preparation, for example, means that retained but previously burned hollow trees are subject to two fires within a short time (two or three years). These processes accelerate hollow tree loss above that caused by conventional clearfelling (Ball *et al.* 1999). Similarly, mechanical seedbed preparation will enhance declines in populations of understorey vegetative resprouters (Ough and Murphy 2004). Seed regenerators that have not yet reached sexual maturity (most) are also likely to decline. For these reasons, seedbed preparation after salvage logging is strongly discouraged.

Timing of salvage logging

There needs to be greater consideration given to the timing of salvage logging operations. Operations that continue for prolonged periods after a wildfire have considerable potential to alter forest recovery (Ough 2002). By logging soon after germination, small germinants will be particularly vulnerable to physical disturbance. Salvage logging at later growth stages (e.g. two to four years after a wildfire) will jeopardise structurally sensitive plant taxa, such as those with easily damaged stems. Conversely, salvage operations that take place too soon after a wildfire may commence before some trees begin to recover through epicormic shooting. Such live standing trees could mistakenly be assumed to be dead, and therefore harvested. Thus, the potential for recovering trees to become part of multi-aged stands is prevented. Studies of fire-damaged stands in the Powelltown region of the Central Highlands of Victoria found many areas supported living trees recovering after the 1983 wildfires (Smith and Woodgate 1985). Many of these trees were cut as part of salvage logging operations – opportunities for the development of multi-aged stands in the area were lost. Areas where montane ash trees have not been killed and are beginning to recover after a wildfire should be added to other places, such as old-growth, that are exempt from salvage logging.

SUMMARY

Mitigating the effects of logging requires multi-scaled conservation and management strategies at the regional and landscape levels through large ecological reserves and meso-scale reserves (Chapter 15) as well as at the stand level. An overall guideline or general principle for mitigating logging effects is to increase levels of congruence between human disturbance and natural disturbance (Hunter 2007). At the stand level, this often translates to greater retention of key components of strand structural complexity at the time of timber harvesting. Within montane ash forests, this means greater retention of existing trees with hollows, trees that can be maintained over time so they will eventually develop cavities, and intact areas of understorey and ground cover vegetation (including coarse woody debris).

Alternative logging methods are needed to facilitate increased levels of structural complexity within some areas of forest that are logged. It is at this smallest spatial scale – the stand level – that the greatest improvements need to be made if the multi-scaled hierarchy of management strategies is applied within montane ash forests. The Cutting Experiment, based on the concept of the variable retention harvest system developed in western North America (Franklin *et al.* 1997), was implemented in montane ash forests to ensure greater retention of stand structure in logged forest (Lindenmayer 2007) and to investigate the impact on biodiversity.

Prior to being damaged by the February 2009 wildfires, the Cutting Experiment was showing considerable promise as a harvesting method suitable for broad adoption across montane ash forests: it better integrates wood production with the conservation of forest biodiversity than does the widespread application of conventional clearfelling (Lindenmayer 2007).

Salvage logging is likely to have a range of effects additional to those of conventional clearfelling. The mitigation of these impacts requires (after Lindenmayer and Ough 2006):

- reducing the intensity of salvage logging operations by increasing the number of biological legacies that are retained, especially in areas which have been only partially burned;
- minimising the mechanical disturbance of the ground layer during logging operations;
- minimising the amount of seedbed preparation, particularly through the application of high-intensity regeneration burns;
- considering the timing and seasonality of salvage logging operations to limit impacts on the recovery of ground-level and understorey plants.

LESSONS LEARNED

Many important lessons have been learned from the past 25 years of work on mitigating logging effects. They include the need to think carefully about the objectives of particular silvicultural systems and to consider where it is appropriate to practise VRHS and in what proportion of coupes. For example, the general recommendation is that ~30% of coupes, and especially those on relatively flat terrain, should be harvested using VRHS where (after Lindenmayer 2007):

- it is safe for forest workers to employ this silvicultural system;
- it is most straightforward to successfully protect retained islands of forest from regeneration burns;
- multi-aged stands that develop are likely to be most suited to target species such as Leadbeater's Possum, especially given the species' aversion to forests on steep slopes (Lindenmayer *et al.* 1991d) (Chapter 4);
- naturally occurring multi-aged stands most often occur, thereby emulating natural disturbance regimes.

Perhaps the most sobering lesson from the work in the Central Highlands of Victoria is that the capacity to implement truly ecologically sustainable forest management is intimately intertwined with the level of commitment (or overcommitment) of forest resources to the timber industry. That is, if sustained yields are set too high, there is a risk of overcutting and limited capacity to accommodate the lower rates of production to meet the needs of other forest values. This problem of overcommitment can become extremely serious if forest losses from stand-replacing wildfires are not factored into sustained yield calculations well before such events take place. This is a significant issue, given the amount of montane ash forest burned in the 2009 wildfires, and highlights a need to recalculate very conservative sustained yields to limit the potential for resource overcommitment – especially given rapid climate change and its effects on fire frequency.

KNOWLEDGE GAPS

Efforts to mitigate the impacts of logging have been at the heart of much of the research conducted in the Central Highlands of Victoria over the past 25 years. Nevertheless, several important knowledge gaps remain. The effectiveness for biodiversity conservation of alternative cutting methods to clearfelling have yet to be quantified. Preliminary results from the Cutting Experiment are promising but it will be some time before convincing answers are obtained.

No work has been conducted on the amount and condition of coarse woody debris that needs to be retained in logged and regenerated montane ash forest. It has been well documented that these forests support very large quantities of fallen timber and that many species are closely associated with this attribute of stand structural complexity (Chapter 8). However, the effects of extensive and prolonged depletion of coarse woody debris remain unknown. Work in other forests around the world, particularly in the northern hemisphere, suggests that the impacts can be substantial (Grove and Hanula 2006; Harmon *et al.* 1986). It should be possible to capture large amounts of fallen timber within retained islands on logged coupes, but the effects of altering the spatial distribution of coarse woody debris on log-dependent taxa are not known.

The effects on biodiversity of accumulating many logging coupes across wood production landscapes remain poorly understood. More work needs to be completed at the landscape scale within montane ash forests. Such work will not be easy, because cumulative landscape-level effects are extremely difficult to quantify.

The effects of thinning operations on biodiversity in montane ash forests are poorly understood. This knowledge gap needs to be urgently addressed because thinning is increasingly applied, particularly in regions such as Toolangi and Powelltown.

FOREST PATTERN, ECOLOGICAL PROCESS AND LINKS TO OTHER CHAPTERS

As outlined in Chapter 13, logging is a major ecological process that influences many other ecological processes and an array of forest patterns. Some of these influences are negative, like those on reduced areas of old-growth stands, accelerated collapse rates of large trees with hollows, impaired large tree recruitment and impaired metapopulation dynamics. Finding ways to mitigate these impacts requires a good understanding of how ecological processes and forest patterns are altered by logging operations. This highlights the strong links between the content of this chapter and almost all the other chapters in this book.

Links to other chapters

This chapter has been written in direct response to Chapters 13 and 14, on logging effects. The links with those chapters are important because it is impossible to mitigate impacts if there is no understanding or quantification of those impacts. Understanding these impacts (and hence mitigating them) is, in turn, based on understanding where key species of interest occur in the landscape, why they occur where they do, and how logging (and other drivers) influences the resources on which they depend. Hence, this chapter is intimately related to Chapter 9, on animal distribution and abundance.

This chapter is linked strongly with Chapter 15, on reserves, because reserve and off-reserve management forms a hierarchical continuum of approaches to mitigate logging effects and better conserve forest biodiversity (Lindenmayer *et al.* 2006).

17 Monitoring

INTRODUCTION

Without rigorous monitoring, there is no way to determine whether management strategies implemented in montane ash forests are effective and whether montane ash forests are being managed in an ecologically sustainable way. Long-term data gathered from monitoring studies are fundamental for many other key activities:

- providing baselines for evaluating environmental change;
- detecting and evaluating changes in forest ecosystem structure and function;
- evaluating responses to disturbance, such as logging and wildfire;
- generating new questions;
- providing empirical data for modelling and for data-mining when exploring new questions (Lindenmayer and Likens 2009).

There is a prolonged worldwide history of poorly planned and unfocused monitoring programs that are either ineffective or fail completely (Norton 1996; Orians 1986; Stankey *et al.* 2003). The monitoring program in the Central Highlands of Victoria is a rare example of a successful monitoring program in Australia, particularly Australian forests. This chapter outlines some of the many reasons why long-term studies and monitoring programs fail. Four key aspects of a good monitoring program are described, to address the problems that typically beset failed monitoring programs. The remainder of the chapter

Table 17.1: Studies focusing on monitoring programs in the montane ash forests of the Central Highlands of Victoria

Study description	Reference
General overview of features of good monitoring programs	Lindenmayer and Likens (2009)
Description of long-term monitoring program in montane ash forests	Lindenmayer *et al.* (2003a)
Linking monitoring data on tree fall and abundance of arboreal marsupials	Lindenmayer *et al.* (unpublished data)
Linking monitoring data and population viability analysis models for arboreal marsupials	Lindenmayer and McCarthy (2006)

describes the long-term monitoring program that has been instigated in the Central Highlands of Victoria. The concluding part touches on some of the challenges associated with maintaining an effective monitoring program.

BACKGROUND DATASETS

The major dataset that has underpinned the monitoring program in the montane ash forests has been a pool of 161 sites established between 1983 and 1997. These sites have been subject to ongoing counts of arboreal marsupials (Lindenmayer *et al.* 2003a) and birds (Lindenmayer *et al.* 2009b) as well as repeated measurements of vegetation structure including the abundance and condition of large trees with hollows (Lindenmayer *et al.* 1997b; Lindenmayer and Wood 2009). Subsets of data have been used to test the accuracy of predictions about the efficacy of management options (Lindenmayer and McCarthy 2006).

WHY MONITORING PROGRAMS OFTEN FAIL

Arguments over what to monitor

One of the reasons that long-term studies and monitoring programs usually fail is that their design is often prefaced by protracted (and usually unresolved) arguments about what to monitor. One solution is to monitor a large number of things (the 'laundry list'), but resource and time constraints frequently mean that this is done badly. An alternative solution is to argue that indicator species or groups should be the targets of monitoring programs. A review by Lindenmayer and Burgman (2005) showed that a vast array of species and over 20 major taxonomic groups have been proposed as indicators, ranging from fungi and bryophytes to invertebrates and virtually all major vertebrate groups. Very rarely was it explicitly stated what these species or groups were actually indicative of, particularly at the ecosystem level, and the circumstances where these species or groups were and were not appropriate indicators.

The best response to the common question, 'What should be monitored?' is 'What is the ecological question?' That is, 'What question do you want to answer by monitoring?' For example, 'Have populations of species X changed between Y and Z years?' 'Have attributes A and B of forest cover influenced the observed changes in populations of species X during Y and Z years?' Driving monitoring efforts or programs by ecological questions is the most efficient and effective strategy to obtain meaningful ecological results, and the best way to avoid inefficient and ineffective monitoring and squandering of limited resources.

Poor design

A major problem with long-term studies and monitoring programs has been that they have often been very poorly designed. Good design is an inherently statistical process, but professional statisticians are often excluded from experimental design phases of monitoring programs. Key issues are overlooked, such as calculations of statistical power to detect trends, the importance of contrasts between treatments (e.g. where there is a human intervention and where there is not) and the value of innovative rotating of sampling methods for increasing inference (Welsh *et al.* 2000).

FEATURES OF A GOOD MONITORING PROGRAM

Monitoring programs have often been driven by a short-term funding opportunity or a political directive (mindless data collection) rather than being underpinned by carefully posed questions and objectives. Indeed, Roberts (1991) argued that too often monitoring has been 'planned backwards on the collect-now (data), think-later (of a useful question) principle'.

A good and effective monitoring program should include four key features:

- it should address well-defined and tractable questions that are specified prior to commencement;
- it should be driven by a human need to know about an ecosystem (e.g. the effects of a pollutant or changes in climate) so that it passes the test of management relevance (Russell-Smith *et al.* 2003);
- it should monitor a small number of things well, rather than many entities badly (Lindenmayer 1999);

- it should be underpinned by rigorous statistical design.

MONITORING IN THE CENTRAL HIGHLANDS OF VICTORIA

The monitoring program in the Central Highlands of Victoria has been carefully crafted to ensure that it includes the four key features outlined above. The two government natural resource agencies responsible for the management of montane ash forests (Parks Victoria and the Victorian Department of Sustainability and Environment) co-sponsor the monitoring program that has been administered from The Australian National University since 1997.

The human need to know about the system

The status of arboreal marsupials, particularly the endangered Leadbeater's Possum, is a major concern to forest and wildlife mangers in Victoria. These concerns have been magnified by the effects of the February 2009 wildfires on extensive areas of montane ash forest. There is a mandatory requirement for monitoring as part of the recovery plan for Leadbeater's Possum (Macfarlane *et al.* 1998), as well as agreements for ecologically sustainable forest management in Victoria. Therefore, there is a management imperative to know about the status of some elements of the biota in montane ash forests.

Targeted subset of taxa

The arboreal marsupial assemblage was targeted for detailed monitoring. The focus was not limited to Leadbeater's Possum because information on the temporal trajectories of commoner species was considered likely to reveal more about the montane ash forest ecosystem than was a rare species for which data can often be difficult to gather in sufficient quantities for robust statistical analyses. Notably, no individual species or set of species was considered to be an indicator taxon. This was because there is no evidence that there are valid indicator species among the taxa in this assemblage (Lindenmayer and Cunningham 1997) (Chapter 11). Nor is there any evidence that this concept is valid in general, not only for montane ash forests, but also in forest ecosystems *per se* (Lindenmayer *et al.* 2000c). Rather, arboreal marsupials were selected because they were an appropriate group for answering key questions (see below), and there is a shared management and public concern about their persistence and conservation.

Forest structure, particularly the abundance of trees with hollows, is a key habitat attribute for almost all species of arboreal marsupials in montane ash forests (Chapters 6 and 9). Considerable effort has been focused on gathering data on the temporal patterns of abundance and condition of trees with hollows on the 161 sites monitored for arboreal marsupials. These data enabled patterns in animal abundance to be linked with abundance patterns in trees with hollows. Data on other key attributes of forest structure for arboreal marsupials, such as the basal area of *Acacia* spp. (Chapters 4 and 10), have been gathered on a repeated basis on all monitoring sites.

In 2002, diurnal birds were added to arboreal marsupials as a target group for monitoring. This has enabled an expansion of inference about the status of the ash forests and their suitability as habitat for avifauna. As for arboreal marsupials, no assumptions were made about a given species of bird or set of bird species being indicator species.

Tractable questions

The monitoring program has been driven by a series of important questions that underpin a need to know about biodiversity conservation in the montane ash forests. Six main questions are:

- What are the temporal trends in populations of arboreal marsupials in montane ash forests?
- What are the temporal trends in populations of trees with hollows in montane ash forests?
- Can animal population trends be tied to trends in the decline of trees with hollows or temporal changes in other stand structural attributes?
- Are species differentially affected by changes in the abundance of trees with hollows and other stand characteristics?
- Is the rate of decline in the abundance of animals similar to that predicted in 1997 (Lindenmayer *et al.* 1997b), based on projections of collapse of large trees with hollows?

- Are there major tenure differences in population trends for arboreal marsupials? That is, do populations in areas managed as closed water catchments or national parks exhibit a population trajectory different from those in forests broadly designated for wood production and where timber and pulpwood harvesting is permitted? Can these changes be tied with landscape-scale changes in forest cover, such as the number and spatial pattern of logging coupes?

Rigorous statistical design

The monitoring program, like all the studies conducted in the Central Highlands of Victoria, was underpinned by detailed guidance from professional statisticians. Data-gathering on trees with hollows and other attributes of stand structure was based on repeated remeasurement of marked trees and vegetation plots on monitoring sites. Some of the results of the longitudinal trends in the abundance of trees with hollows are presented in Chapter 6. The approach for monitoring populations of arboreal marsupials is more complex; it is explained below.

The monitoring program was based on a retrospective approach to resurvey populations of arboreal marsupials on sites first surveyed in 1983. A pool of 161 sites has been established and each site has had repeated field surveys over the past two decades. All 161 sites were resurveyed in 1997–98 and a randomly selected subset (20–50) were resampled for animals each year between 1999 and 2008 (Lindenmayer et al. 2003a). The counting procedure used in all surveys was stag-watching. As described in Chapter 3, this involved all large trees with hollows on a site being carefully observed before and after dusk and the number of each species of arboreal marsupial emerging from their nest sites being counted (Lindenmayer et al. 1991a).

The monitoring program has been designed to provide strong statistical inferences about population trends. Year-to-year partial sampling has allowed short-term temporal fluctuations in population dynamics to be separated from long-term trends such as population declines (Welsh et al. 2000). The probability of site selection was based on the counted number of animals recorded on those sites in previous surveys, including the census year 1997. Overall, the monitoring design allowed for approximately 80% of all sampled sites to remain the same from year to year (Lindenmayer et al. 2003a). A key advantage of the non-standard overlapping and rotating sampling approach was that it enabled a far greater number and range of sites to be included in the monitoring program than if a traditional annual sample and resample method was used. This greatly broadened the kinds of inferences that could be made from the monitoring program. The statistical advantages of non-standard overlapping and rotating is further discussed by Cunningham and Welsh (1996) and Welsh et al. (2000).

The 161 monitoring sites encompassed a wide range of stand conditions that varied in human and natural disturbance history. This ensured the results from the program had broad inference across the study region. The monitoring sites varied in dominant tree species (Mountain Ash, Alpine Ash, Shining Gum), slope angle (inclination 2–38°), aspect, elevation (220–1040 m) and topographic position (gully, midslope, ridge). The number of trees with hollows ranged from two to 30 per site. The monitoring study encompasses sites dominated by different stand age classes including mid 1700s, mid 1850s, early 1900s (1905, 1919, 1926 and 1932), 1939 and 1983. There was a minimum of eight sites in each age cohort, but more stands in 1939-aged forest were examined than other age cohorts (Lindenmayer et al. 2003a). This was because forests dating from the extensive 1939 conflagration dominate the Central Highlands of Victoria and comprise more than 70% of the ash-type eucalypt forest in the region (Chapter 12). Finally, the sites in the monitoring program spanned different land tenures – national parks and wood production forests. This provides an opportunity to quantify landscape-level effects resulting from different land use practices.

Repeated surveys of the vegetation structure on the 161 monitoring sites (Lindenmayer et al. 1997b; Lindenmayer and Wood 2009), coupled with the known history of disturbance patterns, allows testing of hypotheses that attempt to account for observed population declines or increases (if any) at a stand level. That is, the study can relate the patterns of animal abundance at a site with the ecological process of stand structural change over time (e.g. the loss of trees with hollows).

Adaptive monitoring

Aspects of the monitoring program in montane ash forests conform to what has been termed 'adaptive monitoring' (Ringold et al. 1996) – the key questions being addressed and the associated monitoring protocols have been adjusted since the commencement of the program, on the basis of new information. The original protocol for the selection of the subset of sites from the total monitoring pool in any given year was based on targeting sites found to support high numbers of arboreal marsupials, irrespective of species. Analyses of data found limited patterns of persistence of Leadbeater's Possum on such sites from year to year. As Leadbeater's Possum was the species of primary concern for managers, new site selection protocols were adopted based, in part, on ensuring that all sites that supported Leadbeater's Possum in a given year were monitored in the following year. This increased the chance of identifying population trends of Leadbeater's Possum, without jeopardising the quality of data gathered for other species (Lindenmayer et al. unpublished data).

RESULTS OF THE MONITORING PROGRAM

A key part of the monitoring program has been to link patterns of long-term abundance of arboreal marsupials to changes in the abundance of trees with hollows. Prior to the instigation of the monitoring program, it was possible to make predictions about the rate of decline in animal abundance that would accompany predicted declines in the abundance of trees with hollows (Lindenmayer et al. 1997b). This was because of existing data on temporal declines in the abundance of trees with hollows in the Central Highlands region (Lindenmayer et al. 1997b) and significant relationships between the abundance of trees with hollows and the abundance of arboreal marsupials (Lindenmayer et al. 1994b; Lindenmayer et al. 1991d) (Chapter 6).

Figure 17.1 shows the forecast decline in abundance of Leadbeater's Possum, Greater Glider and Mountain Brushtail Possum. All three species demonstrate quite precipitous declines in response to the loss of trees with hollows. These predicted declines were published in the late 1990s and have formed the basis for comparison with the results of ongoing animal population monitoring (Lindenmayer and Wood 2009).

Preliminary analyses of field data gathered during monitoring work over the past two decades indicated that the magnitude of forecast declines in populations of arboreal marsupials have not been realised. Although the collapse of trees with hollows has been substantial, this has not translated to major declines in populations of cavity-dependent animals. There have been reductions in populations of some species, such as the Greater Glider, but others have remained unchanged. Very unexpectedly, the Yellow-bellied

Table 17.2: Estimated abundance of arboreal mammals in each year, standardised to 6.5 trees with hollows per monitoring site

Season	Leadbeater's Possum	Sugar Glider	Yellow-bellied Glider	Mountain Brushtail Possum	Greater Glider	All arboreal marsupials
1997/98	0.54 ± 0.12	0.28 ± 0.06	0.00 ± 0.00	0.66 ± 0.13	1.62 ± 0.19	5.01 ± 0.37
1998/99	0.73 ± 0.14	0.17 ± 0.05	0.19 ± 0.05	0.61 ± 0.12	0.70 ± 0.12	3.95 ± 0.32
1999/2000	0.17 ± 0.08	0.23 ± 0.07	0.07 ± 0.04	0.78 ± 0.18	1.34 ± 0.21	3.68 ± 0.39
2000/01	0.29 ± 0.11	0.11 ± 0.04	0.05 ± 0.03	1.15 ± 0.20	0.94 ± 0.17	3.56 ± 0.36
2001/02	0.34 ± 0.11	0.04 ± 0.03	0.29 ± 0.07	1.26 ± 0.21	0.83 ± 0.16	3.48 ± 0.36
2002/03	0.64 ± 0.21	0.08 ± 0.05	0.12 ± 0.06	1.25 ± 0.21	0.49 ± 0.16	3.55 ± 0.47
2003/04	0.44 ± 0.15	0.03 ± 0.02	0.15 ± 0.06	0.87 ± 0.20	0.51 ± 0.14	2.98 ± 0.37
2006/07	0.80 ± 0.26	0.36 ± 0.12	0.21 ± 0.09	1.00 ± 0.28	0.64 ± 0.21	4.04 ± 0.57
2007/08	0.88 ± 0.30	0.18 ± 0.09	0.24 ± 0.10	1.23 ± 0.33	0.45 ± 0.19	3.08 ± 0.54

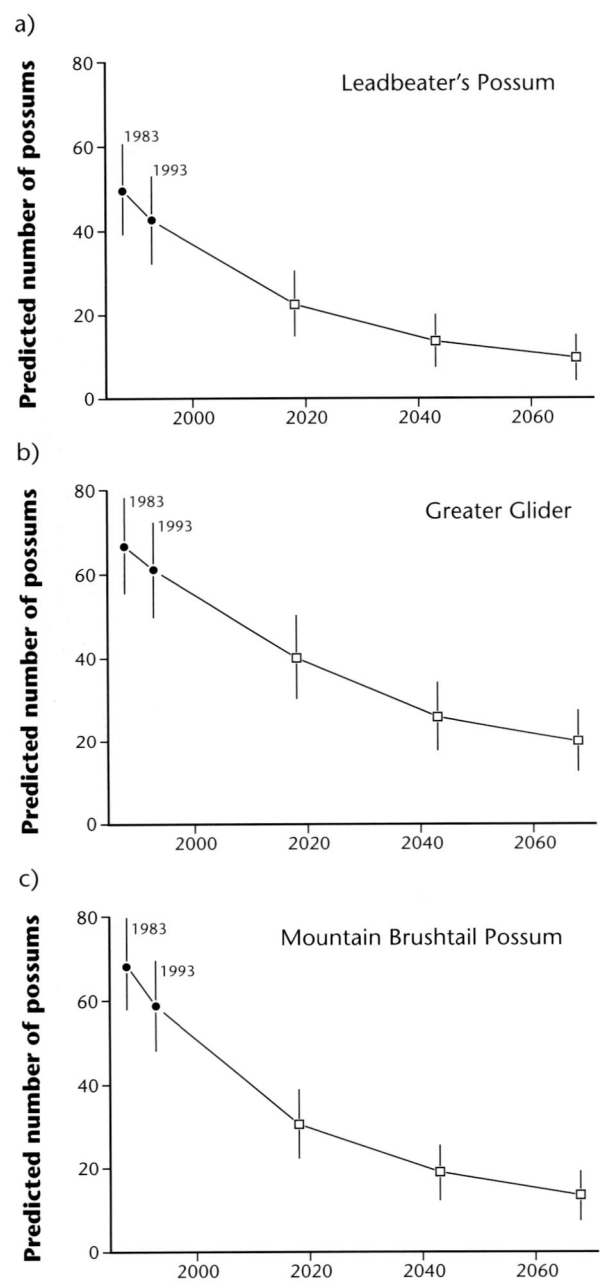

Figure 17.1: Predicted future abundances in response to the decline of trees with hollows in montane ash forests. (Redrawn from Lindenmayer et al. 1997b).

Glider and Mountain Brushtail Possum have actually increased (Table 17.2). It is clear that there are major and often highly significant year-to-year fluctuations in populations of almost all species, except the Mountain Brushtail Possum.

Box 17.1: Adaptive monitoring and the 2009 wildfires

Adaptive monitoring often entails the evolution of a long-term monitoring program through the development of new questions of management relevance. The February 2009 wildfires triggered an urgent need to redirect the long-term monitoring program to address key questions about post-fire ecological recovery. By building on extensive existing knowledge of the biodiversity in montane ash forests and on the long-term (pre-fire) data gathered over the past 25 years, there was a rare but significant chance to establish a powerful rigorous longitudinal study of ecological biodiversity recovery following wildfire. In a true adaptive monitoring approach, key new questions were developed:

1 What are the relationships between pre-fire plant and animal populations and post-fire population trajectories? How does burn severity influence these trajectories, for example, in terms of species richness, community composition and populations of individual species?
2 How are recovery trajectories influenced by initial conditions (stand age at the time of the 2009 fire), the severity of the 2009 fire and the interaction of initial conditions and burn severity?
3 Does burn severity and past fire history differentially influence certain kinds of species, such as functional groups of plants or particular guilds of arboreal marsupials and birds? Are there particular sets of life history attributes common to species which increase rather than decrease over time?
4 Is there a well-defined successional replacement of species within particular groups over time, as predicted by the habitat accommodation model (Fox 1982)?
5 Are there landscape context effects? That is, is the recovery process at a site influenced by the extent of unburned and burned forest surrounding a site?

These key questions set the ongoing basis for the long-term monitoring program and the reasons for ongoing empirical work within it over the coming years.

Figure 17.2: A long-term monitoring site where the new questions in the adaptive monitoring program will focus on ecological recovery following the February 2009 wildfires. The flora and fauna on this and other unburned sites will be compared with that from burned sites. (Photo by David Lindenmayer).

Further analytical work is exploring these long-term patterns and relationships between changes in the availability of trees with hollows and the abundance of different species of arboreal marsupials on the monitoring sites. However, the monitoring program has become more complex given the wildfires in February 2009. It requires refocusing questions towards those associated with post-fire ecological recovery, including the recovery trajectories of arboreal marsupials – adaptive monitoring

THE ULTIMATE CHALLENGE: HOW TO KEEP THE MONITORING GOING?

Countless scientific articles, books, management plans and other documents have been written about the need to conduct long-term studies and monitoring (Canham *et al.* 2003; Franklin *et al.* 1999; Goldsmith 1991; Likens 1989; Shanley *et al.* 2007; Spellerberg 1994; Thompson *et al.* 1998). However, the record of monitoring, both in Australia and globally, is atrocious. It is nearly always the last item funded and the first one cut when funds are limited. Keeping monitoring going is a major challenge that has very rarely been successfully met throughout the chequered history of environmental management anywhere.

There are some broad approaches that can increase the chances that a long-term monitoring study will be successful. The basic attributes of monitoring programs listed above are critical – tractable questions, addressing a problem of management importance, a well-defined set of target responses, and a rigorous and statistically based experimental design (Lindenmayer and Likens 2009). There are other features that the work in Victoria (and elsewhere in south-eastern Australia) suggests are also valuable components of successful monitoring programs.

The first is high-quality staff who can be employed for prolonged periods, to ensure low rates of turnover in personnel. People are actually *the* key infrastructure in a successful monitoring program and are far more important than physical infrastructure such as expensive gadgets. The second is careful attention to databasing and data curation. Good monitoring requires the maintenance of data integrity and data quality over prolonged periods. The importance of high-quality databasing is frequently overlooked in long-term studies. The third is the need for the lead scientists to be productive in formally publishing results of their monitoring project. Publications that have passed the test of peer review have credibility in the scientific fraternity and give a form of scientific status to a monitoring program. This creates pathways to a range of funding for monitoring work and hence diversifies the sources of ongoing financial support. Fourth, lead scientists must continue to pose new questions that are of management relevance. Other studies can be developed which are allied with the monitoring program (e.g. post-graduate student projects); these different strands of research can often be combined to create an overall body of more substantive work that is useful for a given management authority.

The various components of successful monitoring programs require strong personal and professional linkages and institutional arrangements between researchers, resource managers and policy-makers. This is a major challenge for the work in the Victorian forests given the rapid turnover rate of government officials, not only policy-makers but also staff responsible for on-ground management.

The vexed issue of funding

Ongoing funding for monitoring remains a vexed issue. A diversity of funding sources reduces the risk of a monitoring program collapsing if any one funder is unable to continue providing resources. The long-term nature of monitoring programs does not fit with the annual budget cycles of government and grant agencies. Innovative approaches are often necessary to maintain monitoring programs. These can include trust funds or endowments dedicated to monitoring programs, an approach pioneered by Rothamsted Experimental Station in England which has been running for over 175 years (Rothamsted Research 2009). Another possibility is funds provided by levies on timber products (Lindenmayer and Recher 1998) – a strategy that could be linked with the forest certification process.

Given the central role of monitoring as the definitive test of the effectiveness of resource management, as well as its role in assessing compliance with regulatory and market-based goals, solutions to the funding problems associated with the maintenance of long-term monitoring programs must be found.

LINKS BETWEEN THE MONITORING PROGRAM AND OTHER RESEARCH STUDIES

The monitoring program in the montane ash forests is strongly linked with a range of other studies in the forests. The long-term monitoring study has provided a basis on which to establish a new rigorous longitudinal study of ecological biodiversity recovery following the 2009 wildfires, through pre-existing long-term retrospective data and knowledge gained on the 161 permanent research sites established over the past 25 years.

Data from the monitoring program were useful in providing a benchmark for comparison with information generated from the Cutting Experiment (Chapter 16). For example, longitudinal data from the monitoring program highlight expected population trajectories for birds and arboreal marsupials within large areas of 1939 regrowth forest – one of the external controls in the Cutting Experiment.

Data from the monitoring program have been an integral part of the cycle of testing and improving models for population viability analysis (Lindenmayer and McCarthy 2006). As outlined in Chapter 10, linking models and monitoring data led to new insights in the importance of food resources for arboreal marsupials, although this did not substantially change the veracity of management recommendations about the kinds and sizes of reserves needed to promote wildlife conservation within forests broadly designated for wood production (Lindenmayer and McCarthy 2006).

The monitoring program has provided a strong platform on which to build, and/or link to, a range of other important studies in montane ash forests. For example, longitudinal data on the abundance of the Mountain Brushtail Possum, gathered since 1983, are being linked to the ongoing trap–recapture studies of the species at Cambarville that began in 1991 (Banks et al. 2008; Viggers and Lindenmayer 2000). There are plans to instigate a new series of studies on the responses of carnivorous beetles to forest age, stand structure and past disturbance. These studies will add to the overall body of knowledge on montane ash forests and ways to best manage them. They are also ensuring that the financial sponsors are receiving a high level of additional return (in terms of research output) for their investments – a very important consideration, given that taxpayer money is being used to fund the work.

SUMMARY

The long-term monitoring program in the montane ash forests has been underpinned by some fundamental characteristics, particularly the need to answer credible and tractable scientifically based questions that are of management relevance. Many entities could have been monitored but, rather than trying to monitor a laundry list of many things, a small subset of entities was monitored – primarily populations of arboreal marsupials and vegetation structure (particularly the condition and abundance of trees with hollows). These entities were never considered to be indicator species or indicator structures. They were selected to help answer particular questions and because there is a human need to know about them.

The monitoring program has revealed some highly unexpected findings. These include substantial between-species differences in population trajectories over the past 10 years, and changes in populations that do not appear to be strongly linked to reductions in the abundance of trees with hollows.

LESSONS LEARNED

Many valuable lessons have been learned from the long-term monitoring program in the montane ash forests. These have contributed to an understanding of the salient features that underpin a successful monitoring program. There have been other insights, most of which could not have been anticipated when work began in 1983.

First, it is clear that monitoring is as much about statistics as it is about ecology. Monitoring programs are dysfunctional without a rigorous statistical framework. They are also a waste of effort and money, without a rigorous statistical framework. The novel rotating approach to sampling (Lindenmayer *et al.* 2003a) that has been employed would not have been available without the input and collaboration of highly professional statisticians. The vast majority of ecologists are unaware of these novel methods of monitoring programs. Without such a high level of statistical input, a traditional visit–revisit approach would have been used. This would have had substantial limitations, particularly in regard to the numbers of sites in the monitoring program, limited inference across site types and environmental conditions, and lack of capacity to statistically model and predict populations over time.

The second lesson that has been learned (related to the first) is that, with sound statistics, it has been possible to transform the initial cross-sectional studies completed between 1983 and 1993 to a longitudinal investigation which has now become the long-term monitoring program. That has meant inferences could be made over 25 years (1983–2008) rather than 11 years (1997–2008), which is when the formal rotating sampling approach was first implemented (Lindenmayer *et al.* 2003a). The importance of this approach has been strengthened by the redirected focus of the monitoring program on answering key questions about ecological recovery and biodiversity responses following the February 2009 wildfires.

A third important lesson from the monitoring program has been that currently common species, such as the Greater Glider, can be among those at risk of medium-term decline. Conversely, those that are uncommon may show no discernible increasing or decreasing trend. This is a common misconception in ecology – that rare or uncommon species are also declining species. Rare species and declining species may not be the same.

A fourth lesson is a sobering one, about the amount of effort required to maintain a long-term monitoring program. It has proved immensely difficult to garner enthusiasm and financial support from government to maintain the monitoring program. Despite the fact that monitoring is crucial for the assessment of the effectiveness of management actions and the credibility of sustainability schemes like forest certification, the levels of scepticism about the value of the monitoring program remain high within many sectors of government – both state and federal. It is perhaps a sad reality that the value of such kinds of monitoring projects is only recognised after they no longer exist.

Connections between the monitoring program and other key projects such as the Cutting Experiment and the use of PVA models are important. Links to management mean that the work more than qualifies as true active adaptive management. Active adaptive management has been discussed for more than two decades as the best approach to manage natural resources. Despite the number of advocates for the approach, on-ground examples are remarkably rare (Stankey et al. 2003; Wintle and Lindenmayer 2008). The monitoring program and allied projects, such as the Cutting Experiment, need to be maintained for the overall body of work to continue to qualify as true active adaptive management.

KNOWLEDGE GAPS

One of the key knowledge gaps associated with monitoring is not a scientific research question, but a social science and financial one. What is the best way to generate the long-term funding needed to maintain monitoring studies so they best inform natural resource management? The funding issue is a perennial one that has thwarted the establishment of countless good monitoring programs and led to the premature termination of many others. Novel ways to solve the problem have been suggested, such as levies, trust funds and endowments, but innovation through actual funding programs has not been forthcoming in Australia, although there are some useful examples from North America and Europe (Strayer et al. 1986; Rothamsted Research 2009).

FOREST PATTERN, ECOLOGICAL PROCESS AND LINKS TO OTHER CHAPTERS

The monitoring program in montane ash forests has involved efforts to quantify the ecological process of tree fall and hence temporal changes in trees with hollows on sites (Lindenmayer and Wood 2009), temporal changes in the pattern of animal presence and abundance on sites (Lindenmayer et al. 2003a) and relationships linking this process and pattern. The objective is to make inferences about the process of tree fall across the broader forest extent, that is, to extrapolate beyond the sample pool of sites where repeated measurements of tree condition are made (Lindenmayer and Wood 2009), including across different land tenures (e.g. wood production forests versus protection areas). This is also the case for animal populations, hence the need to make broad inferences across the Central Highlands region *per se*. These relationships will be examined even more deeply as part of ongoing monitoring aimed at quantifying the trajectories of post 2009 wildfire recovery, particularly among species of arboreal marsupials.

Links to other chapters

The long-term monitoring program is strongly linked to several chapters in this book. The program's origins and ongoing remit are firmly anchored in an understanding of the distribution and abundance of target species – the primary topic of Chapter 9. However, its aim has been to quantify temporal patterns rather than substituting space for time in the largely observational sets of studies which feature in Chapter 9.

Work linking long-term monitoring data and PVA models was discussed in Chapter 10. The work on communities in Chapter 11 is instructive for some of the discussions in this chapter, because it highlights why the indicator species concept is deeply problematic and therefore an inappropriate way to select target species for monitoring.

Temporal changes in the structure and composition of the forest are measured as part of the monitoring program and the information used to help interpret longitudinal changes in animal populations. This emphasises the links between this chapter and the topics of forest overstorey discussed in Chapters 4 and 6 and the understorey in Chapter 7. Chapters 12 and 14 are also connected to this chapter because natural and human disturbance are important drivers of change in montane ash forests. Finally, attempts to mitigate the impacts of logging through studies such as the Cutting Experiment highlight the levels of cross-fertilisation between the monitoring program and other projects (see Chapter 16) and underscore what is needed for adaptive management.

PART VII

CONCLUSIONS AND FUTURE DIRECTIONS

18 Conclusions and future directions

INTRODUCTION

The final part of this book comprises just one chapter, which discusses a range of broad themes associated with the work to date. The first part of this chapter is dedicated to discussing the role of two fundamentally important groups of people who have been pivotal to the success of the work in the montane ash forests – expert professional statisticians and volunteers. The second part focuses on the importance of long-term ecological research and the challenges of maintaining such work over prolonged periods. The third section of this chapter outlines some emerging issues in the management of montane ash forests. These include the impacts of climate change and the role of ash forests in carbon storage. Two topics appear to be especially important. The first is the importance of improved fire management in a carbon-constrained economy. The second, which is related to the first, is the vast stores of carbon sequestered in montane ash forests. The importance of carbon storage could ultimately mean a major transition from farming trees (logging and wood production) to farming carbon, with substantial implications for the forest industry. Such changes to the forest industry should, in many ways, not be especially surprising, as forest debates will never go away. There will always be conflicts over forest values and forest management. The last part of this chapter sets out some important research themes and research questions that need to be examined in the coming decades of work.

One of the objectives of this book is to underscore the importance of ecological synthesis and how it has underpinned the evolution of the body of work undertaken in montane ash forests over the past 25 years. Two aspects of synthesis are particularly pertinent to the concluding section of this book. The first is the complementarity of single-species and ecosystem-based research, and how they can generate new insights not possible from studying either one in isolation. The second, strongly related to the first, is the importance of studying ecological patterns and ecological processes, and the relationships between them. This greatly strengthens the understanding of how a forest is structured, how it functions and how it responds to processes like disturbance regimes. Indeed, this final section of the book emphasises why the volume was given the title it was, and how the twin foci of forest pattern and ecological process became a useful framework in which to illustrate the links between the various studies completed in montane ash forests over the past 25 years.

THE CRITICAL ROLES OF STATISTICAL SCIENCE

This book has synthesised various ecological research studies completed during the past 25 years. Statistical science has been at the heart of all this work and virtually all projects have been underpinned by collaboration between expert accredited professional statisticians and ecologists. Indeed, statistical science has played key roles in the key phases of each project – experimental design, developing methods for data collection, data organisation, data analysis, data interpretation and writing for scientific publications.

The collaborative approach has been fundamental to the quality of the work for many reasons. First, the discipline of statistics is arguably evolving much faster than that of ecology, making it possible to analyse datasets in ways that were impossible even a handful of years ago. Second, expert professional statistical input has allowed methods of data analysis to be applied that were not previously used in ecology. For example, zero-inflated Poisson (ZIP) regression was originally applied in analysing rare events in fields, such as product defects in production engineering and the risk of shipwrecks. This approach had clear applications to modelling the occurrence of rare species such as Leadbeater's Possum (Cunningham and Lindenmayer 2005; Welsh et al. 1996). Its use may have been overlooked in ecology without the input of professional statisticians. Similarly, the rotating method for site selection from the pool of monitoring sites was a comparatively new statistical approach to monitoring and had rarely been used in ecology. This innovation moved the monitoring program beyond a traditional one of simply revisiting the same set of sites annually (or at another set interval) to include a much larger number of sites in the monitoring program, thereby broadening the range of conditions being monitored and expanding the inferences that could be made (Lindenmayer et al. 2003a) (Chapter 17).

Other reasons why the collaborative approach between ecologists and statisticians has been vital to the quality of the work are:

- it forced a precise and rigorous posing of questions prior to the commencement of field work – something that ecologists are not often good at doing (Peters 1991; Shrader-Frechette and McCoy 1993);
- it facilitated dummy data analyses at various stages during each project to determine if the extent of data collection was sufficient to address the key questions being asked;
- it enabled each project to be statistically 'audited' at the end of each field season. This helped determine, for example, whether it was appropriate to continue field work on a particular project for additional field seasons (in subsequent years) or halt field work and commence detailed statistical analysis.

THE CRITICAL ROLE OF VOLUNTEERS

Volunteers were a second core group of people crucial to the success of the work in the montane ash forests. They continue to be involved in a range of studies, particularly through the support of organisations like Earthwatch Institute, that brings together groups of volunteers from around the world. The stagwatching program is undoubtedly the most important project to which volunteers contribute. Because arboreal marsupials nest in a number of different trees with hollows and swap between those trees on a regular basis (Lindenmayer et al. 1996c) (Chapter 6), it was imperative that all marked trees with hollows be stagwatched simultaneously. Failure to do this would have resulted in over- or undercounting of arboreal marsupials, with implications for subsequent work such as identifying the habitat requirements of particular species. The only way to simultaneously observe all trees with hollows on a field site was to use the assistance of trained volunteer observers. Over 7000 volunteers have participated in stagwatching programs in the past 25 years. In return, it has been possible to offer the stagwatching program as an environmental education opportunity and to communicate the importance of biodiversity conservation as a core part of ecologically sustainable forest management (Lindenmayer et al. 1991a).

There are some important caveats in working with volunteers. First, they clearly do not have (nor should be expected to have) the skills of professional scientists. This is important because data collection by volunteers can affect overall data quality if it is not carefully managed. If researchers are not aware of these problems, perceived trends in data may not be actual trends but artifacts of the data-gathering process. This was always given careful consideration when using volunteers in various projects in montane ash forests. The effectiveness of volunteer work was ensured by providing a well-conceived scientific framework within which volunteers were assigned tasks, and allowing volunteers to participate only in tasks for which they could be quickly trained and for which small levels of observer heterogeneity could be tolerated without breaching overall data integrity.

CHALLENGES OF MAINTAINING LONG-TERM RESEARCH PROJECTS

It is almost a truism to say that without good knowledge and understanding of a natural resource, it is simply not possible to manage that resource in an effective and ecologically sustainable way. For example, the long-term monitoring program outlined in Chapter 17 will be critical to credible forest certification – if that can be obtained for the forest industry in montane ash forests. Yet very few of Australia's (or the world's) natural resources are managed using good scientific knowledge, and the effectiveness of management actions are almost never monitored. The long-term monitoring program will also be crucial for documenting the ecological recovery of montane ash forests and populations of biota in these forests following the February 2009 wildfires (Chapter 17).

There are many challenges in maintaining a long-term forest research and monitoring program. Three of these are outlined below, to illustrate some of the hurdles that can arise and might hopefully be avoided by others embarking on long-term, large-scale research studies.

Data curation and management
The maintenance of high-quality datasets is a pivotal part of any successful long-term ecological project. This sounds like a trivial point to most scientists. But the reality is that the curation of data is often an afterthought in the vast majority of ecological projects. Indeed, the author has personally witnessed several cases where government agencies discarded high-quality datasets that, not long after, were recognised to be extremely important.

For the past five years, a part-time database specialist has been employed to manage the data gathered from various studies in Victoria (and from five other large-scale studies being conducted in south-eastern Australia). Data are added on a three- to six-monthly basis as additional field surveys are completed. The input of data requires careful checking, otherwise error rates average 7–10% of the total imported information.

Data storage remains a serious problem. For example, over the duration of the various projects in Victoria, the evolution of computerised data storage methods has included magnetic tapes, 5½-inch floppy disks, 3½-inch floppy disks, CDs, burnable DVDs and memory sticks. The time to redundancy of these storage methods has been steadily reducing. The development of new data storage methods has required regular updating, a process that can lead to errors being introduced to datasets, or to data loss.

A key revelation has been the importance of primary data in the form of pencil entries on paper copies of field datasheets. Many errors in digital datasets were rectified only through cross-referencing with original field datasheets. The orderly storage of vast quantities of paper records is important, as is its duplication and storage off-site in case of an unforeseen event such as an office fire – a fate not unknown.

Maintenance of a field presence
Many of the studies summarised in this book have been characterised by the collection of high-quality empirical data gathered through repeated surveys from many permanently established field sites. The amount of field work involved in these studies is considerable. For example, the monitoring program in the montane ash forests has involved many visits to each of 161 field sites over many years. Similarly, there have been repeated visits to the 192 nest boxes erected as part of the nest box program (Chapter 6) (Lindenmayer *et al.* 2009c).

One lesson from running these projects has been that the most effective way to maintain them is to station a person permanently within or close to the study region. This has been the only way to complete the extensive field sampling associated with these large studies. This approach also has the advantage of ensuring close interactions with staff from the agencies responsible for land management in the area. This has facilitated the rapid communication of results and can highlight the relevance of science to resource management. Regular contact with agency staff has become increasingly important in recent years because of high rates of staff turnover within those agencies. Recently, communication with government agencies has become even more critical to ensure that monitoring sites within burned forests are not subject to salvage logging and that crucial information on ecological recovery following the Black Saturday 2009 wildfires is not lost.

Funding

Maintaining an ongoing field presence and constantly upgrading databases requires high-quality staff. This requires considerable funding. Securing research funds is the bane of virtually all researchers' lives. Generating funds to maintain the work in the Victorian ash forests has been no exception. More than 85 grant proposals to over 25 different funding bodies have been submitted over the past 25 years. Most were unsuccessful. Funding becomes more difficult to obtain as a large program progresses because the emphasis of most funding bodies is on 'new work'. While new questions can be addressed in a location where research has been underway for some time, few funding bodies are willing to support long-term work that has been focused on the same area for a prolonged period (see Haynes *et al.* 2006; Laurance *et al.* 2007; Likens 1989 for rare exceptions). This is unfortunate, because the most interesting findings often emerge only when a project has been running for a prolonged period.

One approach to tackling the problems associated with the need for new research by funding bodies has involved using the monitoring program and its 161 permanent field sites as a broad frame on which to 'hang' many other kinds of studies, including focused postdoctoral research programs. As outlined in Chapter 17, the long-term monitoring program has been used to answer key new questions about post-fire ecological recovery. This has the advantage that true value-adding can be generated when insights from several strands of research are integrated.

Ongoing funding remains the most significant threat to the maintenance of the work in Victoria. In the unfortunate event that the research program is terminated through a shortage of funds, strict attention to data curation and database management have tried to ensure that field information is stored in a way that allows the work to be recommenced by another researcher in the future if funding should again become available.

The Victorian study region is part of the Australian Long-term Ecological Research (LTER) network, an informal alliance of a small number of research sites nationwide. Nominating the Victorian ash forests for LTER status involves instigating succession planning so that another researcher may take over the work in the future. This is a pivotal point, as the vast majority of long-term studies are terminated when the 'champion' for them is no longer involved (Lindenmayer and Franklin 2002; Norton 1996). Succession planning is often contemplated too late in the career of a senior scientist, leaving major long-term projects without a mid-career researcher to maintain them. Succession planning has commenced for the work in the Victorian ash forests, particularly because the 2009 wildfires have reset the trajectory of the research for the next 25 years.

ECOLOGICALLY SUSTAINABLE FOREST MANAGEMENT

Ecologically sustainable forest management is the holy grail of forest management worldwide and has been a major focus of forest policy-makers in Victoria (Department of Sustainability and Environment 2006, 2007). However, it is an elusive goal because ecological sustainability is a pathway, rather than a readily and precisely definable target (Lindenmayer and Franklin 2003). Despite this, the work in the montane ash forests has the potential to provide the core of ecologically sustainable forest management. This is because of the long-term commitment to forest research and to long-term ecological monitoring. Knowledge of forest resources and an understanding of the effectiveness (or otherwise) of management actions through research and monitoring is fundamental to improving forest management and hence creating the pathway for ecologically sustainable forest management. Maintenance of commitments to research and monitoring should be among the key prerequisites to gaining certification for montane ash forest management through potentially credible bodies such as the Forest Stewardship Council.

Conversely, without continued investments in research and monitoring, claims of ecological sustainability will be just that – claims – without any substantiation. The integrity of the montane ash forest estate may ultimately be at risk, particularly through the impacts of rapid climate change. Without a guarantee of continued investment, certification of forest management practices should not be granted.

ASH FORESTS IN A CHANGING CLIMATE

Few scientists doubt the existence of major and rapid climate change (Rosenzweig *et al.* 2008). Australia will not be immune from these changes. Rather, the continent may well be one of the worst-affected areas on Earth. The wet forests of south-eastern Australia may be especially vulnerable to the effects of rapid climate change, particularly through the impacts on altered fire regimes of higher temperatures and reduced rainfall (Cary 2002; Nitschke *et al.* 2009). Two high-severity stand-replacing fires within 25 years in montane ash forest would eliminate ash trees and result in them being replaced by *Acacia* spp. This would have massive implications not only for the timber industry but for many elements of biota, including the largely ash-dependent Leadbeater's Possum. This possibility is of great concern, particularly given the extent and severity of the February 2009 wildfires and the risks (due to ongoing drought and increased temperatures) of another major conflagration before 2030.

Changes in temperature and rainfall regimes are likely to result in reduced ranges of ash-type forests and of animals such as Leadbeater's Possum (Brereton *et al.* 1995; Lindenmayer *et al.* 1991f). Notably, genetic studies contain strong evidence of past population bottlenecks in species such as Leadbeater's Possum (Hansen *et al.* 2008), most likely as a result of changes in climate conditions that characterised the Pleistocene period, including the glacial maximum approximately 15 000 years ago. It appears likely that

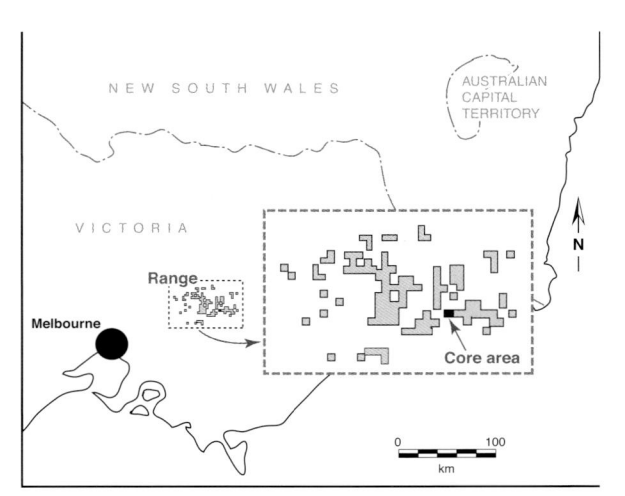

Figure 18.1: Predicted reductions in the distribution of Leadbeater's Possum as a consequence of climate change. (Based on modelling by Lindenmayer *et al.* 1991f).

populations of this and other species will be subject to similar demographic and genetic challenges in the future. Irrespective of the magnitude of these changes, the potential for alterations in species' ranges also underscores the importance of reserve and off-reserve conservation management strategies, as discussed throughout Part V of this book.

Climate change may have other subtle, but nevertheless profound, impacts on montane ash forests. For example, the regeneration niche of montane ash trees may be altered by rising temperatures (Nitschke *et al.* 2009). Ash-type tree species may be highly vulnerable to climate change because the regeneration phase is often a critical phase for plant survival (Bell 1999). Similarly, changes in concentrations of atmospheric carbon dioxide may alter the chemical composition of eucalypt leaves, with knock-on impacts for folivorous animals such as the Greater Glider (Kanowski 2001).

Not all effects of climate change may be negative. For example, rates of decay and collapse of trees with hollows appear to have slowed in montane ash forests, perhaps as a result of lower levels of rainfall in the past decade or so (Lindenmayer and Wood 2009) (Chapter 6). Such findings highlight the need to carefully monitor populations over the coming decades so that pre-emptive conservation and management strategies can be implemented if dramatic collapses in species'

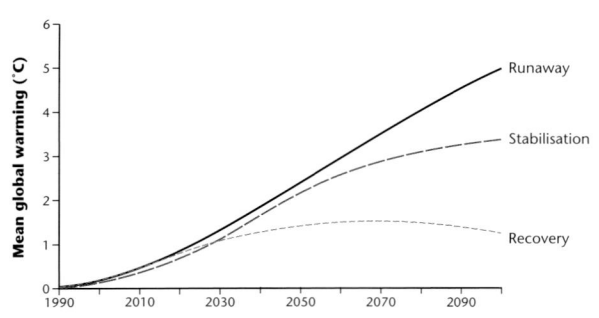

Figure 18.2: Three scenarios of future climate: recovery, stabilisation and runaway. Drawn using the (IPCC 2007) suite of scenarios driven by different assumptions about how humanity responds to the challenge of mitigating CO_2 emissions in the next century. The shorthand names range from the most optimistic 'recovery', through the more realistic 'stabilisation' to the pessimistic 'runaway' (the world's current track). These three scenarios inform risk management against an uncertain future when assessing management options for long-lived biological infrastructure. The scenarios highlight that the Earth is committed to a further warming of at least 0.4°C regardless of human actions, owing to the in-built momentum in the climate system. Thus, the trends for the three scenarios are identical up to approximately 2030. Beyond that, the rate and degree of global mitigation action will determine the alternative possible futures.

ranges begin to take place (see Box 18.1). Indeed, the imperative to embrace rigorous long-term monitoring will become increasingly important as the effects of rapid climate change accumulate. Attempts to best manage the effects of climate change will be largely dependent on the magnitude of the effects of rapid climate change (Figure 18.2).

THE CARBON ECONOMY AND MANAGEMENT OF MONTANE ASH FORESTS

Several strategies have been proposed to tackle the problem of rapid climate change via reducing the net emissions of carbon dioxide and other greenhouse gases to the atmosphere. One of the most prominent is to sequester carbon into the terrestrial biosphere, principally by maintaining and/or planting vegetation (Schulze *et al.* 2000; Strengers *et al.* 2007).

Incentives to sequester carbon through maintaining existing areas of native vegetation and planting new areas of vegetation are likely to increase as the impacts of climate change become more pronounced and intense, and as ecological systems and processes are further modified by climate change (Read 2007). Economic instruments such as carbon taxes and carbon trading markets will become increasingly common (Strengers *et al.* 2007). Income from carbon offsetting and the development of a carbon economy has the potential to lead to major changes in land management (Fargione *et al.* 2008).

The montane ash forests of the Central Highlands of Victoria have the potential to play a very significant role in the carbon economy, because they are some of the most carbon-dense forests on earth (Mackey *et al.* 2008). Indeed, studies completed just before the 2009 wildfires showed that stands of multi-aged and old-growth montane ash forest in the O'Shannassy Water Catchment supported up to 2100 t of above-ground biomass per ha, several times higher than the IPCC (2007) default values for temperate forests. Thus, montane ash forests have an enormous capacity to store vast quantities of carbon.

With emissions trading, a possible scenario is that carbon prices of upwards of $A20/t (at 2009 value) could result in the value of the forest as stored carbon significantly outstripping the royalty returns from timber and pulpwood production. In other words, in a carbon economy, the Victorian Government might receive very large financial returns from *not* logging montane ash forests and instead growing them through to an old-growth stage – with returns far in excess of those from the forest industry. An alternative scenario is that subsets of the Central Highlands region with the highest probability for growing forest through to a carbon-dense old-growth stage might be those targeted for carbon storage. This recognises that some parts of the montane ash forest estate are less likely to be disturbed or subject to high-intensity disturbances than others (Chapter 12). Thus, it might be appropriate to focus carbon storage objectives in parts of montane ash landscapes where disturbance-related analyses indicate they will be most effective (Mackey *et al.* 2002).

Either scenario could have enormous implications for the way montane ash forests are managed. The value of native forests as a carbon storage asset, from both financial and climate change perspectives,

Box 18.1: Climate change scenarios, risk-spreading and conservation strategies

In many ecosystems such as forests and woodlands, long-lived tree species have critical roles as key ecological structures (Lindenmayer and Franklin 2002; Manning et al. 2006). The investments made today in the management of these structures (and hence the large quantities of biodiversity associated with them) can have implications that span centuries, just as our investments in power stations and dams today create infrastructure legacies that last many decades. Yet it is not possible to know what the climatic conditions will be in 100–200 years, because they depend on our mitigation actions in response to the challenge of climate change. It is important to consider management strategies that hedge bets against these different futures. This can be done by identifying the responses that might be taken in response to a variety of future climate scenarios (Figure 18.2). If all scenarios require the same responses, this is a 'no regrets' option and management actions should commence immediately. But if each scenario requires different actions, it is critical to consider implementing multiple actions in the same region.

The montane ash forests provide a useful hypothetical example. As outlined in Chapter 6, montane ash trees do not develop tree hollows suitable for occupancy by more than 40 species of vertebrates for 120–150 years (Lindenmayer et al. 1993b). Exemplar species are the Yellow-bellied Glider and the Yellow-tailed Black Cockatoo (Figure 18.3). A temperature rise of 2–3°C in itself would not prevent the survival of such vertebrate species, as both have extensive distributions covering large parts of eastern Australia. However, they will decline rapidly if they do not have access to tree hollows for nesting and sheltering (Gibbons and Lindenmayer 2002; Nelson and Morris 1994), as would happen if tree species such as Mountain Ash fail to recruit in the face of climate change (Nitschke et al. 2009).

In the recovery scenario, it is important to keep existing populations of tree species alive *in situ*. This is because by 2100 (well within the lifespan of some of the trees germinating today), the climate will return to conditions that the Mountain Ash can tolerate. For this scenario, it is possible to identify key refugia and work hard to keep them free from fire and other disturbances (Mackey et al. 2002). This approach aims to maintain ecosystem resilience.

In the stabilisation scenario, it might be appropriate to establish tree species more typical of lower and warmer elevations, such as Messmate, in areas where Mountain Ash lives today. When the species matures in 100–200 years, trees will provide suitable hollows in areas where Mountain Ash can no longer establish. Under this scenario, it may be appropriate to actively establish Messmate (e.g. through deliberate planting or on-site releases of seed) in areas burned naturally over the coming decades. This approach would encourage directed transformation.

In the runaway scenario, the dominant tree species will continue changing over coming centuries, first to Messmate then to other species characteristic of lower and warmer environments (e.g. Red Stringybark and Narrow-leaved Peppermint). For this scenario, it may be appropriate to encourage the natural movement of species across the landscape to ensure a perpetual supply of large, long-lived trees species that can form hollows. This approach would promote self-organising transformation of the vegetation community.

Although it is not currently possible to know which scenario is actually going to happen, planning needs to occur now. It is possible to plan for such uncertainty by applying all three strategies in different places in the Central Highlands of Victoria. For example, a recommendation might be that one part of the Yarra Ranges National Park follows the recovery strategy, particularly in areas with many good refugia (Mackey *et al.* 2002), another part specifically encourages the recruitment of Messmate, and a third part allows unrestrained recruitment of any long-lived tree species. If an active adaptive management approach is employed (Walters 1986), this multi-faceted decision can be re-examined in 30–50 years as it becomes clearer which scenario is really playing out. Strategies could then be refined.

In contrast to the hypothetical case study for montane ash forests, other vegetation types will face conservation issues where the same management strategy will be needed, irrespective of which future scenario manifests. For example, this may be

> the case for short-lived vegetation types, such as coastal heathlands, where managing disturbance and other threats may be appropriate regardless of the future (Woinarski 1999). In this case, the management strategy is one of 'no regrets' – it is essential to get on with doing the same things in all places with this vegetation type.
>
> This kind of risk management using scenario-based thinking needs to become an imperative part of conservation planning, particularly as the future will not be known until we are living it. It is simply not a realistic option to wait for certainty before making decisions about biodiversity conservation.

means that particular attention would need to be paid to excluding fire from as much forest as possible and for as long as possible. There are good carbon storage-related reasons not to salvage-log large areas of montane ash forests after the February 2009 wildfires, particularly areas that formerly supported old-growth forest. The pre-2009 studies showed that the highest levels of above-ground carbon biomass occurs in places where there has been past wildfire but where salvage logging has not occurred (Keith *et al.* 2009). Another is that fire-scarred or fire-killed trees can remain standing for over 70 years in montane ash forests (Lindenmayer *et al.* 1997b; Lindenmayer and Wood 2009) (Chapter 9).

In summary, a carbon-constrained economy in a world subject to rapid and pronounced climate change could have major implications for how montane ash forests are managed. Such implications may result in a wholesale change in the way different values (e.g. carbon storage) are perceived and assigned economic worth (Mackey *et al.* 2008). Changes in native forest management would have substantial social and economic ramifications and any transitions in forest use would need to be done in a socially just way to ensure that displaced workers can find employment in other sectors.

FUTURE WORK

The final section of almost all the chapters in this book has outlined knowledge gaps which need to be addressed. The aim of this section is not to repeat these gaps as an extended shopping list of additional research tasks. Rather, it will highlight four areas of work that received limited attention in the preceding 17 chapters.

Post-fire ecological recovery

This book was almost completed when the February 2009 wildfires burned large areas of the Central Highlands of Victoria. This was a tragic and monumental event of enormous proportions that will influence human communities and ecological communities for centuries. However, the fires also offered a critical opportunity to study post-fire ecological recovery. Rigorous studies of ecosystem recovery following major catastrophic natural disturbances can radically alter ecological thinking (Franklin and MacMahon 2000), lead to the development of new ecological the-

Figure 18.3: Yellow-tailed Black Cockatoo. (Photo by David Cook/David Cook Wildlife Photography).

Box 18.2: Could the carbon storage values for Mountain Ash forests really be true?

Mackey *et al.* (2008) and Keith *et al.* (2009) published data on carbon storage in native forests in a range of locations in south-eastern Australia. The results of analyses for montane ash forests have proved highly controversial. Values exceeding 2000 t/ha make montane ash forests the highest biomass carbon forests on earth (Keith *et al.* 2009). These values were obtained from detailed surveys of 318 plots on 53 sites within old-growth and multi-aged forests in the O'Shannassy Water Catchment where there has been minimal historical human disturbance, both by Europeans and indigenous people. The results have been questioned by lobbyists from the timber industry. Could the values be true, or is there a major error in the calculations?

The values appear to be valid for several important reasons. First, they are based on actual field-based measurements rather than simulation models of carbon biomass, which is the usual approach to examining carbon dynamics. Second, they include not only the overstorey trees but also the understorey and shrub layers as well as the ground layer – these carbon stores were often ignored in other stand-level analyses of carbon. In fact, past work on logs and coarse woody debris indicate that measured log volumes for many field sites were high (~350 m^3/ha) and at some sites log volumes exceeded 1000 m^3/ha – a value higher than that calculated for many other temperate forest types around the world (Lindenmayer *et al.* 1999d). Third, even seemingly small contributors to carbon storage values in most vegetation types can return very high measured tonnages in montane ash forests. For example, Ashton (1975c) found that litter fall alone averaged 7.66 t/ha annually.

Why are the carbon storage values in Mountain Ash forests higher than in other forests around the world? Nine factors spanning environmental conditions, tree species characteristics, vegetation community dynamics and land use history appear to underpin the extremely high values of biomass carbon for Mountain Ash forests. First, the climatic and soil conditions favour high plant productivity. Mountain Ash grows rapidly in this environment (>1 m/yr for the first 70 years) and eventually becomes the world's tallest flowering plant. Second, decay rates of dead standing trees and coarse woody debris are slow because of cool temperatures. Third, eucalypt foliage is evergreen and trees can grow all year in this environment. Fourth, wood density is high (500–600 g/m^3, 60% greater than some of the world's other large trees such as the Giant Sequoia (*Sequoia gigantea*). Hence, the large tree dimensions in Mountain Ash stems convert to very high amounts of carbon biomass. Fifth, limited crown development in Mountain Ash (through crown shyness) and isolateral leaf form enable high levels of light to penetrate to the forest floor, allowing luxuriant mid- and understoreys to grow. Sixth, stands of Mountain Ash with these high values of biomass carbon density were at least 100 years old, and the species is reproductively fertile for about 400 years. Seventh, past partial stand-replacing wildfires have produced younger cohorts of fast-growing eucalypt trees, left an older cohort of living and dead trees and rejuvenated the understorey of *Acacia* spp. and other tree species. Eighth, forests that have been little influenced by human activities can be multi-aged, and do not necessarily consist exclusively of old trees. The forest ecosystem can have a net uptake of carbon and accumulate carbon stocks over long periods. The final reason for high biomass carbon values is a prolonged absence of direct human disturbance. The O'Shannassy catchment has been closed to public access for over 100 years to provide water for Melbourne and had an almost complete absence of Aboriginal use. Natural disturbances have included wildfire, windstorms and insect attacks. Post-wildfire salvage logging, which removes large amounts of biomass in living and dead trees and thus prevents the development of multiple age cohorts, has been excluded. Therefore, when old-growth stands are disturbed, dead trees remain standing or persist as fallen logs and retain most of their carbon because decomposition is slow – the biological legacies concept (Chapter 12). New cohorts of trees are added to the forest and the understorey is rejuvenated.

Perhaps part of the controversy concerning carbon stocks in montane ash forests lies in confusion between carbon sequestration and carbon storage. There is no doubt that fast-growing young

> stands are sequestering carbon faster than slower-growing old-growth forests. However, the *amount* of carbon stored in old-growth forests appears to be considerably greater than that stored in younger forests. Working through an overall carbon budget for forests requires that all the stocks and flows of carbon be properly examined and quantified, including what happens to forest products once wood is processed. For example, ~80% of all wood removed from montane ash forests is used in paper production and/or becomes export woodchips. This is a short-term form of carbon storage compared to other uses such as the crafting of timber for high-quality furniture. The analyses completed by Mackey *et al.* (2008) and Keith *et al.* (2009) indicate that truly comprehensive climate change mitigation strategies must quantify carbon stocks in native forests accurately and ensure that the value of this carbon is recognised by national and international markets and global regulatory regimes.

ories and concepts as well as reshape existing ones (Connell 1978) and markedly alter human resource management practices (Turner *et al.* 2003; Hutto 2008). This was the case in Yellowstone National Park and at Mount St Helens following the 1988 wildfire and 1984 volcanic eruption, respectively (Franklin and MacMahon 2000; Turner *et al.* 2003; Lindenmayer *et al.* 2008a).

As in the case of the long-term studies at Mount St Helens and in Yellowstone National Park, it is highly likely that many ecological surprises will emerge from the careful study of ecological recovery after the 2009 wildfires in the Central Highlands of Victoria. The work will provide an important platform from which to test many of the ideas summarised in this book, about spatial variation in the severity and effects of natural disturbances (Chapter 12) and relationships between stand structure and wildlife habitat (Chapters 6 and 9). It will also provide an excellent opportunity to test key ecological theories and concepts about biotic responses to major disturbances, including:

- the initial conditions hypothesis (Egler 1954), that response is a function of the state of an environment at the time of disturbance;
- the intermediate disturbance hypothesis (Connell 1978; Shiel and Burslem 2003), that species richness should be highest at intermediate severities and frequencies of disturbance;
- the biological legacies concept (Franklin *et al.* 2000);
- the landscape heterogeneity and mosaic hypothesis (Bennett *et al.* 2006; Parr and Andersen 2006), and the importance of patchiness for influencing species richness;
- the habitat accommodation model of well-defined successional change in the elements of biota which comprise ecological communities (Fox 1982).

In some cases, previous ideas about the effects of natural disturbances and habitat relationships will be proved wrong and new concepts and theories will be needed. This is not an embarrassing outcome but a fundamental part of the scientific process – an evolution of ideas and concepts to improve ecosystem understanding and thus improve understanding about how to best manage forest environments and forest biodiversity.

Poorly known groups and forests

The focus of much of the biodiversity research in montane ash forests has been arboreal marsupials and birds. The best approaches to the conservation of other groups remain largely unknown. Examples include carabid beetles, which are known to be disturbance-sensitive in many other parts of the world (Grove and Hanula 2006), reptiles and bats (Brown and Nelson 1993; Brown *et al.* 1997).

Forests adjacent to montane ash, such as those dominated by dry mixed species, remain almost totally unstudied. A major research effort is needed on them, particularly as very little dry mixed-species old growth remains and they are increasingly targeted for timber and pulpwood production. The limited work conducted in dry mixed-species forests suggests they provide habitat for a range of species including a number of uncommon and/or threatened species (Brown *et al.* 1989; Lumsden *et al.* 1991).

Sambar Deer

An emerging issue in montane ash forests is the prevalence of the introduced Sambar Deer (*Cervus unicolor*). Browsing by the species is now listed as a threatening process under the Victorian *Flora and Fauna Guarantee Act*. The species appears to have increased in abundance in some parts of the Central Highlands of Victoria, particularly the water catchments in the Yarra Ranges National Park.

The invasive potential of populations of Sambar Deer must be carefully examined, as there may be problems similar to those created by large herbivores elsewhere in the world. For example, populations of White-tailed Deer (*Odocoileus virginianus*) have increased dramatically in some parts of North America, resulting in substantial damage to the vegetation cover, including a number of threatened plant taxa (Miller *et al.* 1992; Tilghman 1989). Particular problems have arisen at the edges of logging coupes. Therefore, it is important to understand the potential future risks associated with increased browsing by Sambar Deer, especially in areas where logging is prevalent and many human-created edges exist.

Thinning

The increasing use of machine harvesting in montane ash forests (Flint and Fagg 2007) means that thinning is increasingly prevalent throughout large areas of 20–35-year-old post-logging regrowth forest. What are the impacts of thinning on biodiversity? If there are negative effects, how might they be best mitigated? How might thinning be conducted so that the locations of future islands of forest can be identified well before the application of variable retention harvesting at the end of a timber rotation (Lindenmayer 2007)? The answers to these questions remain unknown, but they are important. First, experience in other ecosystems worldwide demonstrates that thinning applied in appropriate ways can lead to positive outcomes for biodiversity conservation (Carey *et al.* 1999). Second, the logical temporal progression from ecologically sensitive thinning regimes is the application of variable retention harvesting and the maintenance of islands of retained forest throughout thinning and subsequent regeneration harvest phases of a rotation.

Accelerated research adoption

The lag time between new research findings and the implementation of improved forest and biodiversity management practices has been ~20 years. The rate of uptake of new research findings needs to be accelerated, especially given the potential for rapid climate change. Doing this requires a greater commitment to improve communication among policy-makers, researchers and natural resource managers.

Better research adoption is a research topic in its own right and it is not always clear or straightforward how to best accelerate the transfer of knowledge and its subsequent on-ground implementation. The Cutting Experiment (Chapter 16) (Lindenmayer 2007) and the long-term monitoring program (Lindenmayer *et al.* 2003a) have been valuable exercises linking researchers with forest managers and policy-makers, but the true effectiveness of these and other projects lies in including their findings in on-ground management.

CONCLUDING REMARKS

The work in the montane ash forests has been a much longer and more varied journey than could ever have been anticipated at the outset in 1983. The body of work has developed through the continued evolution of ideas and approaches. As outlined at the beginning of this book, the work started as a single-species project

Figure 18.4: Juvenile Sambar Deer. (Photo by Doug Read).

(on Leadbeater's Possum) and over the following two decades expanded to encompass investigations on many other species, multi-species assemblages and ecological processes.

The themes in this body of work are broadly linked – some tightly, others more loosely. Many strongly inform one another, leading to new insights. For example, an understanding of the disturbance ecology of ash forests has indicated why populations of trees with hollows are distributed in particular ways in the landscape and, in turn, why the abundance of cavity-dependent arboreal marsupials is distributed in a patchy and disjunct way (Chapter 6).

One objective of this book has been to weave together the plethora of studies and generate new insights from the synthesis. Two themes particularly demonstrate the value of syntheses from different kinds of work. These are the complementarity of single-species and ecosystem-based research, and links between research on ecological patterns and on ecological processes. This is discussed briefly below, and emphasised by the topic interaction diagrams at the end of most chapters.

The complementarity of single-species and ecosystem-based work

The kinds of studies completed in the montane ash forests emphasise the complementarity between species-oriented and ecosystem-oriented research approaches. As an example, an initial investigation, which commenced in 1983 and focused on the habitat requirements of a single species (Leadbeater's Possum), quickly evolved to include the habitat requirements of other species and ecosystem-oriented investigations of disturbance regimes. Ecosystem-oriented studies focused on wildfire and logging and how they affect stand structural complexity, landscape heterogeneity and the long-term dynamics of critical forest structures (Lindenmayer et al. 1997b). For example, relationships were quantified between wildfires and the location of old-growth patches (Mackey et al. 2002) (Chapters 4 and 12), the structural composition of multi-aged stands (Lindenmayer et al. 1999b) and the spatial distribution and abundance of large trees, which influenced the composition and abundance of animal communities (Lindenmayer et al. 1991d) (Chapter 6). This ecosystem-oriented forest research led to the recognition that disturbances were major factors significantly influencing the abundance and population dynamics not only of Leadbeater's Possum but of entire assemblages of forest-dependent fauna (Lindenmayer and Franklin 2002). Detailed understanding of the habitat requirements of individual species, coupled with an ecosystem-level understanding of natural disturbance dynamics, has been critical for forest management (Chapter 16) and has contributed to the development of new, more ecologically sensitive silvicultural systems (Lindenmayer 2007). Neither the changes in conservation management, nor the insights that underpinned them, would have been possible using one research approach alone (Lindenmayer et al. 2007b).

Forest patterns and ecological processes

The work in montane ash forests has examined not only individual species and ecosystem-level issues, but forest patterns and ecological processes and the links between them. This is important because a lot of ecological research has quantified ecological patterns but has not explored the ecological processes which give rise to those patterns. Simply quantifying patterns, without an understanding of ecological processes, can make it difficult to understand why some kinds of changes occur and hence to predict future responses to changes in an ecosystem.

Links between ecological patterns and ecological processes can often be established only through a marriage of work on both processes and patterns and, hence, by threading together strands of information and insights from a range of different sources. The body of work accumulated from 25 years of empirical study in montane ash forests has fostered valuable new perspectives on links between ecological patterns and ecological processes. For example, the extensive range of studies on trees with hollows has emphasised the links between several ecological patterns and the process of cavity tree recruitment, development, decay and collapse (Figure 18.5).

Other work on ecological processes and patterns highlighted the fact that links between processes and patterns are often more complex than previously rec-

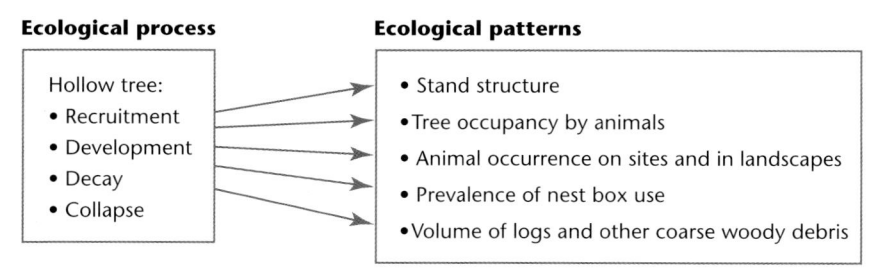

```
Ecological process              Ecological patterns

Hollow tree:                    • Stand structure
  • Recruitment                 • Tree occupancy by animals
  • Development                 • Animal occurrence on sites and in landscapes
  • Decay                       • Prevalence of nest box use
  • Collapse                    • Volume of logs and other coarse woody debris
```

Figure 18.5: Links between the ecological process of the recruitment, development, decay and collapse of trees with hollows, and a range of ecological patterns.

ognised. Relationships between disturbance and stand structure are a useful example. Natural disturbances in montane ash forests were examined in detail in Chapter 12. The material in that chapter and several others in this book (e.g. Chapters 4, 7 and 8) indicated that natural disturbances influence many spatial patterns in montane ash forests (Figure 18.6). Such links between ecological processes and forest patterns have taken on greater significance as a result of the February 2009 wildfires and will be pivotal to quantifying and understanding ecological recovery.

The diagram in Figure 18.6 is a simple one showing unidirectional links between natural disturbance and a range of ecological patterns. However, deeper thought indicates that this is overly simplistic, because some ecological patterns can have major impacts on ecological processes. For example, natural disturbances can influence the development of old-growth and multi-aged forests (Lindenmayer et al. 2000a) as well as the occurrence of rainforest (Lindenmayer et al. 2000b). Likewise, some of the patterns can influence natural disturbance. Large-diameter old-growth trees with thick bark at the base are considered less likely to be killed by wildfires. It is also possible that stands of old-growth forest and wet rainforest may limit the spread of wildfires across landscapes. Hence, there may be two-way relationships between some ecological patterns and processes, such as natural disturbance. This kind of understanding matters – for example, it highlights why multi-aged stands can develop and, given the complex stand structural features that result, why places where such kinds of stand structure occur are important for the diversity of animal groups such as arboreal marsupials.

Some final words

Despite the major body of work already completed, the journey is far from over as the existing challenges of ecologically sustainable forest management remain and new ones associated with rapid climate change emerge. This is underscored by the 2009 Black Saturday wildfires and the associated need to understand:

- what the damage to large areas of native forest will mean for sustained yields of timber and pulpwood, particularly following major salvage logging operations;
- the impacts of wildfire (and subsequent salvage logging) on carbon emissions and carbon storage;
- the process of ecological and biodiversity recovery;
- the long-term future of montane ash forest itself under fire regimes altered by climate-change, in which fires may become more frequent and more severe.

Seeking new knowledge and answering key questions has always been at the heart of the work in the montane ash forests. Guiding research by questions and continuing to develop and pose new questions is

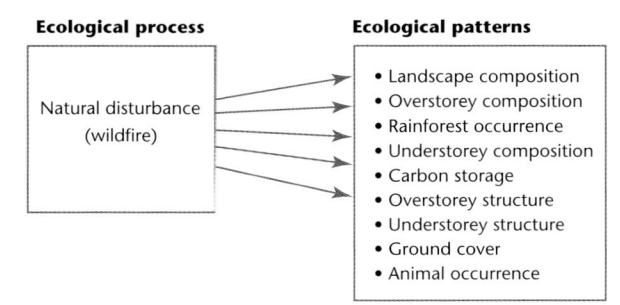

```
Ecological process         Ecological patterns

Natural disturbance        • Landscape composition
    (wildfire)             • Overstorey composition
                           • Rainforest occurrence
                           • Understorey composition
                           • Carbon storage
                           • Overstorey structure
                           • Understorey structure
                           • Ground cover
                           • Animal occurrence
```

Figure 18.6: Links between the ecological process of natural disturbance and a range of ecological patterns.

widely considered to be a fundamental part of successful, high-quality long-term research (Strayer *et al.* 1986; Likens 1992; Lovett *et al.* 2007). High-quality data are pivotal to best-practice natural resource management and best-practice biodiversity conservation. Without that, natural resource management is based on guesses, and typically poorly informed guesses at that. Changing societal demands on forests, together with the future challenges arising from rapid climate change, has made the gathering and analysis of high-quality data more important than ever. The hope is that the body of work in the montane ash forests can continue to grow and that, as a result, the quality of management and conservation efforts will be worthy of these truly extraordinary forests.

Bibliography

Aakala T, Kuuluvainen T, Gauthier S and Grandpre DL (2008) Standing dead trees and their decay-class dynamics in the north-eastern boreal old growth forests of Quebec. *Forest Ecology and Management* **255**, 410–420.

Adams MH and Attiwill PM (1984) The role of *Acacia* spp. in nutrient balance and cycling in regenerating *Eucalyptus regnans* F. Muell forests 1. Temporal changes in biomass and nutrient content. *Australian Journal of Botany* **32**, 205–215.

Ajani J (2007) *The Forest Wars*. University of Melbourne Press, Melbourne.

Akçakaya HR and Ferson S (1990) RAMAS/Space User Manual. Spatially Structured Population Models for Conservation Biology. Exeter Software, Setauket, New York.

Allee WC, Emerson AE, Park O, Park T and Schmidt KP (1949) *Principles of Animal Ecology*. Saunders, Philadelphia.

Allen TFH and Starr TB (1988) *Hierarchy Perspectives for Ecological Complexity*. University of Chicago Press, Chicago.

Amaranthus MP and Perry DA (1994) The functioning of ectomycorrhizal fungi in the field: linkages in space and time. *Plant and Soil* **159**(1), 133–140.

Ambrose GJ (1982) An ecological and behavioural study of vertebrates using hollows in eucalypt branches. PhD thesis. La Trobe University, Melbourne.

Andersen AN, Cook GD, Corbett LK, Douglas MM, Eager RW, Russell-Smith J, Setterfield SA, Williams RJ and Woinarski JCZ (2005) Fire frequency and biodiversity conservation in Australian tropical savannas: implications from the Kapalga fire experiment. *Austral Ecology* **30**(2), 155–167.

Anstis M (2002) *Tadpoles of South-eastern Australia: A Guide with Keys*. Reed New Holland, Sydney.

Appleby MW (1998) The incidence of exotic species following clearfelling of *Eucalyptus regnans* forest in the Central Highlands, Victoria. *Australian Journal of Ecology* **23**, 457–463.

Ashton D (1962) Some aspects of root competition in *E. regnans*. In *Proceedings of the 3rd General Conference. Institute of Foresters of Australia*. Institute of Foresters of Australia, Melbourne.

Ashton DH (1975a) The root and shoot development of *Eucalyptus regnans* F. Muell. *Australian Journal of Botany* **23**, 867–887.

Ashton DH (1975b) The seasonal growth of *Eucalyptus regnans* F. Muell. *Australian Journal of Botany* **23**, 239–252.

Ashton D (1975c) Studies of litter in *Eucalyptus regnans* forests. *Australian Journal of Botany* **24**, 397–414.

Ashton DH (1976) The development of even-aged stands of *Eucalyptus regnans* F. Muell. in central Victoria. *Australian Journal of Botany* **24**, 397–414.

Ashton D (1979) Seed harvesting by ants in forests of *Eucalyptus regnans* F. Muell. in Central Victoria. *Australian Journal of Ecology* **4**, 265–277.

Ashton DH (1981a) The ecology of the boundary between *Eucalyptus regnans* F. Muell. and *E. obliqua* L. Herit in Victoria. *Proceedings of the Ecological Society of Australia* **11**, 75–94.

Ashton DH (1981b) Fire in tall open forests (wet sclerophyll forests). In *Fire and the Australian Biota*. (Eds AM Gill, RH Groves and IR Noble) pp. 339–366. Australian Academy of Science, Canberra.

Ashton DH (1981c) Tall open forests. In *Australian Vegetation*. (Ed. RH Groves) pp. 121–151. Cambridge University Press, Cambridge.

Ashton DH (1986) Ecology of bryophytic communities in mature *Eucalyptus regnans* F. Muell. forest at Wallaby Creek, Victoria. *Australian Journal of Botany* **34**, 107–129.

Ashton DH (2000) The Big Ash forest, Wallaby Creek, Victoria: changes during one lifetime. *Australian Journal of Botany* **48**, 1–26.

Ashton DA and Attiwill P (1994) Tall open forests. In *Australian Vegetation*. (Ed. RH Groves) pp. 157–196. Cambridge University Press, Melbourne.

Ashton DH and Bassett OD (1997) The effects of foraging by the superb lyrebird (*Menura novaehollandiae*) in *Eucalyptus regnans* forests at Beenak,

Victoria. *Australian Journal of Ecology* **22**, 383–394.

Ashton DH and Sandiford EM (1988) Natural hybridization between *Eucalyptus regnans* F. Muell and *Eucalyptus macrorhyncha* F. Muell in the Cathedral Range, Victoria. *Australian Journal of Botany* **36**(1), 1–22.

Ashton D and Willis EJ (1982) Antagonisms in the regeneration of *Eucalyptus regnans* in the mature forest. In *The Plant Community as a Working Mechanism*. (Ed. EI Newman) pp. 113–128. Blackwell, Oxford.

Attiwill PM (1994a) The disturbance of forest ecosystems: the ecological basis for conservative management. *Forest Ecology and Management* **63**(2–3), 247–300.

Attiwill PM (1994b) Ecological disturbance and the conservative management of eucalypt forests in Australia. *Forest Ecology and Management* **63**, 301–346.

Austin MP, Nicholls AO and Margules CR (1990) Measurement of the realized qualitative niche: environmental niches of five *Eucalyptus* species. *Ecological Monographs* **60**, 161–177.

Backhouse F and Manning T (1996) 'Predator habitat enhancement as a strategy for managing small mammal damage in BC forests: a problem analysis'. Guidelines prepared for BC Environment. British Columbia Forest Service, Vancouver.

Ball IR, Lindenmayer DB and Possingham HP (1999) A tree hollow dynamics simulation model. *Forest Ecology and Management* **123**(2–3), 179–194.

Ball SJ, Lindenmayer DB and Possingham HP (2003) The predictive accuracy of viability analysis: a test using data from two small mammal species in a fragmented landscape. *Biodiversity and Conservation* **12**(12), 2393–2413.

Banks JC (1993) 'Tree-ring analysis of two mountain ash trees *Eucalyptus regnans* F. Muell from the Watts and O'Shannassy catchments, Central Highlands, Victoria'. Report to the Central Highlands Old Growth Forest Project, August 1993. Department of Conservation and Natural Resources, Melbourne.

Banks SC, Hoyle SD, Horsup A, Sunnicks P, Witton A and Taylor AC (2003) Demographic monitoring of an entire species by genetic analysis of non-invasively collected material. *Animal Conservation* **6**, 101–108.

Banks SC, Knight E, Dubach JE and Lindenmayer DB (2008) Microscale heterogeneity influences offspring sex ratio and spatial kin structure in possums. *Journal of Animal Ecology*. doi: 10.1111/j.1365-2656.01448.x

Banks SC, Dubach JE, Viggers KL and Lindenmayer DB (2009) Heterozygosity at MHC and microsatellites increases adult survival in possums. *Oecologia*. In press.

Barker P and Kirkpatrick JB (1994) *Phyllocladus aspleniifolius*: variability in the population structure, the regeneration niche and dispersion patterns in Tasmanian forest. *Australian Journal of Botany* **42**, 163–190.

Barker J, Grigg G and Tyler M (1995) *A Field Guide to Australian Frogs*. Surrey Beatty & Sons, Sydney.

Barrett GW, Silcocks A, Barry S, Cunningham R and Poulter R (2003) *The New Atlas of Australian Birds*. Birds Australia, Melbourne.

Baum JK and Myers RA (2004) Shifting baselines and the decline of pelagic sharks in the Gulf of Mexico. *Ecology Letters* **7**, 135–145.

Baumgartner LL (1939) Fox squirrel dens. *Journal of Mammalogy* **20**, 456–465.

Beale R (2007) *If Trees Could Speak. Stories of Australia's Greatest Trees*. Allen & Unwin, Sydney.

Beese WJ, Dunsworth BG, Zielke K and Bancroft B (2003) Maintaining attributes of old growth forests in coastal BC through variable retention. *Forestry Chronicle* **79**, 570–578.

Beezobs T and Sanson G (1997) The effects of plant and tooth structure on intake and digestibility in two small mammalian herbivores. *Physiological Zoology* **70**, 338–351.

Beissinger SR and McCullough DR (2002) *Population Viability Analysis*. University of Chicago Press, Chicago.

Bell DT (1999) Turner Review No. 1: the process of germination in Australian species. *Australian Journal of Botany* **47**, 475–517.

Belyea LR and Lancaster J (1999) Assembly rules within a contingent ecology. *Oikos* **86**(3), 402–416.

Bender DJ, Tischendorf L and Fahrig L (2003) Evaluation of patch isolation metrics for predicting

Bennett AF, Lumsden LF, Alexander JSA, Duncan PE, Johnson PG, Robertson P and Silveira CE (1991) Habitat use by arboreal marsupials along an environmental gradient in north-eastern Victoria. *Wildlife Research* **18**, 125–146.

Bennett AF, Radford JQ and Haslem A (2006) Properties of land mosaics: implications for nature conservation in agricultural landscapes. *Biological Conservation* **133**, 250–264.

Berg A, Ehnstrom B, Gustafsson L, Hallingback T, Jonsell M and Weslien J (1994) Threatened plant, animal, and fungus species in Swedish forests: distribution and habitat associations. *Conservation Biology* **8**(3), 718–731.

Bergeron Y, Bradshaw R and Engelmark O (1993) *Disturbance Dynamics in Boreal Forest*. Opulus Press, Uppsala.

Bergeron Y, Harvey B, Leduc A and Gauthier S (1999) Forest management guidelines based on natural disturbance dynamics: stand- and forest-level considerations. *Forestry Chronicle* **75**(1), 49–54.

Beschta R, Rhodes JJ, Kauffman JB, Gresswell RE, Minshall GW, Karr JR, Perry DA, Hauer FR and Frissell CA (2004) Postfire management on forested public lands of the western United States. *Conservation Biology* **18**, 957–967.

Beyer GL and Goldingay RL (2006) The value of nest boxes in the research and management of Australian hollow-using arboreal marsupials. *Wildlife Research* **33**, 161–174.

Bibby CJ, Jones M and Marsden S (1998) *Expedition Field Techniques*. Royal Geographical Society, London.

Boland DJ, Brooker MI, Chippendale GM, Hall N, Hyland BP, Johnson RD, Kleinig DA, McDonald MW and Turner JD (2006) *Forest Trees of Australia*. 5th edn. CSIRO Publishing, Melbourne.

Boose EP, Serrano MI and Foster DF (2004) Landscape and regional impacts of hurricanes in Puerto Rico. *Ecological Monographs* **74**, 335–352.

Bowman DM (1999) *Australian Rainforests. Islands of Green in a Land of Fire*. Cambridge University Press, Melbourne.

Bowman D and Kirkpatrick JB (1984) Geographic variation in the demographic structure of stands of *Eucalyptus delegatensis Baker, RT* on dolerite in Tasmania. *Journal of Biogeography* **11**, 427–437.

Boyce MS (1992) Population viability analysis. *Annual Review of Ecology and Systematics* **23**, 481–506.

Bradstock RA, Williams JE and Gill AM (Eds) (2002) *Flammable Australia: The Fire Regimes and Biodiversity of a Continent*. Cambridge University Press, Melbourne.

Braithwaite LW (1984) The identification of conservation areas for possums and gliders in the Eden woodpulp concession district. In *Possums and Gliders*. (Eds AP Smith and ID Hume) pp. 501–508. Surrey Beatty & Sons, Sydney.

Brereton R, Bennett S and Mansergh I (1995) Enhanced greenhouse climate change and its potential effect on selected fauna of south-eastern Australia: a trend analysis. *Biological Conservation* **72**, 339–354.

Brook BW (2000) Pessimistic and optimistic bias in population viability analysis. *Conservation Biology* **14**, 564–566.

Brook BW, O'Grady JJ, Chapman AP, Burgman MA, Akçakaya HR and Frankham R (2000) Predictive accuracy of population viability analysis in conservation biology. *Nature* **404**(6776), 385–387.

Brooker L and Brooker M (2002) Dispersal and population dynamics of the blue-breasted fairy-wren, *Malurus pulcherrimus*, in fragmented habitat in the Western Australian wheatbelt. *Wildlife Research* **29**, 225–233.

Brown MJ and Hickey J (1990) Tasmanian forest: genes or wilderness. *Search* **21**, 86–87.

Brown GW and Howley ST (1990) The bat fauna (Chioptera: Verpertilionidae) of the Acheron Valley, Victoria. *Australian Mammalogy* **13**, 65–70.

Brown GW and Nelson JL (1993) Influence of successional stage on *Eucalyptus regnans* (mountain ash) on habitat use by reptiles in the Central Highlands of Victoria. *Australian Journal of Ecology* **18**, 405–418.

Brown GW, Earl GE, Griffiths RC, Horrocks GF and Williams LM (1989) Flora and fauna of the Acheron Forest Block, Central Highlands, Victoria. In *Ecological Survey Report No. 30*. Department of Conservation, Forests and Lands, Melbourne.

Brown GW, Nelson JL and Cherry K (1997) The influence of habitat structure on insectivorous bat activity in montane ash forests of the Central Highlands of Victoria. *Australian Forestry* **60**, 138–146.

Bruns H (1960) The economic importance of birds in forests. *Bird Study* **7**, 193–208.

Bull EL and Partridge AD (1981) Creating snags with explosives. In *USDA Forest Service PNW 393*. Pacific Northwest Forest and Range Station, Portland, Oregon.

Bull EL and Partridge AD (1986) Methods of killing trees for use by cavity nesters. *Wildlife Society Bulletin* **14**, 142–146.

Bunnell F (1995) Forest-dwelling fauna and natural fire regimes in British Columbia: patterns and implications for conservation. *Conservation Biology* **9**, 636–644.

Bunnell F, Dunsworth G, Huggard D and Kremsater L (2003) Learning to sustain biological diversity on Weyerhauser's coastal tenure. Weyerhauser Co., Vancouver, BC.

Bunnell FL, Squires KA and Houde I (2004) Evaluating effects of large-scale salvage logging for mountain pine beetle on terrestrial and aquatic vertebrates. In *Mountain Pine Beetle Initiative Working Paper 2004-2*. Pacific Forestry Centre, Canadian Forest Service, Victoria, BC.

Burgman MA (1996) Characterisation and delineation of the eucalypt old-growth forest estate in Australia: a review. *Forest Ecology and Management* **83**(3), 149–161.

Burgman MA and Ferguson IS (1995) 'Rainforest in Victoria: a review of the scientific basis of current and proposed protection measures'. Report to Victorian Department of Conservation and Natural Resources, *Forest Services Technical Reports 95-4*. Department of Conservation and Natural Resources, Melbourne.

Burgman MA, Ferson S and Akçakaya HR (1993) *Risk Assessment in Conservation Biology*. Chapman & Hall, New York.

Burgman M, Ferson S and Lindenmayer DB (1995) The effect of initial distribution on extinction risks: implications for the reintroduction of Leadbeater's possum. In *Reintroduction Biology of Australasian Fauna*. (Ed. M Serena) pp. 15–19. Surrey Beatty & Sons, Sydney.

Burgman MA, Lindenmayer DB and Elith J (2005) Managing landscapes for conservation under uncertainty. *Ecology* **86**(8), 2007–2017.

Burton PJ, Messier C, Smith DW and Adamowicz WL (2003) *Towards Sustainable Management of the Boreal Forest*. National Research Council of Canada, Ottawa.

Busby JR (1986) A biogeoclimatic analysis of *Nothofagus cunninghamii* Hook Oesrt in south-eastern Australia. *Australian Journal of Ecology* **11**, 1–7.

Busby J and Brown M (1994) Southern rainforests. In *Vegetation of Australia*. (Ed. RH Groves) pp. 131–155. Cambridge University Press, Melbourne.

Cai W and Cowan T (2008) Dynamics of late autumn rainfall reduction over south-eastern Australia. *Geophysical Research Letters*, **35**. L09708, doi:10.1029/2008GL033727.

Calhoun AJ and de Maynadier PG (Eds) (2008) *Science and Conservation of Vernal Pools in Northeastern North America*. CRC Press, New York.

Campbell RG (1984) The eucalypt forests. In *Silvicultural and Environmental Aspects of Harvesting Some Major Commercial Eucalypt Forests in Victoria: A Review*. (Eds RG Campbell, EA Chesterfield, FG Craig, PC Fagg, PW Farrell, GR Featherstone, DW Flinn, P Hopmans, JD Kellas, CJ Leitch, RH Loyn, MA Macfarlane, LA Pederick, RO Squire, HT Stewart and GC Suckling) pp. 1–12. Forests Commission Victoria, Melbourne.

Campbell RG (1997) Evaluation and development of sustainable silvicultural systems for mountain ash forests. Discussion Paper. In *VSP Technical Report No. 28*. Forests Service, Department of Natural Resources and Environment, Melbourne.

Canham CD, Cole JJ and Lauenroth WK (Eds) (2003) *Models in Ecosystem Science*. Princeton University Press, Princeton, NJ.

Carey AB, Lippke BR and Sessions J (1999) Intentional systems management: managing forests for biodiversity. *Journal of Sustainable Forestry* **9**, 83–125.

Carron PL, Happold DCD and Bubela TM (1990) Diet of two sympatric Australian subalpine rodents,

Mastacomys fuscus and *Rattus fuscipes*. *Australian Wildlife Research* **17**, 479–489.

Cary G (2002) Importance of a changing climate for fire regimes in Australia. In *Flammable Australia*. (Eds RA Bradstock, JE Williams and AM Gill) pp. 26–48. Cambridge University Press, Melbourne.

Cary G, Lindenmayer DB and Dovers S (Eds) (2003) *Australia Burning: Fire Ecology, Policy and Management Issues*. CSIRO Publishing, Melbourne.

Caughley G (1994) Directions in conservation biology. *Journal of Animal Ecology* **63**, 215–244.

Caughley GC and Gunn A (1996) *Conservation Biology in Theory and Practice*. Blackwell Science, Cambridge, MA.

Chambers CL, Carrigan T, Sabin T, Tappeiner J and McComb WC (1997) Use of artificially created Douglas fir snags by cavity-nesting birds. *Western Journal of Applied Forestry* **12**, 93–97.

Chazdon RL (2008) Beyond deforestation: restoring forests and ecosystem services on degraded lands. *Science* **320**, 1458–1460.

Chesterfield EA, McCormick J and Hepworth G (1991) The effect of low root temperatures on the growth of mountain forest eucalypts in relation to the ecology of *Eucalyptus nitens*. *Proceedings of the Royal Society of Victoria* **103**, 67–76.

Chippendale GM and Wolfe L (1985) EUCALIST: a computerized data retrieval system for *Eucalyptus* (Myrtaceae). *Australian Forest Research* **14**, 147–152.

Christensen NL, Bartuska A, Brown JH, Carpenter S, Dantonio C, Francis R, Franklin JF, MacMahon JA, Noss RF, Parsons DJ, Peterson CH, Turner MG and Woodmansee RG (1996) The report of the Ecological Society of America committee on the scientific basis for ecosystem management. *Ecological Applications* **6**(3), 665–691.

Churchill S (2008) *Australian Bats*. Reed New Holland, Sydney.

Claridge AW and Lindenmayer DB (1993) The mountain brushtail possum (*Trichosurus caninus* Ogilby): disseminator of fungi in the mountain ash forests of the Central Highlands of Victoria? *Victorian Naturalist* **110**(2), 91–95.

Claridge AW and Lindenmayer DB (1994) The need for a more sophisticated approach toward wildlife corridor design in the multiple-use forests of south-eastern Australia: the case for mammals. *Pacific Conservation Biology* **1**, 301–307.

Claridge AW and Lindenmayer DB (1998) Consumption of hypogeous fungi by the mountain brushtail possum (*Trichosurus caninus*) in eastern Australia. *Mycological Research* **102**(3), 269–272.

Claridge AW, Paull D, Dawson J, Mifsud G, Murray AJ, Poore R and Saxon MJ (2005) Home range of the Spotted-tailed Quoll (*Dasyurus maculatus*), a marsupial carnivore, in a rainshadow woodland. *Wildlife Research* **32,** 7–14.

Cline SP, Berg AB and Wight HM (1980) Snag characteristics and dynamics in Douglas fir forests, western Oregon. *Journal of Wildlife Management* **44**, 773–786.

Cockburn A and Lee AK (1988) Marsupial femmes fatales. *Natural History* **97**(3), 40–46.

Cockburn A, Scott MP and Scotts DJ (1985) Inbreeding avoidance and male-biased natal dispersal in *Antechinus* spp (Marsupialia, Dasyuridae). *Animal Behaviour* **33**, 908–915.

Cogger H (2000) *Reptiles and Amphibians of Australia*. 6th edn. Reed New Holland, Sydney.

Commonwealth of Australia (1992) *National Forest Policy Statement*. Australian Government Publishing Service, Canberra.

Commonwealth of Australia and Department of Natural Resources and Environment (1997) *Comprehensive Regional Assessment: Biodiversity. Central Highlands of Victoria*. Commonwealth of Australia/Department of Natural Resources and Environment, Canberra.

Connell JH (1978) Diversity in tropical forests and coral reefs. *Science* **199**, 1302–1310.

Costermans L (1994) *Native Trees and Shrubs of South-eastern Australia*. 2nd edn. Rigby, Sydney.

Coulson T, Mace GM, Hudson E and Possingham H (2001) The use and abuse of population viability analysis. *Trends in Ecology and Evolution* **16**, 219–221.

Coxson DS and Stevenson SK (2005) Retention of canopy lichens after partial-cut harvesting in wet-belt interior cedar-hemlock forests, British Columbia, Canada. *Forest Ecology and Management* **204**, 97–112.

Coy R and Burgess J (1994) Psyllids in mountain ash forests. In *Research and Development Note No. 27*. Department of Conservation and Natural Resources, Melbourne.

Craig SA (1985) Social organisation, reproduction and feeding behaviour of a population of yellow-bellied gliders, *Petaurus australis* (Marsupialia: Petauridae). *Australian Wildlife Research* **12**, 1–18.

Cremer KW (1975) Temperature and other climatic influences on shoot development and growth of *Eucalyptus regnans*. *Australian Journal of Botany* **23**, 27–44.

Crisafulli CM, Swanson FJ and Dale VH (2005) Overview of ecological responses to the eruption of Mount St Helens: 1980–2005. In *Ecological Responses to the Eruption of Mount St Helens*. (Eds CM Crisafulli, FJ Swanson and VH Dale) pp. 287–299. Springer, New York.

Crowe MP, Paxton J and Tyers G (1984) Felling dead trees with explosives. *Australian Forestry* **47**, 84–87.

Cunningham TM (1960) The natural regeneration of *Eucalyptus regnans*. *Bulletin of the School of Forestry, University of Melbourne* **1**, 1–158.

Cunningham RB and Lindenmayer DB (2005) Modeling count data of rare species: some statistical issues. *Ecology* **86**(5), 1135–1142.

Cunningham RB and Welsh AH (1996) 'Coral sea reserves: monitoring seabird populations 1992–1996. Survey design and analysis'. Report to the Australian Nature Conservation Agency. Australian Nature Conservation Agency, Canberra.

Cunningham RB, Lindenmayer DB, Nix HA and Lindenmayer BD (1999) Quantifying observer heterogeneity in bird counts. *Australian Journal of Ecology* **24**(3), 270–277.

Cunningham RB, Lindenmayer DB, MacGregor C, Welsh AW and Barry S (2005) Small mammal populations in a wet eucalypt forest: factors influencing the probability of capture. *Wildlife Research* **32**, 657–671.

Cunningham RB, Lindenmayer DB, MacGregor C, Crane M and Michael D (2008) What factors influence bird biota on farms? Putting restored vegetation into context. *Conservation Biology* **22**, 742–752.

Dargarvel J (1995) *Fashioning Australia's Forests*. Oxford University Press, Melbourne.

Davey SM (1989) The environmental relationships of arboreal marsupials in a *Eucalypt* forest: a basis for Australian wildlife management. PhD thesis. Australian National University, Canberra.

Dawson TJ (1995) *Kangaroos: Biology of the Largest Marsupials*. University of New South Wales Press, Sydney.

Deans AM, Malcolm JR, Smith SM and Bellocq MI (2005) Edge effects and the responses of aerial insect assemblages to structural retention harvesting in Canadian boreal peatland forests. *Forest Ecology and Management* **204**, 249–266.

DeLong SC and Kessler WB (2000) Ecological characteristics of mature forest remnants left by wildfire. *Forest Ecology and Management* **131**, 93–106.

Denslow JS (1987) Tropical rainforest gaps and tree species diversity. *Annual Review of Ecology and Systematics* **18**, 431–451.

Department of Conservation, Forests and Lands (1989) *Code of Practice. Code of Forest Practices for Timber Production. Revision No. 1*. Department of Conservation, Forests and Lands, Melbourne.

Department of Natural Resources and Environment (1996) *Code of Practice. Code of Forest Practice for Timber Production. Revision No. 2, November 1996*. Department of Natural Resources and Environment, Melbourne.

Department of Natural Resources and Environment (2002) *Central Forest Management Area Estimate of Sawlog Resource*. Department of Natural Resources and Environment, Melbourne.

Department of Sustainability and Environment (2003) *Proposed Salvage Logging Prescriptions for the 2003 Eastern Victoria Fire*. Department of Sustainability and Environment, Melbourne.

Department of Sustainability and Environment (2004) *Baw Baw Frog. Action Statement No. 55*. Revised 2004. Department of Sustainability and Environment, Melbourne.

Department of Sustainability and Environment (2007) *Salvage Harvesting Prescriptions Table*. Department of Sustainability and Environment, Melbourne.

Department of Sustainability and Environment (2008) *Leadbeater's Possum Recovery Team. Meeting 3: 2 September 2008*. Department of Sustainability and Environment, Melbourne.

Diamond JM (1975) Assembly of species communities. In *Ecology and Evolution of Communities*. (Eds ML Cody and J Diamond) pp. 342–444. Harvard University Press, Cambridge, MA.

Dick RS (1975) A map of Australia according to Koppen's principles of definition. *Queensland Geographic Journal (3rd Series)* **3**, 33–69.

Dickman CR (1995a) Agile antechinus, *Antechinus agilis*. In *Mammals of Australia* (Ed. R Strahan) pp. 99–101. Reed Books, Sydney.

Dickman CR (1995b) Dusky antechinus, *Antechinus swainsonii*. In *Mammals of Australia*. (Ed. R Strahan) pp. 97–98. Reed Books, Sydney.

Dickman CR, Parnaby HE, Crowther MS and King DH (1998) *Antechinus agilis* (Marsupialia : Dasyuridae): a new species from the A-stuartii complex in south-eastern Australia. *Australian Journal of Zoology* **46**(1), 1–26.

Doeg TJ and Joehn JD (1990) A review of Australian studies on the effects of forestry practices on aquatic values. In *Silvicultural Systems Project Technical Report No. 5*. Department of Conservation and Environment, Melbourne.

Driscoll D and Lindenmayer DB (2009) Assembly rules for bird communities in south-eastern Australia. *Journal of Biogeography*. In review.

Duncan BD and Isaac G (1986) *Ferns and Allied Plants of Victoria, Tasmania and South Australia*. Melbourne University Press, Melbourne.

Dunlop M and Brown PR (2008) 'Implications of climate change for Australia's National Reserve System: a preliminary assessment'. Report to Department of Climate Change and Department of the Environment, Water, Heritage and the Arts. CSIRO Sustainable Ecosystems, Canberra.

Dunsworth G and Beese B (2000) New approaches in managing temperate rainforests. In *Mountain Forests and Sustainable Development*, pp. 24–25. Prepared by Mountain Agenda for the Commission on Sustainable Development 2000 spring session. Swiss Agency for Development and Cooperation, Berne, Switzerland.

Eberhart KE and Woodard PM (1987) Distribution of residual vegetation associated with large fires in Alberta. *Canadian Journal of Forest Research* **117**, 1207–1222.

Eggler WA (1948) Plant communities in the vicinity of the volcano El Parícutin, Mexico, after two and a half years of eruption. *Ecology* **29**, 415–436.

Egler FE (1954) Vegetation science concepts. I. Initial floristic composition: a factor in old field vegetation development. *Vegetation* **4**, 412–417.

Elliott HJ, Ohmart CP and Wylie FR (1998) *Insect Pests of Australian Forests*. Inkata Press, Melbourne.

Elliott KJ, Hitchcock SL and Krueger L (2002) Vegetation response to large-scale disturbance in a southern Appalachian forest: Hurricane Opal and salvage logging. *Journal of the Torrey Botanical Society* **129**, 48–59.

Ellis RC (1971) Mobilization of iron by extracts of *Eucalyptus* leaf litter. *Journal of Soil Science* **22**, 8–22.

Ellner SP, Fieberg J, Ludwig D and Wilcox C (2002) Precision of population viability analysis. *Conservation Biology* **16**, 258–261.

Elton CS (1927) *Animal Ecology*. Sidgwick & Jackson, London.

Enoksson B, Angelstam P and Larsson K (1995) Deciduous forests and resident birds: the problem of fragmentation within a coniferous forest landscape. *Landscape Ecology* **10**, 267–275.

Epting J, Verbyla D and Sorbel B (2005) Evaluation of remotely sensed indices for assessing burn severity in interior Alaska using Landsat TM and ETM+. *Remote Sensing of Environment* **96**, 328–339.

Fagan WF, Meir E, Prendergast J, Folarin A and Karieva P (2001) Characterizing population vulnerability for 758 species. *Ecology Letters* **4**, 132–138.

Fargione J, Hill J, Tilman D, Polasky S and Hawthorne P (2008) Land clearing and the biofuel carbon debt. *Science* **319**, 1235–1238.

Fazey I, Fischer J and Lindenmayer DB (2005) What do conservation biologists publish? *Biological Conservation* **124**(1), 63–73.

Fieberg J and Ellner SP (2001) Stochastic matrix models for conservation and management: a comparative review of methods. *Ecology Letters* **4**, 244–266.

Field SA, Tyre AJ and Possingham HP (2002) Estimating bird species richness: how should repeat surveys be organized in time? *Austral Ecology* **27**, 624–629.

Fischer J and Lindenmayer DB (2000) A review of relocation as a conservation management tool. *Biological Conservation* **96**(1), 1–11.

Fischer WC and McClelland BR (1983) Cavity-nesting bird bibiography including related titles on forest snags, fire, insects, diseases and decay. Intermountain Forest and Range Experiment Station, Ogden, Utah.

Fischer J, Lindenmayer DB, Nix HA and Stein J (2001) The bioclimatic domains of the mountain brushtail possum. *Journal of Biogeography* **28**(3), 293–304.

Fischer J, Lindenmayer DB, Blomberg S, Montague-Drake R and Felton A (2007) Functional richness and relative resilience of bird communities in regions with different land use intensities. *Ecosystems* **10**, 964–974.

Fischer J, Lindenmayer DB and Montague-Drake R (2008) Avian body mass is correlated with vegetation structure at the regional scale: a test of the landscape texture hypothesis. *Biodiversity and Distributions* **14**, 38–46.

Fitzpatrick EA and Nix HA (1970) The climatic factor in grassland ecology. In *Australian Grasslands*. (Ed. RM Moore) pp. 3–26. Australian National University Press, Canberra.

Flannigan MD, Logan KA, Amiro BD, Skinner WR and Stocks BJ (2005) Future area burned in Canada. *Climatic Change* **72**, 1–16.

Flint A and Fagg P (2007) *Mountain Ash in Victoria's State Forests. Silviculture Reference Manual No. 1.* Department of Sustainability and Environment, Melbourne.

Florence RG (1969) The application of ecology to forest management with particular reference to eucalypt forests. *Proceedings of the Ecological Society of Australia* **4**, 82–100.

Florence RG (1996) *Ecology and Silviculture of Eucalypt Forests.* CSIRO Publishing, Melbourne.

Folke C, Carpenter S, Walker B, Scheffer M, Elmqvist T, Gunderson L and Holling CS (2004) Regime shifts, resilience, and biodiversity in ecosystem management. *Annual Review of Ecology and Systematics* **35**, 557–581.

Ford HA, Barrett GW, Saunders D and Recher HF (2001) Why have birds in the woodlands of southern Australia declined? *Biological Conservation* **97**, 71–88.

Forestry Tasmania (2004) 'Towards new silviculture in Tasmania's old growth forests. 2. Sustaining wood yields'. Forestry Tasmania Draft Report. Forestry Tasmania, Hobart.

Forman RT (1964) Growth under controlled conditions to explain the hierarchical distributions of a moss, *Tetraphis pellucida. Ecological Monographs* **34**, 1–25.

Forman RTT (1995) *Land Mosaics: The Ecology of Landscapes and Regions.* Cambridge University Press, New York.

Foster DR and Boose ER (1992) Patterns of forest damage resulting from catastrophic wind in central New England, USA. *Journal of Ecology* **80**, 79–98.

Foster DR and Orwig DA (2006) Preemptive and salvage harvesting of New England forests: when doing nothing is a viable alternative. *Conservation Biology* **20**, 959–970.

Fox BJ (1982) Fire and mammalian secondary succession in an Australian coastal heath. *Ecology* **63**, 1332–1341.

Fox BJ and Brown JH (1993) Assembly rules for functional groups in North American desert rodent communities. *Oikos* **67**, 358–370.

Franklin JF and Agee JK (2003) Forging a science-based national forest fires policy. *Issues in Science and Technology* **20**, 59–66.

Franklin JF and Fites-Kaufman JA (1996) Assessment of late-successional forests of the Sierra Nevada. In *Sierra Nevada Ecosystem Project. Final Report to Congress. Vol. II. Assessment and Scientific Basis for Management Options.* pp. 627–656. University of California, Centers for Water and Wildland Resources, Davies, California.

Franklin JF and Forman RT (1987) Creating landscape patterns by forest cutting: ecological consequences and principles. *Landscape Ecology* **1**, 5–18.

Franklin JF and MacMahon JA (2000) Messages from a mountain. *Science* **288**, 1183–1185.

Franklin JF, Cromack K Jr, Denison W, McKee A, Maser C, Sedell J, Swanson F and Juday G (1981) Ecological attributes of old-growth Douglas fir forests. In *USDA Forest Service General Technical*

Report PNW-118. Pacific Northwest Forest and Range Experimental Station, Portland, Oregon.

Franklin JF, Swanson FJ, Harmon ME, Perry DA, Spies TA, Dale VH, McKee A, Ferrell WK, Means JE, Gregory SV, Lattin JD, Schowalter TD and Larsen D (1991) Effects of global climate change on forests in north-western America. *Northwest Environmental Journal* **7**, 233–254.

Franklin JF, Berg DE, Thornburgh DA and Tappeiner JC (1997) Alternative silvicultural approaches to timber harvest: variable retention harvest systems. In *Creating a Forestry for the 21st Century*. (Eds KA Kohm and JF Franklin) pp. 111–139. Island Press, Covelo, CA.

Franklin JF, Harmon ME and Swanson FJ (1999) Complementary roles of research and monitoring: lessons from the US LTER Program and Tierra del Fuego. Paper presented at *Toward a Unified Framework for Inventorying and Monitoring Forest Ecosystem Resources Symposium*. Guadalajara, Mexico.

Franklin JF, Lindenmayer DB, MacMahon JA, McKee A, Magnuson J, Perry DA, Waide R and Foster DR (2000) Threads of continuity: ecosystem disturbances, biological legacies and ecosystem recovery. *Conservation Biology in Practice* **1**(1), 8–16.

Frelich LE (2005) *Forest Dynamics and Disturbance Regimes: Studies from Temperate Evergreen-Deciduous Forests*. Cambridge University Press, Cambridge.

Friend GR (1993) Impact of fire on small vertebrates in mallee woodlands and heathlands of temperate Australia: a review. *Biological Conservation* **65**, 99–114.

Fries C, Johansson O, Petterson B and Simonsson P (1997) Silvicultural models to maintain and restore natural stand structures in Swedish boreal forests. *Forest Ecology and Management* **94**, 89–103.

Garber SM, Brown JP, Wilson DS, Maguire DA and Heath LS (2005) Snag longevity under alternative silvicultural systems in mixed-species forests of Central Maine. *Canadian Journal of Forest Research* **35**, 787–796.

Gaston KJ and Spicer JI (2004) *Biodiversity: An Introduction*. 2nd edn. Blackwell Publishing, Oxford.

Geertsema M and Pojar JJ (2007) Influence of landslides on biophysical diversity: a perspective from British Columbia. *Geomorphology* **89**, 55–69.

Gibbons P and Lindenmayer DB (1997a) Developing tree retention strategies for hollow-dependent arboreal marsupials in the wood production eucalypt forests of eastern Australia. *Australian Forestry* **60**(1), 29–45.

Gibbons P and Lindenmayer DB (1997b) The performance of prescriptions for the conservation of hollow-dependent fauna: implications for the Comprehensive Regional Forest Agreement process. In *CRES Working Paper 1997/2*. Centre for Resource and Environmental Studies, Australian National University, Canberra.

Gibbons P and Lindenmayer DB (2002) *Tree Hollows and Wildlife Conservation in Australia*. CSIRO Publishing, Melbourne.

Gibbons P, Lindenmayer DB, Fischer J, Manning AD, Weinberg A, Sedden J, Ryan P and Barrett G (2008) The future of scattered trees in agricultural landscapes. *Conservation Biology* **22**, 1309–1319.

Gilbert JM (1959) Forest succession in the Florentine Valley, Tasmania. *Papers and Proceedings of Royal Society Tasmania* **93**, 129–151.

Gill AM (1975) Fire and the Australian flora: a review. *Australian Forestry* **38**, 4–25.

Gill AM (1981) Past settlement fire history in Victorian landscapes. In *Fire and the Australian Biota*. (Eds AM Gill, RH Groves and IR Noble) pp. 77–98. Australian Academy of Sciences, Canberra.

Gill AM (1999) Biodiversity and bushfires: an Australia-wide perspective on plant-species changes after a fire event. In *Australia's Biodiversity: Responses to Fire. Plants, Birds and Invertebrates*. (Eds AM Gill, J Woinarski and A York) pp. 9–53. Environment Australia, Canberra.

Gilpin ME and Soulé ME (1986) Minimum viable populations: processes of extinction. In *Conservation Biology: The Science of Scarcity and Diversity*. (Ed. ME Soule) pp. 19–34. Sinauer Associates, Sunderland, MA.

Gippel CJ, Finlayson BL and O'Neill IC (1996) Disturbance and hydraulic significance of large woody debris in a lowland Australian river. *Hydrobiologica* **318**, 179–194.

Goldingay RL (1986) Feeding behaviour of the yellow-bellied glider, *Petaurus australis* Marsupilia:

Petauridae at Bombala, New South Wales. *Australian Mammalogy* **9**, 17–25.

Goldingay R and Jackson S (2004) *The Biology of Australian Possums and Gliders*. Surrey Beatty & Sons, Sydney.

Goldingay RG and Kavanagh RP (1991) The yellow-bellied glider: a review of its ecology and management considerations. In *Conservation of Australia's Forest Fauna*. (Ed. D Lunney) pp. 365–375. Surrey Beatty & Sons, Sydney.

Goldingay RL and Kavanagh RP (1993) Home range estimates and habitat of the yellow-bellied glider (*Petaurus australis*) at Waratah Creek, NSW. *Wildlife Research* **20**, 387–404.

Goldingay RL, Carthew SM and Whelan RJ (1991) The importance of non-flying mammals in pollination. *Oikos* **61**, 79–87.

Goldsmith B (1991) *Monitoring for Conservation and Ecology*. Chapman & Hall, London.

Gooday P, Whish-Wilson P and Weston L (1997) Regional Forest Agreements. Central Highlands of Victoria. In *Australian Forest Products Statistics*, pp. 1–11. Australian Bureau of Agricultural and Resource Economics, Canberra.

Gordon G, Brown AS and Pulsford T (1988) A koala (*Phascolarctos cinereus* Goldfuss) population crash during drought and heatwave conditions in south-western Queensland. *Australian Journal of Ecology* **13**, 451–461.

Government of British Columbia (2008) Mountain pine beetles in British Columbia.

Government of Victoria (1986) *Timber Industry Strategy. Government Statement No. 9*. Government Printer, Melbourne.

Green K and Osborne W (1994) *Wildlife of the Australian Snow Country*. Reed Books, Sydney.

Green K and Osborne W (2003) The distribution and status of the broad-toothed rat *Mastacomys fuscus* (Rodentia: Muridae) in New South Wales and the Australian Capital Territory. *Australian Zoologist* **32**, 229–237.

Greenacre MJ (2007) *Theory and Applications of Correspondence Analysis*. 2nd edn. Academic Press, Orlando, FL.

Greenberg CH, Harris LD and Neary DG (1995) A comparison of bird communities in burned and salvaged-logged clearcut, and forested Florida sand pine scrub. *Wilson Bulletin* **107**, 40–54.

Greenstein BJ, Curran HA and Pandolfi JM (1998) Shifting ecological baselines and the demise of *Acropora cervicornis* in the western North Atlantic and Caribbean province: a Pleistocene perspective. *Coral Reefs* **17**, 249–261.

Gregory SV (1997) Riparian management in the 21st century. In *Creating a Forestry for the 21st Century*. (Eds KA Kohm and JF Franklin) pp. 69–85. Island Press, Washington, DC.

Grove SJ and Hanula JL (Eds) (2006) *Insect Biodiversity and Dead Wood: Proceedings of a Symposium for the 22nd International Congress of Entomology*. USDA, Forest Service, Southern Research Station, Asheville, North Carolina.

Grove SJ, Meggs JF and Goodwin A (2002) 'A review of biodiversity conservation issues relating to coarse woody debris management in the wet eucalypt production forests of Tasmania'. Report to Forestry Tasmania. Forestry Tasmania, Hobart.

Gustaffson L, de Jong J and Noren M (1999) Evaluation of Swedish woodland key habitats using red-listed bryophytes and lichens. *Biodiversity and Conservation* **8**, 1101–1114.

Hall S (1980) The diets of two co-existing species of *Antechinus* (Marsupialia: Dasyuridae). *Australian Wildlife Research* **7**, 365–378.

Halpern CB and Raphael MG (1999) Special issue on retention harvests in north-western forest ecosystems. Demonstration of Ecosystem Management Options (DEMO) study. *Northwest Science* **73**, 1–125.

Hansen BD (2008) Population genetic structure of Leadbeater's possum *Gynobelideus leadbeateri*, and its implications for species conservation. PhD thesis. Monash University, Melbourne.

Hansen A and Rotella J (1999) Abiotic factors. In *Managing Biodiversity in Forest Ecosystems*. (Ed. M Hunter III) pp. 161–209. Cambridge University Press, Cambridge.

Hansen B, Harley D, Lindenmayer DB and Taylor AC (2008) Population genetic analysis reveals long-term decline of a threatened endemic Australian marsupial. *Molecular Ecology*. In press.

Hanski I (1994) A practical model of metapopulation dynamics. *Journal of Animal Ecology* **63**, 151–162.

Hanski I (1998) Metapopulation dynamics. *Nature* **396**, 41–49.

Haramis GM and Thompson DQ (1985) Density-production characteristics of box-nesting wood ducks in a northern green tree impoundment. *Journal of Wildlife Management* **49**, 429–436.

Harley DK (2004) Patterns of nest box use by Leadbeater's possum *Gymnobelideus leadbeateri*: applications to research and conservation. In *The Biology of Australian Possums and Gliders*. (Eds RL Goldingay and SM Jackson) pp. 318–329. Surrey Beatty & Sons, Sydney.

Harmon ME and Franklin JF (1989) Tree seedlings on logs in *Picea-Tsuga* forests of Oregon and Washington. *Ecology* **70**, 48–59.

Harmon M, Franklin JF, Swanson F, Sollins P, Gregory SV, Lattin JD, Anderson NH, Cline SP, Aumen NG, Sedell JR, Lienkaemper GW, Cromack K and Cummins K (1986) Ecology of coarse woody debris in temperate ecosystems. *Advances in Ecological Research* **15**, 133–302.

Harper MJ, McCarthy MA and van der Ree R (2005) The use of nest boxes in urban natural vegetation remnants by vertebrate fauna. *Wildlife Research* **32**, 509–516.

Harris LD (1984) *The Fragmented Forest: Island Biogeography Theory and the Preservation of Biotic Diversity*. University of Chicago Press, Chicago.

Harris SG (2004) *Regeneration of Flora Following Timber Harvesting in the Wet Forests of the Otway Ranges*. Department of Sustainability and Environment, Melbourne.

Hautala H, Jalonen J, Laak-Lindberg S and Vanha-Majamaa I (2004) Impacts of retention felling on coarse woody debris (CWD) in mature boreal spruce forests in Finland. *Biodiversity and Conservation* **13**, 1541–1554.

Haynes RW, Bormann BT, Lee DC and Martin JR (2006) *Northwest Forest Plan – The First 10 Years (1993–2003): Synthesis of Monitoring and Research Results*. US Department of Agriculture, Pacific Northwest Research Station, Portland, Oregon.

Helms JA (1998) *Dictionary of Forestry*. Society of American Foresters, Bethesda, Maryland.

Hennon PE and Loopstra EM (1991) Persistence of western hemlock and western red cedar trees 38 years after girdling at Cat Island in south-east Alaska. In *USDA Forest Service Research Note PNW-RN-507*, p. 4. USDA Forest Service.

Henry SR (1984) Social organisation of the greater glider (*Petauroides volans*) in Victoria. In *Possums and Gliders*. (Eds AP Smith and ID Hume) pp. 222–228. Surrey Beatty & Sons, Sydney.

Henry SR and Craig SA (1984) Diet ranging behaviour and social organisation of the yellow-bellied glider in Victoria. In *Possums and Gliders*. (Eds AP Smith and ID Hume) pp. 331–341. Surrey Beatty & Sons, Sydney.

Hickey JE, Neyland MG, Edwards LG and Dingle JK (1999) Testing alternative silvicultural systems for wet eucalypt forests in Tasmania. In *Practising Forestry Today. 18th Biennial Conference of the Institute of Foresters of Australia*. (Eds RC Ellis and PJ Smethurst) pp. 136–141. Hobart, Tasmania.

Hobson KA and Schieck J (1999) Changes in bird communities in boreal mixed wood forest: harvest and wildfire effects over 30 years. *Ecological Applications* **9**, 849–863.

Holling CS (Ed.) (1978) *Adaptive Environmental Assessment and Management*. John Wiley & Sons, Toronto.

Holling CS (1992) Cross-scale morphology, geometry, and dynamics of ecosystems. *Ecological Monographs* **62**, 447–502.

Hollis G (2003) *Recovery Plan for the Baw Baw Frog (Philoria frosti). Stage II (2002–2006)*. Envionment Australia, Canberra.

Hollis G (2004) Ecology and conservation biology of the Baw Baw Frog *Philoria frosti* (Anura: Myobatrachidae). PhD thesis. University of Melbourne, Melbourne.

Hollstedt C and Vyse A (1997) Sicamous Creek silvicultural systems project. In *Sicamous Creek Silvicultural Systems Project: Workshop Proceedings*. (Eds C Hollstedt and A Vyse). Working Paper 24/1997. Research Branch, British Columbia Ministry of Forests, Kamloops, BC.

Holtam BW (1971) Windblow of Scottish forests in January 1968. In *Forestry Commission Bulletin No. 45*. Her Majesty's Stationery Office, Edinburgh.

How RA (1972) The ecology and management of *Trichosurus* spp. (Marsupialia) in NSW. University of New England, Armidale.

How RA, Barnett JL, Bradley AJ, Humphreys WF and Martin R (1984) The population biology of *Pseudocheirus peregrinus* in a *Leptospermum laevigatum* thicket. In *Possums and Gliders*. (Eds AP Smith and ID Hume) pp. 261–268. Surrey Beatty & Sons, Sydney.

Howard TM (1973) Studies in the ecology of *Nothofagus cunninghamii* Oerst. I. Natural regeneration on the Mt Donna Buang massif, Victoria. *Australian Journal of Botany* **21**, 67–78.

Howard TM and Ashton DH (1973) The distribution of *Nothofagus cunninghamii* rainforest. *Proceedings of the Royal Society of Victoria* **86**, 47–75.

Huang C, Ward SM and Lee AK (1987) Comparison of the diet of the feathertail glider, *Acrobates pygmaeus*, and the eastern pygmy possum, *Cercartetus nanus* (Marsupialia: Burramyidae) in sympatry. *Australian Mammalogy* **10**, 47–50.

Huggett R and Cheeseman J (2002) *Topography and the Environment*. Prentice Hall, London.

Hunter ML (1993) Natural fire regimes as spatial models for managing boreal forests. *Biological Conservation* **65**(2), 115–120.

Hunter ML (2007) Core principles for using natural disturbance regimes to inform landscape management. In *Managing and Designing Landscapes for Conservation*. (Eds DB Lindenmayer and RJ Hobbs) pp. 408–422. Blackwell Publishing, Oxford.

Hutto RL (1995) Composition of bird communities following stand-replacement fires in northern Rocky Mountain (USA) conifer forests. *Conservation Biology* **10**, 1041–1058.

Hutto R (2008) The ecological importance of severe wildfires: some like it hot. *Ecological Applications* **18**, 1827–1834.

Incoll WD (1979) Effects of overwood trees on growth of young stands of *Eucalyptus sieberi*. *Australian Forestry* **42**, 110–116.

Incoll RD, Loyn RH, Ward SJ, Cunningham RB and Donnelly CF (2000) The occurrence of gliding possums in old-growth patches of mountain ash (*Eucalyptus regnans*) in the Central Highlands of Victoria. *Biological Conservation* **98**, 77–88.

Inions G, Tanton MT and Davey SM (1989) Effects of fire on the availability of hollows in trees used by the common brushtail possum, *Trichosurus vulpecula* Kerra 1792, and the ringtail possum, *Pseudocheirus peregrinus* Boddaerts 1785. *Australian Wildlife Research* **16**, 449–458.

IPCC (2007) *Climate Change 2007: Mitigation of Climate Change*. IPCC Secretariate, Geneva.

Jackson WD (1968) Fire, air, water and earth: an elemental ecology of Tasmania. *Proceedings of Ecological Society of Australia* **3**, 9–16.

Jackson RB and Schlesinger WH (2004) Curbing the US carbon deficit. *Proceedings of National Academy of Sciences* **101**, 15827–15829.

Jacobs MR (1955) *Growth Habits of Eucalypts*. Forestry and Timber Bureau, Canberra.

James IL and Norton DA (2002) Helicopter-based natural forest management for New Zealand's rimu (*Dacrydium cupressinum*, Podocarpaceae) forests. *Forest Ecology and Management* **155**, 337–346.

Jarrett PH and Petrie AH (1929) The vegetation of the Black Spur region: a study in the ecology of some Australian *Eucalyptus* forests. II. Pyric succession. *Journal of Ecology* **17**, 269–281.

Jarvis PG and McNaughton KG (1986) Stomatal control of transpiration: scaling up from leaf to region. *Advances in Ecological Research* **15**, 1–49.

Jelinek A, Cameron D, Belcher C and Turner I (1995) New perspectives on the ecology of Lake Mountain: the discovery of Leadbeater's possum *Gymnobelideus leadbeateri* McCoy in sub-alpine woodland. *Victorian Naturalist* **112**, 112–115.

Jenkins R and Bartell R (1980) *A Field Guide to Reptiles of the Australian High Country*. Inkata Press, Melbourne.

Kafka V, Gauthier S and Bergeron Y (2001) Fire impacts and crowning in the boreal forest: study of a large wildfire in western Quebec. *International Journal of Wildland Fire* **10**, 119–127.

Kaila L, Martikainen P and Punttila P (1997) Dead trees left in clear-cuts benefit saproxylic *Coleoptera* adapted to natural disturbances in boreal forests. *Biodiversity and Conservation* **6**, 1–18.

Kanowski J (2001) Effects of elevated CO_2 on the foliar chemistry of seedlings of two rainforest from north-east Australia: implications for folivorous marsupials. *Australian Ecology* **26**, 165–172.

Karr JR, Rhodes JJ, Minshall GW, Hauer FR, Beschta RL, Frissell CA and Perry DA (2004) The effects of postfire salvage logging on aquatic ecosystems in the American west. *BioScience* **54**, 1029–1033.

Kavanagh RP and Lambert MJ (1990) Food selection by the greater glider, *Petauroides volans*: is foliar nitrogen a determinant of habitat quality? *Australian Wildlife Research* **17**, 285–299.

Keen FP (1955) The role of natural falling of beetle-killed ponderosa pine snags. *Journal of Forestry* **27**, 720–723.

Keith D, McCaw WL and Whelan RJ (2002) Fire regimes in Australian heathlands and their effects on plants. In *Flammable Australia: The Fire Regimes and Biodiversity of a Continent*. (Eds R Bradstock, J Williams and AM Gill) pp. 199–237. Cambridge University Press, Cambridge.

Keith H, Mackey B and Lindenmayer DB (2009) The world's most carbon-dense forests. *Proceedings of the National Academy of Sciences*. In press.

Kerle A (2001) *Possums: The Brushtails, Ringtails and Greater Glider*. UNSW Press, Sydney.

Knight EH and Fox BJ (2000) Does habitat structure mediate the effects of forest fragmentation and human-induced disturbance on the abundance of *Antechinus stuartii*? *Australian Journal of Zoology* **48**, 577–595.

Koehn J (1993) Fish need trees. *Victorian Naturalist* **110**, 255–257.

Korpilahti E and Kuuluvainen TE (Eds) (2002) Disturbance dynamics in boreal forests: defining the ecological basis of restoration and management of biodiversity. *Silva Fennica* **36**, 1–447.

Kotliar NB, Kennedy PL and Ferree K (2007) Avifaunal responses to fire in south-western montane forests along a burn severity gradient. *Ecological Applications* **17**, 491–507.

Kraaijeveld-Smit FJL, Ward SJ and Temple-Smith PD (2002a) Multiple paternity in a field population of a small carnivorous marsupial, the agile antechinus, *Antechinus agilis*. *Behavioral Ecology and Sociobiology* **52**(1), 84–91.

Kraaijeveld-Smit FL, Lindenmayer DB and Taylor AC (2002b) Dispersal patterns and population structure in a small marsupial, *Antechinus agilis*, from two forests analysed using microsatellite markers. *Australian Journal of Zoology* **50**(4), 325–338.

Kraaijeveld-Smit FJL, Ward SJ and Temple-Smith PD (2003) Paternity success and the direction of sexual selection in a field population of a semelparous marsupial, *Antechinus agilis*. *Molecular Ecology* **12**(2), 475–484.

Kraaijeveld-Smit FJ, Lindenmayer DB, Taylor AC, MacGregor C and Wertheim B (2007) Comparative genetic structure reflects underlying life histories of three sympatric small mammal species in continuous forest of south-eastern Australia. *Oikos* **116**, 1819–1830.

Krebs C (2008) *The Ecological World View*. CSIRO Publishing, Melbourne.

Kulakowski D and Veblen TT (2007) Effect of prior disturbance on the extent and severity of wildfire in Colorado subalpine forests. *Ecology* **88**, 759–769.

Lacy RC (2000) Structure of the VORTEX simulation model for population viability analysis. *Ecological Bulletin* **48**, 191–203.

Lacy RC and Lindenmayer DB (1995) Using population viability analysis (PVA) to explore the impacts of population sub-division on the mountain brushtail possum *Trichosurus caninus* Ogilby (Phalangeridae, Marsupialia) in south-eastern Australia. II. Changes in genetic variability in sub-divided populations. *Biological Conservation* **73**(2), 131–142.

Lacy RC, Flesness NR and Seal US (1989) 'Puerto Rican parrot population viability analysis'. Report to the US Fish and Wildlife Service. Captive Breeding Specialist Group, Species Survival Commission, IUCN, Apple Valley, Minnesota.

Land Conservation Council (1994) *Final Recommendations: Melbourne Area. District 2 Review*. Land Conservation Council, Melbourne.

Langford KJ (1976) Change in yield of water following a bushfire in a forest of *Eucalyptus regnans*. *Journal of Hydrology* **29**, 87–114.

Landres PB, Verner J and Thomas JW (1988) Ecological uses of vertebrate indicator species: a critique. *Conservation Biology* **2**(4), 316–328.

Laurance WF, Nascimento HE, Laurance SG, Andrade A, Ewers RM, Harms KE, Luizao RC and Ribeiro JE (2007) Habitat fragmentation, variable edge effects, and the landscape-divergence hypothesis. *PLoS One* **2**(10), e1017 doi:10.371/journal.pone.0001017.

Law B and Dickman CR (1998) The use of habitat mosaics by terrestrial vertebrate fauna: implications for conservation and management. *Biodiversity and Conservation* **7**, 323–333.

Lazaruk LW, Kernaghan G, Macdonald SE and Khasa D (2005) Effects of partial cutting on the ectomycorrhizae of *Picea glauca* forests in north-western Alberta. *Canadian Journal of Forest Research* **35**, 1442–1454.

Lazenby-Cohen KA and Cockburn A (1991) Social and foraging components of the home range in *Antechinus stuartii* (Dasyuridae, Marsupialia). *Australian Journal of Ecology* **16**(3), 301–307.

Lebreton JD, Burnham KP, Clobert J and Anderson DR (1992) Modeling survival and testing biological hypotheses using marked animals: a unified approach with case studies. *Ecological Monographs* **62**, 67–118.

Lecomte N, Simard M and Bergeron Y (2006) Effects of fire severity and initial tree composition on stand structural development in the coniferous boreal forest of north-western Quebec, Canada. *EcoScience* **13**, 152–163.

Lee AK and Cockburn A (1985) *Evolutionary Ecology of Marsupials*. Cambridge University Press, Sydney.

Lenihan JM, Drapek R, Bachelet D and Neilson RP (2003) Climate change effect on vegetation distribution, carbon, and fire in California. *Ecological Applications* **13**, 1667–1681.

Lesslie R and Maslen M (1995) *National Wilderness Inventory Handbook*. 2nd edn. Australian Government Publishing Service, Canberra.

Likens GE (Ed.) (1989) *Long-term Studies in Ecology: Approaches and Alternatives*. Springer-Verlag, New York.

Likens GE (1992) *The Ecosystem Approach: Its Use and Abuse*. Ecology Institute, Oldendorf/Luhe, Germany.

Lill A (1996) The foraging behaviour of the Superb Lyrebird. *Corella* **20**, 77–87.

Lindenmayer DB (1989) *The Ecology and Habitat Requirements of Leadbeater's Possum*. Australian National University Press, Canberra.

Lindenmayer DB (1991) A note on the occupancy of nest trees by Leadbeater's possum at Cambarville in the montane ash forests of the Central Highlands of Victoria. *Victorian Naturalist* **108**(6), 128–129.

Lindenmayer DB (1992) The ecology and habitat requirements of arboreal marsupials in the montane ash forests of the Central Highlands of Victoria: a summary of studies. In *Value Adding and Silvicultural Systems Program, VSP Internal Report No. 6*. Department of Conservation and Natural Resources, Melbourne.

Lindenmayer DB (1994a) The impacts of timber harvesting on arboreal marsupials at different spatial scales and its implications for ecologically sustainable forest use and nature conservation. *Australian Journal of Environmental Management* **1**, 56–68.

Lindenmayer DB (1994b) Timber harvesting in the montane ash forests of the Central Highlands of Victoria: impacts at different spatial scales on arboreal marsupials and the implications for ecologically sustainable forest use. In *Ecology and Sustainability of Southern Temperate Ecosystems*. (Eds TW Norton and SR Dovers) pp. 31–50. CSIRO Publishing, Melbourne.

Lindenmayer DB (1996) *Wildlife and Woodchips: Leadbeater's Possum as a Testcase of Sustainable Forestry*. UNSW Press, Sydney.

Lindenmayer DB (1997) Differences in the biology and ecology of arboreal marsupials in forests of south-eastern Australia. *Journal of Mammalogy* **78**(4), 1117–1127.

Lindenmayer DB (1999) Future directions for biodiversity conservation in managed forests: indicator species, impact studies and monitoring programs. *Forest Ecology and Management* **115**(2–3), 277–287.

Lindenmayer DB (2000) Factors at multiple scales affecting distribution patterns and their implications for animal conservation: Leadbeater's possum as a case study. *Biodiversity and Conservation* **9**(1), 15–35.

Lindenmayer DB (2002) *Gliders of Australia: A Natural History*. UNSW Press, Sydney.

Lindenmayer DB (2007) *The Variable Harvest Retention System and its Implications in the Mountain Ash Forests of the Central Highlands of Victoria*.

Fenner School of Environment and Society, Australian National University, Canberra.

Lindenmayer DB (2009) Old forests, new perspectives: insights from the wet ash forests of the Central Highlands of Victoria. *Forest Ecology and Management*. In press.

Lindenmayer DB and Burgman MA (2005) *Practical Conservation Biology*. CSIRO Publishing, Melbourne.

Lindenmayer DB and Cunningham RB (1996a) A habitat-based microscale forest classification system for zoning wood production areas to conserve a rare species threatened by logging operations in south-eastern Australia. *Environmental Monitoring and Assessment* **39**(1–3), 543–557.

Lindenmayer DB and Cunningham RB (1996b) Microscale forest classification for zoning wood production areas to conserve a rare species threatened by logging operations in south-eastern Australia. In *Global to Local: Ecological Land Classification*. (Eds RA Sims, IGW Corns and K Klinka) pp. 543–557. Kluwer Academic, London.

Lindenmayer DB and Cunningham RB (1997) Patterns of co-occurrence among arboreal marsupials in the forests of central Victoria, south-eastern Australia. *Australian Journal of Ecology* **22**(3), 340–346.

Lindenmayer DB and Dixon JM (1992) An additional historical record of Leadbeater's possum, *Gymnobelideus leadbeateri* McCoy, prior to the 1961 rediscovery of the species. *Victorian Naturalist* **109**(6), 217–218.

Lindenmayer DB and Franklin JF (1996) The importance of stand structure for the conservation of wildlife in logged forests: a case study from Victoria. In *CRES Working Paper 96/1*. Centre for Resource and Environmental Studies, Australian National University, Canberra.

Lindenmayer DB and Franklin JF (1997) Managing stand structure as part of ecologically sustainable forest management in Australian mountain ash forests. *Conservation Biology* **11**(5), 1053–1068.

Lindenmayer DB and Franklin JF (2002) *Conserving Forest Biodiversity: A Comprehensive Multiscaled Approach*. Island Press, Washington, DC.

Lindenmayer DB and Franklin JF (Eds) (2003) *Towards Forest Sustainability*. CSIRO Publishing, Melbourne.

Lindenmayer DB and Lacy RC (1995a) Metapopulation viability of arboreal marsupials in fragmented old-growth forests: comparison among species. *Ecological Applications* **5**(1), 183–199.

Lindenmayer DB and Lacy RC (1995b) A simulation study of the impacts of population subdivision on the mountain brushtail possum *Trichosurus caninus* Ogilby (Phalangeridae, Marsupialia) in south-eastern Australia. I. Demographic stability and population persistence. *Biological Conservation* **73**(2), 119–129.

Lindenmayer DB and Lacy RC (1995c) Metapopulation viability of Leadbeater's possum, *Gymnobelideus leadbeateri*, in fragmented old-growth forests. *Ecological Applications* **5**(1), 164–182.

Lindenmayer DB and Likens G (2009) Adaptive monitoring: a new paradigm for long-term studies and monitoring. *Trends in Ecology and Evolution*. In press.

Lindenmayer DB and McCarthy MA (2002) Congruence between natural and human forest disturbance: a case study from Australian montane ash forests. *Forest Ecology and Management* **155**(1–3), 319–335.

Lindenmayer DB and McCarthy MA (2006) Evaluation of PVA models of arboreal marsupials. *Biodiversity and Conservation* **15**, 4079–4096.

Lindenmayer DB and Meggs RA (1996) Use of den trees by Leadbeater's possum (*Gymnobelideus leadbeateri*). *Australian Journal of Zoology* **44**(6), 625–638.

Lindenmayer DB and Nix HA (1993) Ecological principles for the design of wildlife corridors. *Conservation Biology* **7**(3), 627–630.

Lindenmayer DB and Noss RF (2006) Salvage logging, ecosystem processes, and biodiversity conservation. *Conservation Biology* **20**(4), 949–958.

Lindenmayer DB and Ough K (2006) Salvage harvesting in the montane ash forests of the Central Highlands of Victoria, south-eastern Australia. *Conservation Biology* **20**, 1005–1015.

Lindenmayer DB and Possingham HP (1995a) Modelling the impacts of wildfire on the viability of metapopulations of the endangered Australian species of arboreal marsupial, Leadbeater's possum. *Forest Ecology and Management* **74**(1–3), 197–222.

Lindenmayer DB and Possingham HP (1995b) The conservation of arboreal marsupials in the montane ash forests of the Central Highlands of Victoria, south-eastern Australia.VII. Modeling the persistence of Leadbeater's possum in response to modified timber harvesting practices. *Biological Conservation* **73**(3), 239–257.

Lindenmayer DB and Possingham HP (1995c) *The Risk of Extinction: Ranking Management Options for Leadbeater's Possum*. Centre for Resource and Environmental Studies, Australian National University, Canberra.

Lindenmayer DB and Possingham HP (1995d) Modelling the viability of metapopulations of the endangered Leadbeater's possum in south-eastern Australia. *Biodiversity and Conservation* **4**(9), 984–1018.

Lindenmayer DB and Possingham HP (1996a) Modeling the relationships between habitat connectivity, corridor design and wildlife conservation within intensively logged wood production forests of south-eastern Australia. *Landscape Ecology* **11**, 79–105.

Lindenmayer DB and Possingham HP (1996b) Modelling the inter-relationships between habitat patchiness, dispersal capability and metapopulation persistence of the endangered species, Leadbeater's possum, in south-eastern Australia. *Landscape Ecology* **11**(2), 79–105.

Lindenmayer DB and Possingham HP (1996c) Ranking conservation and timber management options for Leadbeater's possum in south-eastern Australia using population viability analysis. *Conservation Biology* **10**(1), 235–251.

Lindenmayer DB and Press K (1989) *Spotlighting Manual for Rangers and Group Leaders*. ACT Parks and Conservation Service, Canberra.

Lindenmayer DB and Recher HF (1998) Aspects of ecologically sustainable forestry in temperate eucalypt forests: beyond an expanded reserve system. *Pacific Conservation Biology* **4**, 4–10.

Lindenmayer DB and Wood J (2009) Long-term tree fall patterns in Australian wet forests. *Canadian Journal of Forest Research*. In review.

Lindenmayer DB, Cunningham RB, Tanton MT, Smith AP and Nix HA (1990a) The conservation of arboreal marsupials in the montane ash forests of the Central Highlands of Victoria, south-eastern Australia. I. Factors influencing the occupancy of trees with hollows. *Biological Conservation* **54**(2), 111–131.

Lindenmayer DB, Cunningham RB, Tanton MT and Smith AP (1990b) The conservation of arboreal marsupials in the montane ash forests of the Central Highlands of Victoria, south-eastern Australia. II. The loss of trees with hollows and its implications for the conservation of Leadbeater's possum *Gymnobelideus leadbeateri* McCoy (Marsupialia, Petauridae). *Biological Conservation* **54**(2), 133–145.

Lindenmayer DB, Cunningham RB, Tanton MT, Smith AP and Nix HA (1990c) The habitat requirements of the mountain brushtail possum and the greater glider in the montane ash-type eucalypt forests of the Central Highlands of Victoria. *Australian Wildlife Research* **17**, 467–478.

Lindenmayer DB, Craig SA, Linga T and Tanton MT (1991a) Public participation in stagwatching surveys for a rare mammal: applications for environmental education. *Australian Journal of Environmental Education* **7**, 63–70.

Lindenmayer DB, Cunningham RB, Nix HA, Tanton MT and Smith AP (1991b) Predicting the abundance of hollow-bearing trees in montane ash forests of south-eastern Australia. *Australian Journal of Ecology* **16**, 91–98.

Lindenmayer DB, Cunningham RB, Tanton MT and Nix HA (1991c) Aspects of the use of den trees by arboreal and scansorial marsupials inhabiting montane ash forests in Victoria. *Australian Journal of Zoology* **39**, 57–65.

Lindenmayer DB, Cunningham RB, Tanton MT, Nix HA and Smith AP (1991d) The conservation of arboreal marsupials in the montane ash forests of the Central Highlands of Victoria, south-eastern Australia. III. The habitat requirements of Leadbeater's possum *Gymnobelideus leadbeateri* and models of the diversity and abundance of arboreal marsupials. *Biological Conservation* **56**(3), 295–315.

Lindenmayer DB, Cunningham RB, Tanton MT, Smith AP and Nix HA (1991e) Characteristics of hollow-bearing trees occupied by arboreal marsupials in the montane ash forests of the Central Highlands of Victoria, south-east Australia. *Forest Ecology and Management* **40**(3–4), 289–308.

Lindenmayer DB, Nix HA, McMahon JP, Hutchinson MF and Tanton MT (1991f) The conservation of Leadbeater's possum, *Gymnobelideus leadbeateri* McCoy: a case study of the use of bioclimatic modelling. *Journal of Biogeography* **18**, 371–383.

Lindenmayer DB, Norton TW and Tanton MT (1991g) Differences between the effects of wildfire and clearfelling in montane ash forests of Victoria and its implications for fauna dependent on tree hollows. *Australian Forestry* **53**, 61–68.

Lindenmayer DB, Cunningham RB and Donnelly CF (1993a) The conservation of arboreal marsupials in the montane ash forests of the Central Highlands of Victoria, south-east Australia. IV. The distribution and abundance of arboreal marsupials in retained linear strips (wildlife corridors) in timber production forests. *Biological Conservation* **66**(3), 207–221.

Lindenmayer DB, Cunningham RB, Donnelly CF, Tanton MT and Nix HA (1993b) The abundance and development of cavities in *Eucalyptus* trees: a case study in the montane forests of Victoria, south-eastern Australia. *Forest Ecology and Management* **60**(1–2), 77–104.

Lindenmayer DB, Lacy RC, Thomas VC and Clark TW (1993c) Predictions of the impacts of changes in population-size and environmental variability on Leadbeater's possum, *Gymnobelideus leadbeateri* McCoy (Marsupialia, Petauridae) using population viability analysis : an application of the computer program Vortex. *Wildlife Research* **20**(1), 67–86.

Lindenmayer DB, Boyle S, Burgman MA, McDonald D and Tomkins B (1994a) The sugar and nitrogen content of the gums of *Acacia* spp. in the mountain ash and alpine ash forests of Central Victoria and its potential implications for exudivorous arboreal marsupials. *Australian Journal of Ecology* **19**, 169–177.

Lindenmayer DB, Cunningham RB and Donnelly CF (1994b) The conservation of arboreal marsupials in the montane ash forests of the Central Highlands of Victoria, south-eastern Australia. VI. The performance of statistical models of the nest tree and habitat requirements of arboreal marsupials applied to new survey data. *Biological Conservation* **70**(2), 143–147.

Lindenmayer DB, Cunningham RB, Donnelly CF, Triggs BE and Belvedere M (1994c) Factors influencing the occurrence of mammals in retained linear strips (wildlife corridors) and contiguous stands of montane ash forest in the Central Highlands of Victoria, south-eastern Australia. *Forest Ecology and Management* **67**(1–3), 113–133.

Lindenmayer DB, Cunningham RB, Donnelly CF, Triggs BJ and Belvedere M (1994d) The conservation of arboreal marsupials in the montane ash forests of the Central Highlands of Victoria, south-eastern Australia. V. Patterns of use and the microhabitat requirements of the mountain brushtail possum *Trichosurus caninus* Ogilby in retained linear habitats (wildlife corridors). *Biological Conservation* **68**(1), 43–51.

Lindenmayer DB, Tanton MT and Viggers KL (1994e) The fur-inhabiting ectoparasites of Leadbeater's possum, *Gymnobelideus leadbeateri* McCoy Marsupialia: Petauridae. *Australian Mammalogy* **17**, 109–111.

Lindenmayer DB, Burgman MA, Akçakaya HR, Lacy RC and Possingham HP (1995a) A review of the generic computer programs Alex, Ramas/Space and Vortex for modelling the viability of wildlife metapopulations. *Ecological Modelling* **82**(2), 161–174.

Lindenmayer DB, Ritman K, Cunningham RB, Smith JDB and Horvath D (1995b) A method for predicting the spatial distribution of arboreal marsupials. *Wildlife Research* **22**(4), 445–456.

Lindenmayer DB, Viggers KL, Cunningham RB and Donnelly CF (1995c) Morphological variation among populations of the mountain brushtail possum, *Trichosurus caninus* Ogilby (Phalangeridae, Marsupialia). *Australian Journal of Zoology* **43**(5), 449–458.

Lindenmayer DB, Mackey B and Nix HA (1996a) Climatic analyses of the distribution of four commercially important wood production eucalypt trees from south-eastern Australia. *Australian Forestry* **59**, 11–26.

Lindenmayer DB, Mackey BG and Nix HA (1996b) The bioclimatic domains of four species of commercially important eucalypts from south-eastern Australia. *Australian Forestry* **59**(2), 74–89.

Lindenmayer DB, Welsh A, Donnelly CF and Meggs RA (1996c) Use of nest trees by the mountain brushtail possum (*Trichosurus caninus*) (Phalangeridae, Marsupialia). I. Number of occupied trees and frequency of tree use. *Wildlife Research* **23**(3), 343–361.

Lindenmayer DB, Welsh A, Donnelly CF and Cunningham RB (1996d) Use of nest trees by the mountain brushtail possum (*Trichosurus caninus*) (Phalangeridae, Marsupialia). II. Characteristics of occupied trees. *Wildlife Research* **23**(5), 531–545.

Lindenmayer DB, Wong A and Triggs BE (1996e) A comparison of the detection of small mammals by hairtubing and by scat analysis. *Australian Mammalogy* **18**, 91–92.

Lindenmayer DB, Welsh A and Donnelly CF (1997a) Use of nest trees by the mountain brushtail possum (*Trichosurus caninus*) (Phalangeridae, Marsupialia). III. Spatial configuration and co-occupancy of nest trees. *Wildlife Research* **24**(6), 661–677.

Lindenmayer DB, Cunningham RB and Donnelly CF (1997b) Decay and collapse of trees with hollows in eastern Australian forests: impacts on arboreal marsupials. *Ecological Applications* **7**(2), 625–641.

Lindenmayer DB, Lacy RC and Viggers KL (1998) Modelling survival and capture probabilities of the mountain brushtail possum (*Trichosurus caninus*) in the forests of south-eastern Australia using trap-recapture data. *Journal of Zoology* **245**(1), 1–13.

Lindenmayer DB, Cunningham RB and McCarthy MA (1999a) The conservation of arboreal marsupials in the montane ash forests of the central highlands of Victoria, south-eastern Australia. VIII. Landscape analysis of the occurrence of arboreal marsupials in the montane ash forests. *Biological Conservation* **89**(1), 83–92.

Lindenmayer DB, Cunningham RB and Pope ML (1999b) A large-scale 'experiment' to examine the effects of landscape context and habitat fragmentation on mammals. *Biological Conservation* **88**(3), 387–403.

Lindenmayer DB, Cunningham RB, Pope ML and Donnelly CF (1999c) The response of arboreal marsupials to landscape context: a large-scale fragmentation study. *Ecological Applications* **9**(2), 594–611.

Lindenmayer DB, Incoll RD, Cunningham RB and Donnelly CF (1999d) Attributes of logs on the floor of Australian mountain ash (*Eucalyptus regnans*) forests of different ages. *Forest Ecology and Management* **123**(2–3), 195–203.

Lindenmayer DB, Incoll RD, Cunningham RB, Pope ML, Donnelly CF, MacGregor CI, Tribolet C and Triggs BE (1999e) Comparison of hairtube types for the detection of mammals. *Wildlife Research* **26**(6), 745–753.

Lindenmayer DB, Mackey BG, Mullen IC, McCarthy MA, Gill AM, Cunningham RB and Donnelly CF (1999f) Factors affecting stand structure in forests: are there climatic and topographic determinants? *Forest Ecology and Management* **123**(1), 55–63.

Lindenmayer DB, Cunningham RB, Donnelly CF and Franklin JF (2000a) Structural features of old growth Australian montane ash forests. *Forest Ecology and Management* **134**(1–3), 189–204.

Lindenmayer DB, Mackey BG, Cunningham RB, Donnelly CF, Mullen IC, McCarthy MA and Gill AM (2000b) Factors affecting the presence of the cool temperate rainforest tree myrtle beech (*Nothofagus cunninghamii*) in southern Australia: integrating climatic, terrain and disturbance predictors of distribution patterns. *Journal of Biogeography* **27**(4), 1001–1009.

Lindenmayer DB, Margules CR and Botkin DB (2000c) Indicators of biodiversity for ecologically sustainable forest management. *Conservation Biology* **14**(4), 941–950.

Lindenmayer DB, Cunningham RB, Donnelly CF, Incoll RD, Pope ML, Tribolet CR, Viggers KL and Welsh AH (2001) How effective is spotlighting for detecting the greater glider (*Petauroides volans*)? *Wildlife Research* **28**(1), 105–109.

Lindenmayer DB, Claridge AW, Gilmore AM, Michael DR and Lindenmayer BD (2002a) The ecological roles of logs in Australian forests and the potential impacts of harvesting intensification on log-using biota. *Pacific Conservation Biology* **8**(2), 121–140.

Lindenmayer DB, Cunningham RB, Donnelly CF and Lesslie R (2002b) On the use of landscape indices as ecological indicators in fragmented forests. *Forest Ecology and Management* **159**(3), 203–216.

Lindenmayer DB, Cunningham RB, Donnelly CF, Nix HA and Lindenmayer BD (2002c) The distribution of birds in a novel landscape context. *Ecological Monographs* **72**(1), 1–18.

Lindenmayer DB, Dubach J and Viggers KL (2002d) Geographic dimorphism in the mountain brushtail possum (*Trichosurus caninus*): the case for a new species. *Australian Journal of Zoology* **50**(4), 369–393.

Lindenmayer DB, Cunningham RB, MacGregor C and Incoll RD (2003a) A long-term monitoring study of the population dynamics of arboreal marsupials in the Central Highlands of Victoria. *Biological Conservation* **110**(1), 161–167.

Lindenmayer DB, Possingham HP, Lacy RC, McCarthy MA and Pope ML (2003b) How accurate are population models? Lessons from landscape-scale tests in a fragmented system. *Ecology Letters* **6**(1), 41–47.

Lindenmayer DB, MacGregor CI, Cunningham RB, Incoll RD, Crane M, Rawlins D and Michael DR (2003c) The use of nest boxes by arboreal marsupials in the forests of the Central Highlands of Victoria. *Wildlife Research* **30**(3), 259–264.

Lindenmayer DB, Franklin JF, Angelstam P, Bunnell F, Brown M, Dovers S, Hickey J, Kremsater L, Niemela J, Norton D, Perry D and Soulé M (2004) The Victorian Forestry Roundtable Meeting: discussing transitions to sustainability in Victorian forests. *Australian Forestry* **67**, 1–4.

Lindenmayer DB, Cunningham RB and Peakall R (2005) The recovery of populations of bush rat *Rattus fuscipes* in forest fragments following major population reduction. *Journal of Applied Ecology* **42**, 649–658.

Lindenmayer DB, Franklin JF and Fischer J (2006) General management principles and a checklist of strategies to guide forest biodiversity conservation. *Biological Conservation* **131**(3), 433–445.

Lindenmayer DB, Hobbs R, Montague-Drake R, Alexandra J, Bennett A, Burgman M, Cale P, Calhoun A, Cramer V, Cullen P, Driscoll D, Fahrig L, Fischer J, Franklin J, Haila Y, Hunter M, Gibbons P, Lake S, Luck G, McIntyre S, MacNally R, Manning A, Miller J, Mooney H, Noss R, Possingham H, Saunders D, Schmiegelow F, Scott M, Simberloff D, Sisk T, Walker B, Wiens J, Woinarski J and Zavaleta E (2007a) A checklist for ecological management of landscapes for conservation. *Ecology Letters* **10**, 1–14.

Lindenmayer DB, Fischer J, Felton A, Montague-Drake R, Manning A, Simberloff D, Youngentob K, Saunders D, Blomberg S, Wilson D, Felton AM, Blackmore C, Lowe A and Elliott CP (2007b) The complementarity of single-species and ecosystem-oriented research in conservation research. *Oikos* **116**, 1220–1226.

Lindenmayer DB, Burton PJ and Franklin JF (2008a) *Salvage Logging and its Ecological Consequences*. Island Press, Washington, DC.

Lindenmayer DB, MacGregor C, Wood JT, Cunningham RB, Crane M, Michael D, Montague-Drake R and Brown D (2008b) Testing hypotheses associated with bird recovery after wildfire. *Ecological Applications* **18**, 1967–1983.

Lindenmayer DB, Welsh AW, Donnelly CF, Cunningham RB, MacGregor C, Crane M and Michael D (2008c) Nesting biology of the common ringtail possum. *Australian Journal of Zoology* **56**, 1–11.

Lindenmayer DB, Wood J and MacGregor C (2009a) Do observer differences in bird detection significantly influence inferences about environmental impacts?. *Emu*. In press.

Lindenmayer DB, Wood J, Cunningham RB, McBurney L, Crane M, Montague-Drake R, Michael D and MacGregor C (2009b) Are gullies best for biodiversity? An empirical examination of Australian wet forest types: a case study from the wet ash forests of Victoria, south-eastern Australia. *Forest Ecology and Management*. In press.

Lindenmayer DB, Welsh A, Donnelly CF, Crane M, Michael D, McBurney L, MacGregor C and Cunningham RB (2009c) Are nest boxes a viable alternative source of cavities for hollow-dependent animals? Long-term monitoring of nest box occupancy, pest use and attrition. *Biological Conservation* **142**, 33–42.

Linder P and Östlund L (1998) Structural changes in three mid-boreal Swedish forest landscapes, 1885–1996. *Biological Conservation* **85**, 9–19.

Lorimer CG and Frelich LE (1994) Natural disturbance regimes in old-growth northern hardwoods. *Journal of Forestry* **January 1994**, 33–38.

Lovett GM, Burns DA, Driscoll CT, Jemkins JC, Mitchell MJ, Rustad L, Shanley JB, Likens GE and Haeuber R (2007) Who needs environmental monitoring? *Frontiers in Ecology and the Environment* **5**, 253–260.

Loyn RH (1985) Bird populations in successional forests of mountain ash *Eucalyptus regnans* in Central Victoria. *Emu* **85**, 213–230.

Loyn RH (1998) Birds in patches of old-growth ash forest, in a matrix of younger forest. *Pacific Conservation Biology* **4**, 111–121.

Luke RH and McArthur AG (1977) *Bushfires in Australia*. Australian Government Publishing Service, Canberra.

Lumsden LF, Alexander JSA, Hill FAR, Krasna SP and Silveira CE (1991) The vertebrate fauna of the Land Conservation Council Melbourne 2 study area. In *Arthur Rylah Institute for Environmental Research Technical Report Series No. 115*. Department of Conservation and Environment, Melbourne.

Lumsden LF, Bennett AF, Silins J and Krasna S (1994) 'Fauna in a remnant vegetation-farmland mosaic: movements, roosts and foraging ecology of bats'. Report to Australian Nature Conservation Agency 'Save the Bush' Program. Flora and Fauna Branch, Department of Conservation and Natural Resources, Melbourne.

Lunney D (1983) Bush rat. In *Complete Book of Australian Mammals*. (Ed. R Strahan) pp. 443–445. Angus & Robertson, Sydney.

Lunney D (Ed.) (2004) *Forest Fauna II*. Suurey Beatty & Sons, Sydney.

Lutze MT, Campbell RG and Fagg PC (1999) Development of silviculture in the native state forests of Victoria. *Australian Forestry* **62**, 236–244.

MacArthur RH and MacArthur JW (1961) On bird species diversity. *Ecology* **42**, 594–598.

Macfarlane MA (1988) Mammal populations in mountain ash (*Eucalyptus regnans*) forests of various ages in the Central Highlands of Victoria. *Australian Forestry* **51**, 14–27.

Macfarlane MA and Seebeck JH (1991) Draft management strategies for the conservation of Leadbeater's possum, *Gymnobelideus leadbeateri*, in Victoria. In *Arthur Rylah Institute for Environmental Research Technical Report Series No. 111*. Department of Conservation and Environment, Melbourne.

Macfarlane MA, Smith J and Lowe K (1998) *Leadbeater's Possum Recovery Plan, 1998–2002*. Department of Natural Resources and Environment, Melbourne.

Mackey BG (1993) A spatial analysis of the environmental relations of rainforest structural types. *Journal of Biogeography* **20**, 303–336.

Mackey BG and Lindenmayer DB (2001) Towards a hierarchical framework for modelling the spatial distribution of animals. *Journal of Biogeography* **28**(9), 1147–1166.

Mackey BG, Lesslie RG, Lindenmayer DB and Nix HA (1998) Wilderness and its place in conservation in Australia. *Pacific Conservation Biology* **4**, 182–185.

Mackey B, Lindenmayer DB, Gill AM, McCarthy MA and Lindesay JA (2002) *Wildlife, Fire and Future Climate: A Forest Ecosystem Analysis*. CSIRO Publishing, Melbourne.

Mackey B, Keith H, Berry SL and Lindenmayer DB (2008) *Green Carbon: The Role of Natural Forests in Carbon Storage*. ANU E-Press, Canberra.

Mackowski CM (1987) *Wildlife Hollows and Timber Management in Blackbutt Forest*. University of New England, Armidale.

MacNally R (1994) Habitat-specific guild structure of forest birds in south-eastern Australia: a regional-scale perspective. *Journal of Animal Ecology* **63**(4), 988–1001.

Majer JD, Recher HF and Postle AC (1994) Comparison of arthropod species richness in eastern and western Australian canopies: a contribution to the species number debate. *Memoirs of the Queensland Museum* **36**, 121–131.

Manning AD, Fischer J and Lindenmayer DB (2006) Scattered trees are keystone structures: implications for conservation. *Biological Conservation* **132**(3), 311–321.

Margules CR and Pressey RL (2000) Systematic conservation planning. *Nature* **405**, 243–253.

Margules CR, Milkovits GA and Smith GT (1994) Contrasting effects of habitat fragmentation on the scorpion *Cercophonius squama* and an amphipod. *Ecology* **75**(7), 2033–2042.

Maser C and Trappe JM (1984) The seen and unseen world of the fallen tree. *USDA Forest Service General Technical Report PNW-GTR-164*. US Department of Agriculture.

Maser C, Trappe JM and Ure DC (1977) Implications of small mammal mycophagy to the management of western coniferous forests. In *Transactions of the North American Wildlife and Natural Resources Conference, 78-88*.

May SA and Norton TW (1996) Influence of fragmentation and disturbance on the potential impact of feral predators on native fauna in Australian forest ecosystems. *Wildlife Research* **23**, 387–400.

McAlpine CA, Syktus J, Deo RC, Lawrence PJ, McGowan HA, Watterson IG and Phinn SR (2007) Modeling the impact of historical land cover change on Australia's regional climate. *Geophysical Research Letters* **34**, L22711.1–L11.6.

McCallum H, Timmers P and Hoyle S (1995) Modelling the impact of predation on reintroductions of bridled nailtail wallabies. *Wildlife Research* **22**, 163–171.

McCarthy MA and Lindenmayer DB (1998) Multi-aged mountain ash forest, wildlife conservation and timber harvesting. *Forest Ecology and Management* **104**(1–3), 43–56.

McCarthy MA and Lindenmayer DB (1999) Incorporating metapopulation dynamics of greater gliders into reserve design in disturbed landscapes. *Ecology* **80**(2), 651–667.

McCarthy MA and Lindenmayer DB (2000) Spatially correlated extinction in a metapopulation model of Leadbeater's possum. *Biodiversity and Conservation* **9**(1), 47–63.

McCarthy MA, Gill AM and Lindenmayer DB (1999) Fire regimes in mountain ash forest: evidence from forest age structure, extinction models and wildlife habitat. *Forest Ecology and Management* **124**(2–3), 193–203.

McCay TS (2000) Use of woody debris by cotton mice (*Peromyscus gossypinus*) in a south-eastern pine forest. *Journal of Mammalogy* **81**, 527–535.

McHugh PJ (1991) *Statement of Resources, Uses and Values: Dandenong Yarra Forests Management Area*. Department of Conservation and Environment, Melbourne.

McIver JD and Starr L (2000) Environmental effects of postfire logging: literature review and annotated bibliography. *PNW-GTR-486*. Pacific Northwest Research Station, USDA Forest Service, Portland, Oregon.

McIver JD and Starr L (2001) A literature review on the environmental effects of postfire logging. *Western Journal of Applied Forestry* **16**, 159–168.

McKay GM (1983) The greater glider. In *Complete Book of Australian Mammals*. (Ed. R Strahan) pp. 134–135. Angus & Robertson, Sydney.

McKenney DW and Lindenmayer DB (1994) An economic assessment of a nest-box strategy for the conservation of an endangered species. *Canadian Journal of Forest Research* **24**(10), 2012–2019.

Meggs J (1996) 'Pilot study of the effects of modern logging practices on the decaying-log habitat in wet eucalypt forest in south-east Tasmania'. Report to the Tasmanian RFA Environment and Heritage Technical Committee. Commonwealth of Australia/Government of Tasmania.

Meggs JM (1997) *Simsons Stage Beetle, Hoplogonus simsoni, in North-east Tasmania: Distribution, Habitat Characteristics and Conservation Requirements*. Forest Practices Board of Tasmania, Hobart.

Meggs RA, Lindenmayer DB, Linga T and Morris BJ (1991) An improved design for trap brackets used in tree trapping. *Wildlife Research* **18**, 589–591.

Menkhorst P (Ed.) (1995) *Mammals of Victoria: Distribution, Ecology and Conservation*. Oxford University Press, Melbourne.

Menkhorst P and Knight F (2001) *A Field Guide to the Mammals of Australia*. Oxford University Press, Melbourne.

Milledge DR, Palmer CL and Nelson JL (1991) 'Barometers of change': the distribution of large owls and gliders in mountain ash forests of the Victorian Central Highlands and their potential as management indicators. In *Conservation of Aus-*

tralia's Forest Fauna. (Ed. D Lunney) pp. 53–65. Royal Zoological Society of NSW, Sydney.

Miller SG, Bratton SP and Hadidian J (1992) Impacts of white-tailed deer on endangered and threatened vascular plants. *Natural Areas Journal* **12**, 67–74.

Milne J and Short M (1999) Invertebrates associated with the moss *Dicranoloma* Ren. In *The Other 99%: Conservation and Biodiversity of Invertebrates*. (Eds W Ponder and D Lunney) pp. 129–132. Surrey Beatty & Sons, Sydney.

Mirande C, Lacy RC and Seal US (1991) *Whooping Crane Population Viability Analysis and Species Survival Plan*. Captive Breeding Specialist Group, Species Survival Commission, IUCN, Apple Valley, Minnesota.

Mitchell RJ, Neel WL, Hiers JK, Cole FT and Atkinson JB (2000) *A Model Management Plan for Conservation Easements in Longleaf Pine-dominated Landscapes*. Joseph E Jones Ecological Research Center, Newton, Georgia.

Mittermeier RA, Mittermeier CG, Brooks TM, Pilgrim JD, Konstant WR, da Fonseca GAB and Kormos C (2003) Wilderness and biodiversity conservation. *Proceedings of the National Academy of Sciences* **100**, 10309–10313.

Moore S, Wallington T, Hobbs R, Ehrlich P, Holling CS, Levin S, Lindenmayer DB, Pahl-Wostl C, Possingham HP, Turner M and Westoby M (2008) Diversity of current ecological thinking: implications for environmental management. *Environmental Management* **43**, 17–27.

Morgan P, Hardy CC, Swetnam TW, Rollins MG and Long DG (2001) Mapping fire regimes across time and space: understanding coarse and fine-scale fire patterns. *International Journal of Wildland Fire* **10**, 329–342.

Moriarty JJ and McComb WC (1983) The long-term effect of timber stand improvement on snag and cavity densities in the Central Appalachians. In *Snag Habitat Management. Proceedings of the Symposium 7–9 June 1983*. Northern Arizona University, Flagstaff. (Eds JW Davis, GA Goodwin and RA Ockenfels) pp. 40–44.

Morissette JL, Cobb TP, Brigham RM and James PC (2002) The response of boreal forest songbird communities to fire and post-fire harvesting. *Canadian Journal of Forest Research* **12**, 2169–2183.

Morrison PH and Swanson FJ (1990) Fire history and pattern in a Cascade mountain landscape. *General Technical Report PNW-GTR-254*. USDA Forest Service, Portland, Oregon.

Morrison ML, Marcot BG and Mannan RW (2006) *Wildlife–habitat Relationships. Concepts and Applications*. Island Press, Washington, DC.

Mueck SG (1990) The floristic composition of mountain ash and alpine ash forests in Victoria. *SSP Technical Report No. 4*. Department of Conservation and Environment, Melbourne.

Mueck SG, Ough K and Banks JC (1996) How old are wet forest understoreys? *Australian Journal of Ecology* **21**, 345–348.

Munks SA, Mooney N, Pemberton D and Gales R (2004) An update on the distribution and status of possums and gliders in Tasmania, including offshore islands. In *The Biology of Australian Possums and Gliders*. (Eds RL Goldingay and SM Jackson) pp. 111–129. Surrey Beatty & Sons, Sydney.

Myroniuk P (Ed.) (1995) *International Studbook for Leadbeater's possum, Gymnobelideus leadbeateri*. Royal Melbourne Zoo, Melbourne.

Nagy KA (1987) Field metabolic rate and food requirement scaling in mammals and birds. *Ecological Monographs* **57**, 111–128.

Naiman RJ and Bilby RE (Eds) (1998) *River Ecology and Management: Lessons from the Pacific Coastal Ecoregion*. Springer-Verlag, New York.

Naiman RJ and Turner MG (2000) A future perspective on North America's freshwater ecosystems. *Ecological Applications* **10**, 958–970.

National Association of Forest Industries (1989) *Wood Production and the Environment: Working in Harmony with Nature*. National Association of Forest Industries, Canberra.

Nelson CR and Halpern CB (2005) Edge-related responses of understorey plants to aggregated retention harvest in the Pacific Northwest. *Ecological Applications* **15**, 196–209.

Nelson JL and Morris BJ (1994) Nesting requirements of the yellow-tailed black cockatoo *Calyptorynchus funereus* in mountain ash forest *Eucalyptus regnans* and implications for forest management. *Wildlife Research* **21**, 257–268.

Neumann FG (1991) Responses of litter arthropods to major natural or artificial ecological disturbances in mountain ash forest. *Australian Journal of Ecology* **16**, 19–32.

Neumann FG (1992) Responses of foraging ant populations to high-intensity wildfire, salvage logging and natural regeneration processes in *Eucalyptus regnans* regrowth forest of the Victorian Central Highlands. *Australian Forestry* **55**, 29–38.

Neumann FG and Marks GV (1976) A synopsis of important pests and diseases in Australian forests and nurseries. *Australian Forestry* **39**, 83–102.

Neumann FG, Harris JA and Wood CH (1977) The phasmatid problem in mountain ash forests of the Central Highlands of Victoria. *Bulletin No. 25*. Forests Commission of Victoria, Melbourne.

New TR (1995) Onychophora in invertebrate conservation: priorities, practice and prospects. *Onychophora: Past and Present* **114**(1), 77–89.

Newton I (1994) The role of nest sites in limiting the numbers of hole-nesting birds: a review. *Biological Conservation* **70**(3), 265–276.

Newton I (1998) *Population Limitation in Birds*. Academic Press, London.

Niemela J, Langor D and Spence JR (1993) Effects of clear-cut harvesting on boreal ground-beetle assemblages (Coleoptera: Carabidae) in western Canada. *Conservation Biology* **7**, 551–561.

Niemela J, Haila Y and Punttila P (1996) The importance of small-scale heterogeneity in boreal forests: variation in diversity in forest-floor invertebrates across the succession gradient. *Ecography* **19**(3), 352–368.

Nitschke C, Hickey G, Kennan R and Arndt S (2009) Assessing the vulnerability of forest ecosystems to climate change in south-east Australia. *Global Ecology and Biogeography*.

Nix HA (1986) A biogeographic analysis of the Australian elapid snakes. In *Atlas of Elapid Snakes. Australian Flora and Fauna Series, No. 7*. (Ed. R Longmore) pp. 4–15. Australian Government Publishing Service, Canberra.

Noble WS (1977) *Ordeal by Fire. The Week a State Burned Up*. Hawthorn Press, Melbourne.

Noble IR and Slatyer RO (1980) The use of vital attributes to predict successional changes in plant communities subject to recurrent disturbances. *Plant Ecology* **43**, 5–21.

Norton TW (1988) *Ecology of Greater Gliders in Different Eucalypt Forests in South-eastern New South Wales*. Australian National University, Canberra.

Norton DA (1996) Monitoring biodiversity in New Zealand's terrestrial ecosystems. In *Papers from a Seminar Series on Biodiversity*. (Eds B McFadgen and S Simpson) pp. 19–41. New Zealand Department of Conservation, Wellington.

Noss RF, Beier P, Covington W, Grumbine E, Lindenmayer DB, Prather J, Schmiegelow F, Sisk T and Vosick D (2006) Integrating ecological restoration and conservation biology: a case study for ponderosa pine ecosystems of the south-west. *Restoration Ecology* **14**, 4–10.

O'Dowd DJ and Gill AM (1984) Predator satiation and site alteration following fire: mass reproduction of alpine ash (*Eucalyptus delegatensis*) in south-eastern Australia. *Ecology* **65**, 1052–1066.

O'Grady JJ, Reed DH, Brook BW and Frankham R (2004) What are the best correlates of predicted extinction risk? *Biological Conservation* **118**, 513–520.

Opdam P and Wascher D (2004) Climate change meets habitat fragmentation: linking landscape and biogeographical scale levels in research and conservation. *Biological Conservation* **117**, 285–297.

Orians GH (1986) The place of science in environmental problem-solving. *Environment* **28**, 12–17, 38–41.

O'Shaughnessy P and Jayasuriya J (1991) Managing the ash-type forest for water production in Victoria. In *Forest Management in Australia*. (Eds FH McKinnell, ER Hopkins and JED Fox) pp. 341–363. Surrey Beatty & Sons, Sydney.

Ough K (2002) Regeneration of wet forest flora a decade after clearfelling or wildfire: is there a difference? *Australian Journal of Botany* **49**, 645–664.

Ough K and Murphy A (1996) The effect of clearfell logging on tree-ferns in Victorian wet forest. *Australian Forestry* **59**, 178–188.

Ough K and Murphy A (1998) Understorey islands: a method of protecting understorey flora during clearfelling operations. *Internal VSP Report No. 29.* Department of Natural Resources and Environment, Melbourne.

Ough K and Murphy A (2004) Decline in tree-fern abundance after clearfell harvesting. *Forest Ecology and Management* **199**, 153–163.

Ough K and Ross J (1992) Floristics, fire and clearfelling in wet forests of the Central Highlands of Victoria. *Silvicultural Systems Project Technical Report No. 11.* Department of Conservation and Environment, Melbourne.

Outerbridge RA and Trofymow JA (2004) Diversity of ectomycorrhizae on experimental planted Douglas fir seedlings in variable retention forestry sites on Vancouver Island. *Canadian Journal of Botany* **82**, 1671–1681.

Owen WH (1964) Studies in mammalian ecology. MSc thesis. University of Melbourne, Melbourne.

Pahl L (1984) Diet preference, diet composition and population density of the ringtail possum (*Pseudocheirus peregrinus cooki*) in several plant communities in southern Victoria. In *Possums and Gliders.* (Eds AP Smith and ID Hume) pp. 253–260. Australian Mammal Society, Sydney.

Pahl L (1987) Survival, age determination and population structure of the common ringtail possum in a *Eucalyptus* woodlands and *Leptospermum* thickets in southern Victoria. *Australian Journal of Zoology* **35**, 487–506.

Palmer GC and Bennett AF (2006) Riparian zones provide for distinct bird assemblages in forest mosaics of south-east Australia. *Biological Conservation* **130**, 447–457.

Parks Victoria (2000) *State of the Parks 2000: Park Profiles.* Parks Victoria, Melbourne.

Parks Victoria (2002) *Yarra Ranges National Park Management Plan.* Parks Victoria, Melbourne.

Parmesan C (2006) Ecological and evolutionary responses to recent climate change. *Annual Review of Ecology Evolution and Systematics* **37**, 637–669.

Parmesan C and Yohe G (2003) A globally coherent fingerprint of climate change impacts across natural systems. *Nature* **421**, 37–42.

Parminter J (1998) Natural disturbance ecology. In *Conservation Biology Principles for Forested Landscapes.* (Eds J Voller and S Harrison) pp. 3–41. UBC Press, Vancouver, BC.

Parr CL and Andersen AN (2006) Patch mosaic burning for biodiversity conservation: a critique of the pyrodiversity paradigm. *Conservation Biology* **20**(6), 1610–1619.

Parry BB (1997) Abiotic edge effects in wet sclerophyll forest in the Central Highlands of Victoria. MSc thesis. University of Melbourne, Melbourne.

Peakall R, Ruibal M and Lindenmayer DB (2003) Spatial autocorrelation analysis offers new insights into gene flow in the Australian bush rat, *Rattus fuscipes. Evolution* **57**(5), 1182–1195.

Perry DA (1994) *Forest Ecosystems.* Johns Hopkins Press, Baltimore.

Peters RH (1991) *A Critique for Ecology.* Cambridge University Press, Cambridge.

Pickett STA (1989) Space-for-time substitution as an alternative to long-term studies. In *Longterm Studies in Ecology: Approaches and Alternatives.* (Ed. GE Likens) pp. 110–135. Springer-Verlag, New York.

Pickett ST and Thompson JH (1978) Patch dynamics and the design of nature reserves. *Biological Conservation* **13**, 27–37.

Pittock AB (2005) *Climate Change: Turning Up the Heat.* CSIRO Publishing, Melbourne.

Polglase PJ and Attiwill PM (1992) Nitrogen and phosphorus cycling in relation to stand age of *Eucalyptus regnans* F. Muell. I. Return from plant to soil in litterfall. *Plant and Soil* **142**(2), 151–166.

Pope ML, Lindenmayer DB and Cunningham RB (2004) Patch use by the greater glider (*Petauroides volans*) in a fragmented forest ecosystem. I. Home range size and movements. *Wildlife Research* **31**(6), 559–568.

Possingham HP and Davies I (1995) ALEX: a model for the viability analysis of spatially structured populations. *Biological Conservation* **73**(2), 143–150.

Possingham HP, Lindenmayer DB and Norton TW (1993) A framework for improved threatened species management using population viability analysis. *Pacific Conservation Biology* **1**, 39–45.

Possingham HP, Lindenmayer DB, Norton TW and Davies I (1994) Metapopulation viability analysis of the greater glider *Petauroides volans* in a wood production area. *Biological Conservation* **70**(3), 227–236.

Possingham HP, Lindenmayer DB and McCarthy MA (2001) Population viability analysis. In *Encyclopedia of Biodiversity*. (Ed. SA Levin) Vol. 4, pp. 831–843. Academic Press, San Diego.

Predavec M (1990) Predation and prey community structure: a study of the impact of *Antechinus stuartii* (Marsupialia: Dasyuridae) on leaf-litter invertebrate communities. BSc Hons thesis. University of Sydney, Sydney.

Price O, Woinarski JC, Liddle DL and Russel-Smith J (1995) Patterns of species composition and reserve design for a fragmented estate monsoon rainforests in the Northern Territory. *Biological Conservation* **74**, 9–19.

Pryor LD (1976) *The Biology of Eucalypts*. Edward Arnold, London.

Puettmann KJ, Coates KD and Messier C (2009) *A Critique of Silviculture: Managing for Complexity*. Island Press, Washington, DC.

Pulliam HR, Dunning JB and Liu J (1992) Population dynamics in complex landscapes: a case study. *Ecology Applications* **2**, 165–177.

Pyke GH and Recher HF (1983) Censusing Australian birds: a summary of procedures and a scheme for standardisation of data presentation and storage. In *Methods of Censusing Birds in Australia. Proceedings of a Symposium Organised by the Zoology Section of ANZAAS and Western Australian Group of the Royal Australasian Ornithologists Union*. (Ed. SJ Davies) pp. 55–63. Department of Conservation and Environment, Perth.

Radeloff VC, Mladenoff DJ and Boyce MS (2000) Effects of interacting disturbances on landscape patterns: budworm defoliation and salvage logging. *Ecological Applications* **10**, 233–247.

Read P (2007) Biosphere carbon stock management: addressing the threat of abrupt climate change in the next few decades: an editorial essay. *Climatic Change*.

Read J and Brown M (1996) Ecology of Australian *Nothofagus* forests. In *The Ecology and Biogeography of Nothofagus Forests*. (Eds TT Veblen, RS Hill and J Read) pp. 131–181. Yale University Press, New Haven.

Rebertus AJ, Kitzberger T, Veblen TT and Roovers LM (1997) Blowdown history and landscape patterns in the Andes of Tierra del Fuego, Argentina. *Ecology* **78**, 678–692.

Recher HF (1969) Bird species diversity and habitat diversity in Australia and North America. *American Naturalist* **103**, 75–121.

Recher HF (1996) Conservation and management of eucalypt forest vertebrates. In *Conservation of Faunal Diversity in Forested Landscapes*. (Eds R DeGraff and I Miller) pp. 339–88. Chapman & Hall, London.

Reed JM, Mills LS, Dunning JB, Menges ES, McKelvey KS, Frye R, Beissenger SR, Antett M and Miller P (2002) Emerging issues in population viability analysis. *Conservation Biology* **16**, 7–19.

Resource Assessment Commission (1991) 'Forest and timber inquiry'. Draft Report, Vol. 1. Australian Government Publishing Service, Canberra.

Resource Assessment Commission (1992) 'Forest and timber inquiry'. Draft Report, Vol. 2. Australian Government Publishing Service, Canberra.

Ringold PL, Alegria J, Czaplwski RL, Mulder BS, Tolle T and Burnett K (1996) Adaptive monitoring design for ecosystem management. *Ecological Applications* **6**, 745–747.

Roberts KA (1991) Field monitoring: confessions of an addict. In *Monitoring for Conservation and Ecology*. (Ed. FB Goldsmith) pp. 179–212. Chapman & Hall, London.

Robinson AC (1987) The ecology of the bush rat, *Rattus fuscipes* (Rodentia: Muridae), in Sherbrooke Forest, Victoria. *Australian Mammalogy* **11**, 35–49.

Robinson G and Zappieri J (1999) Conservation policy in time and space: lessons from divergent approaches to salvage logging on public lands. *Conservation Ecology* **3**(1).

Roe JH and Ruseink A (2000) *Natural Dynamics Silviculture: A Discussion of Natural Community-based Forestry Practices*. Nature Conservancy, Washington, DC.

Rose C, Marcot BG, Mellen TK, Ohmann JL, Waddell K, Lindley D and Schreiber B (2001) Decaying wood in Pacific north-west forests: concepts and tools for habitat management. In *Wildlife–habitat Relationships in Oregon and Washington*. (Eds D

Johnson and T O'Neil) pp. 580–623. Oregon State University Press, Corvallis.

Rosenweig ML (1995) *Species Diversity in Space and Time*. Cambridge University Press, Cambridge.

Rosenzweig C, Karoly D, Vicarelli M, Neofotis P, We QC, Menzel AG, Root TL, Estrella N, Seguin, B, Tryjanowski P, Liu C, Eawlins S and Imeson A (2008) Attributing physical and biological impacts to anthropogenic climate change. *Nature* **45**, 353–357.

Rothamsted Research (2009) *Making a Difference: The Past and Future Economic and Societal Impact of Rothamsted Research*. Rothamsted Research, Harpenden, England.

Rotherham I (1983) Suppression of surrounding vegetation by veteran trees in karri *Eucalyptus diversicolor*. *Australian Forestry* **46**, 8–13.

Routley R and Routley V (1975) *The Fight for the Forests: The Takeover of Australian Forests for Pines, Woodchips and Intensive Forestry*. Research School of Social Sciences, Australian National University, Canberra.

Rubsamen K, Hume ID, Foley WJ and Rubsamen U (1984) Implications of the large surface area to body mass ratio on the heat balance of the greater glider (*Petaroides volans:* Marsupialia). *Journal of Comparative Physiology B*, **154**, 105–111.

Rülcker C, Angelstam P and Rosenberg P (1994) Natural forest-fire dynamics can guide conservation and silviculture in boreal forests. *SkogForsk* **2**, 1–4.

Runkle JR (1982) Patterns of disturbances in some old-growth mesic forests of eastern North America. *Ecology* **63**, 1533–1546.

Russell RE, Saab VA, Dudley JG and Rotella JJ (2006) Snag longevity in relation to wildfire and postfire salvage logging. *Forest Ecology Management* **232**, 179–187.

Russell-Smith J, Whitehead PJ, Cook GD and Hoare JL (2003) Response of eucalyptus-dominated savanna to frequent fires: lessons from Munmarlary 1973–1996. *Ecological Monographs* **73**, 349–375.

Saint-Germain M, Drapeau P and Héebert C (2004) Comparison of *Coleoptera* assemblages from recently burned and unburned black spruce forests of north-eastern North America. *Biological Conservation* **118**, 583–592.

Sanderson HR (1975) Den-tree management for gray squirrels. *Wildlife Society Bulletin* **3**, 125–131.

Sanecki GM, Green K, Wood H, Lindenmayer D and Sanecki KL (2006) The influence of snow cover on home range and activity of the Bush Rat (*Rattus fuscipes*) and the Dusky Antechinus (*Antechinus swainsonii*). *Wildlife Research* **33**, 489–496.

Saveneh AG and Dignan P (1998) The use of shelterwood in *Eucalyptus regnans* forest: the effect of overwood removal at three years on regeneration stocking and health. *Australian Forestry* **4**, 252–259.

Savolainen P, Leitner T, Wilton AN, Matisoo-Smith E and Lundeberg J (2004) A detailed picture of the origin of the Australian dingo, obtained from the study of mitochondrial DNA. *Proceedings of the National Academy of Sciences* **101**, 12387–12390.

Schmiegelow FKA, Machtans CS and Hannon SJ (1997) Are boreal birds resilient to forest fragmentation? An experimental study of short-term community responses. *Ecology* **78**(6), 1914–1932.

Schmiegelow FKA, Stepnisky DP, Stambaugh CA and Koivula M (2006) Reconciling salvage logging of boreal forests with a natural-disturbance management model. *Conservation Biology* **20**(4), 971–983.

Schoener TW, Spiller DA and Losos JB (2004) Variable ecological effects of hurricanes: the importance of seasonal timing for survival of lizards on Bahamian islands. *Proceedings of the National Academy of Sciences* **101**, 177–181.

Schowalter TD, Zhang YL and Progar RA (2005) Canopy arthropod response to density and distribution of green trees retained after partial harvest. *Ecological Applications* **15**, 1594–1603.

Schulte LA and Mladenoff DJ (2005) Severe wind and fire regimes in northern forests: historical variability at the regional scale. *Ecology* **86**, 431–445.

Schulze E-D, Wirth C and Heimann M (2000) Climate change: managing forests after Kyoto. *Science* **289**, 2058–2059.

Scott LK, Hume ID and Dickman CR (1999) Ecology and population biology of Long-nosed Bandicoots (*Perameles nasuta*) at North Head, Sydney Harbour National Park. *Wildlife Research* **26**, 805–821.

Scott JM, Abbitt RJF and Groves CR (2001) What are we protecting? *Conservation Biology in Practice* **2**, 18–19.

Scotts DJ (1991) Old-growth forests: their ecological characteristics and value to forest-dependent vertebrate fauna of south-east Australia. In *Conservation of Australia's Forest Fauna*. (Ed. D Lunney) pp. 147–159. Royal Zoological Society of NSW, Sydney.

Scotts DJ and Craig SA (1988) Improved hair-sampling tube for the detection of small mammals. *Australian Wildlife Research* **15**, 469–472. <http://www.publish.csiro.au/nid/144/paper/WR9880469A.htm>

Sedell JR, Bisson PA, Swanson FJ and Gregory SV (1988) What we know about large trees that fall into streams and rivers. In *From the Forest to the Sea: A Story of Fallen Trees*. (Eds C Maser, RF Tarrant, JM Trappe and JF Franklin) pp. 47–81. Technical Report PNW-GTR-229, USDA Forest Service General, Portland, Oregon.

Seebeck JH (1971) Distribution and habitat of the broad-toothed rat *Mastacomys fuscus* Thomas (Rodentia, Muriade) in Victoria. *Victorian Naturalist* **88**, 310–323.

Seebeck JH, Suckling GC and Macfarlane MA (1983) Leadbeater's possum: survey by stagwatching. *Victorian Naturalist* **100**, 92–97.

Seebeck JH, Warneke RM and Baxter BJ (1984) Diet of the bobuck, *Trichosurus caninus* Ogilby (Marsupialia: Phalangeridae) in a mountain forest in Victoria. In *Possums and Gliders*. (Eds AP Smith and ID Hume) pp. 145–154. Surrey Beatty & Sons, Sydney.

Seymour RS, White AS and deMaynadier PG (2002) Natural disturbance regimes in north-eastern North America: evaluating silvicultural systems using natural scales and frequencies. *Forest Ecology and Management* **155**, 357–367.

Shaffer ML (1990) Population viability analysis. *Conservation Biology* **4**, 39–40.

Shakesby RA, Boakes DJ and Coelho C (1996) Limiting the soil degradational impacts of wildfire in pine and *Eucalyptus* forests in Portugal: a comparison of alternative post-fire management practices. *Applied Geography* **16**, 337–355.

Shanley JB, Likens GE and Haeuber R (2007) Who needs environmental monitoring? *Frontiers in Ecology and the Environment* **5**, 253–260.

Shea SR, Abbott JA, Armstrong JA and McNamara KJ (1997) Sustainable conservation: a new integrated approach to nature conservation in Australia. In *Conservation Outside Nature Reserves*. (Eds P Hale and D Lamb) pp. 39–48. University of Queensland Press, Brisbane.

Shiel D and Burslem FR (2003) Disturbing hypotheses in tropical forests. *Trends in Ecology and Evolution* **18**, 18–26.

Shine R (1991) *Australian Snakes: A Natural History*. Reed Books, Sydney.

Shore TL, Brooks J and Stone JE (2003) Mountain pine beetle symposium: challenges and solutions. *Information Report BC-X-399*. Canadian Forest Service/Pacific Forestry Centre, Victoria, BC.

Shrader-Frechette KS and McCoy ED (1993) *Method in Ecology: Strategies for Conservation*. Cambridge University Press, Cambridge.

Sicamma TG, Fahey TJ, Johnson CE, Sherry TW, Denny EG, Girdler EB, Likens GE and Schwartz PA (2007) Population and biomass dynamics of trees in a northern hardwood forest at Hubbard Brook. *Canadian Journal of Forest Research* **37**, 737–749.

Smith AP (1980) The diet and ecology of Leadbeater's possum and the sugar glider. PhD thesis. Monash University, Melbourne.

Smith AP (1982) Diet and feeding strategies of the marsupial sugar glider in temperate Australia. *Journal of Animal Ecology* **51**, 149–166.

Smith AP (1984a) Demographic consequences of reproduction, dispersal and social interaction in a population of Leadbeater's possum *Gymnobelideus leadbeateri*. In *Possums and Gliders*. (Eds AP Smith and ID Hume) pp. 359–373. Surrey Beatty & Sons, Sydney.

Smith AP (1984b) Diet of Leadbeater's possum *Gymnobelideus leadbeateri* (Marsupialia). *Australian Wildlife Research* **11**, 265–273.

Smith GC and Agnew G (2002) The value of 'bat boxes' for attracting hollow-dependent fauna to farm forestry plantations in south-east Queensland. *Ecological Management and Restoration* **3**, 37–46.

Smith AP and Winter JW (1984) A key and field guide to the Australian possums, gliders and koala. In *Possums and Gliders*. (Eds AP Smith and ID Hume) pp. 579–594. Surrey Beatty & Sons, Sydney.

Smith RB and Woodgate P (1985) Appraisal of fire damage for timber salvage by remote sensing in mountain ash forests. *Australian Forestry* **48**, 252–263.

Smith AP, Nagy KA, Fleming MR and Green B (1982) Energy requirements and turnover in free-living Leadbeater's possums, *Gymnobelideus leadbeateri* (Marsupialia: Petauridae). *Australian Journal of Zoology* **30**, 737–749.

Smith AP, Lindenmayer DB and Suckling GC (1985) The ecology and management of Leadbeater's possum. *Research Report to World Wildlife Fund.* University of New England, Armidale, NSW.

Smith AP, Lindenmayer DB, Begg RJ, Macfarlane MA, Seebeck JH and Suckling GC (1989a) Evaluation of the stag-watching technique for census of possums and gliders in tall open forest. *Australian Wildlife Research* **16**(5), 575–580.

Smith AP, Wellham GS and Green SW (1989b) Seasonal foraging activity and microhabitat selection by echidnas *Tachyglossus aculeatus* on the New England Tablelands. *Australian Journal of Ecology* **14**, 457–466.

Smith B, Augee M and Rose S (2003) Radio-tracking studies of common ringtail possums, *Pseudocheirus peregrinus*, in Manly Dam Reserve, Sydney. *Proceedings of Linnean Society of New South Wales* **124**, 183–194.

Sollins P, Cline SP, Verhoeven T, Sachs D and Spycher G (1987) Patterns of log decay in old-growth Douglas fir forests. *Canadian Journal of Forest Research* **17**, 1585–1595.

Sousa WP (1984) The role of disturbance in natural communities. *Annual Review of Ecology and Systematics* **15**, 353–391.

Specht RL (1981) Major vegetation formations in Australia. In *Ecological Biogeography of Australia*. (Ed. A Keast) pp. 81–103. W. Junk, The Hague.

Specht RL and Specht A (1999) *Australian Plant Communities: Dynamics of Structure, Growth and Biodiversity*. Oxford University Press, Melbourne.

Spellerberg IF (1994) *Monitoring Ecological Change*. 2nd edn. Cambridge University Press, Cambridge.

Spies TA and Turner MG (1999) Dynamic forest mosaics. In *Managing Biodiversity in Forest Ecosystems*. (Ed. M Hunter III) pp. 95–160. Cambridge University Press, Cambridge.

Spies TA, Hemstrom MA, Youngblood A and Hummel S (2004) Conserving old-growth forest diversity in disturbance-prone landscapes. *Conservation Biology in Practice* **20**, 351–362.

Spies TA, McComb BC, Kennedy RSH, McGrath MT, Olsen K and Pabst RJ (2007) Potential effects of forest policies on terrestrial biodiversity in a multi-ownership province. *Ecological Applications* **17**(1), 48–65.

Squire RO (1987) *Silvicultural Systems for Victoria's Commercially Important Mountain Eucalypt Forests: Project Brief.* Public Lands and Forest Division, Melbourne.

Squire RO (1990) *Report on the Progress of the Silvicultural Systems Project July 1986–June 1989*. Department of Conservation and Environment, Melbourne.

Squire RO (1993) The professional challenge of balancing sustained wood production and ecosystem conservation in the native forests of south-eastern Australia. *Australian Forestry* **56**, 237–248.

Squire RO, Campbell RG, Wareing KJ and Featherston GR (1991) The mountain ash forests of Victoria: ecology, silviculture and management for wood production. In *Forest Management in Australia*. (Eds FH McKinnell, ER Hopkins and JED Fox) pp. 38–57. Surrey Beatty & Sons, Sydney.

Stankey GH, Bormann BT, Ryan C, Shindler B, Sturtevant V, Clark RN and Philpot C (2003) Adaptive management and the Northwest Forest Plan: rhetoric and reality. *Journal of Forestry* **101**(1), 40–46.

Starfield AM and Bleloch AL (1992) *Building Models for Conservation and Wildlife Management*. Burgess International Group, Edina.

Stocks BJ, Mason JA, Todd JB, Bosch EM, Wotton BM, Amiro BD, Flannigan MD, Hirsch KG, Logan KA, Martell DL and Skinner WR (2002) Large forest fires in Canada, 1959–1997. *Journal of Geophysical Research* **108**(D1:FFR5), 1–12.

Strahan R (Ed.) (1995) *Complete Book of Australian Mammals*. Angus & Robertson, Sydney.

Strayer DL, Glitzenstein JS, Jones C, Kolasa J, Likens GE, McDonnell M, Parker GG and Pickett STA (1986) Long-term ecological studies: an illustrated account of their design, operation, and importance to ecology. In *Occasional Publication of the Institute* of Ecosystem Studies, Vol. 2, pp. 1–38. Institute of Ecosystem Studies, Millbrook, New York.

Strengers BJ, van Minnen JG and Eickhout B (2007) The role of carbon plantations in mitigating climate change: potentials and costs. *Climate Change* **88**, 343–366.

Suckling GC (1978) A hair sampling tube for the detection of small mammals in trees. *Australian Wildlife Research* **5**, 249–252.

Suckling GC (1982) Value of reserved habitat for mammal conservation in plantations. *Australian Forestry* **45**, 19–27.

Suckling GC (1984) Population ecology of the sugar glider in a system of fragmented habitats. *Australian Wildlife Research* **11**, 49–75.

Sullivan TP and Sullivan DS (2001) Influence of variable retention harvests on forest ecosystems. II. Diversity and population dynamics of small mammals. *Journal of Applied Ecology* **38**, 1234–1252.

Sutherland W (1996) *Ecological Census Techniques*. Cambridge University Press, Cambridge.

Sverdrup-Thygeson A and Lindenmayer DB (2003) Ecological continuity and assumed indicator fungi in boreal forest: the importance of the landscape matrix. *Forest Ecology and Management* **174**(1–3), 353–363.

Swanson FJ, Jones JA, Wallin DO and Cissel JH (1994) Natural variability: implications for ecosystem management. In *Eastside Forest Ecosystem Healthy Assessment. Vol. II. Ecosystem Management: Principles and Applications.* (Eds ME Jensen and PS Bourgeron) pp. 80–94. General Technical Report PNW-GTR-318. USDA Forest Service, Portland, Oregon.

Swift K (2006) Variable retention forestry science forum: overview and key messages. *BC Journal of Ecosystems and Management* **7**, 3–8.

Syrjanen K, Kalliola R, Puolasmaa A and Mattsson J (1994) Landscape structure and forest dynamics in subcontinental Russian European taiga. *Annales Zoologici Fennici* **31**(1), 19–34.

Tallmon D and Mills LS (1994) Use of logs within home ranges of Californian red-backed voles on a remnant of forest. *Journal of Mammalogy* **75**, 97–101.

Tang SM and Gustafson EJ (1997) Perception of scale in forest management planning: challenges and implications. *Landscape and Urban Planning* **39**, 1–9.

Taulman JF, Smith KG and Thill RE (1998) Demographic and behavioural responses of southern flying squirrels to experimental logging in Arkansas. *Ecological Applications* **8**, 1144–1155.

Taylor R (1990) Occurrence of log-dwelling invertebrates in regeneration and old-growth wet sclerophyll forest in southern Tasmania. *Papers and Proceedings of Royal Society of Tasmania* **119**, 7–15.

Taylor BL (1995) The reliability of using population viability analysis for risk classification of species. *Conservation Biology* **9**, 551–558.

Taylor AC, Tyndale-Biscoe H and Lindenmayer DB (2007) Unexpected persistence on habitat islands: genetic signatures reveal dispersal of a eucalypt-dependent marsupial through a hostile pine matrix. *Molecular Ecology* **16**, 2655–2666.

Temple SA and Cary JR (1988) Modelling dynamics of habitat interior bird populations in fragmented landscapes. *Conservation Biology* **2**, 340–347.

Thomas JW (1979) Wildlife habitats in managed forests in the Blue Mountains of Oregon and Washington. In *USDA Agricultural Handbook 553*, p. 512. US Government Printing Office, Washington, DC.

Thompson WL, White GC and Gowan C (1998) *Monitoring Vertebrate Populations*. Academic Press, London.

Thompson JR, Spies TA and Ganio LM (2007) Reburn severity in managed and unmanaged vegetation in a large wildfire. *Proceedings of National Academy of Sciences.*

Thrower J (2005) Earth Island Institute versus United States Forest Service: salvage logging plans in Star Fire region undermine Sierra Nevada framework. *Ecology Law Quarterly* **32**, 721–728.

Tilghman NG (1989) Impacts of white-tailed deer on forest regeneration in north-western Pennsylvania. *Journal of Wildlife Management* **53**, 524–532.

Travis JMJ (2003) Climate change and habitat destruction: a deadly anthropogenic cocktail. *Proceedings of Royal Society of London, B Series* **270**, 467–473.

Triggs B (1996) *Tracks, Scats and Other Traces*. Oxford University Press, Melbourne.

Troy S and Coulson G (1993) Home range of the Swamp Wallaby *Wallabia bicolor*. *Wildlife Research* **20,** 571–577.

Tunbridge BR and Glenane TJ (1983) *Fisheries Value and Classification of Fresh and Estuarine Waters in Victoria*. Government Printer, Melbourne.

Turner V (1983) Eastern pygmy possum. In *Complete Book of Australian Mammals*. (Ed. R Strahan) pp. 160–161. Angus & Robertson, Sydney.

Turner MG, Romme WH, Gardner RH and Hargrove WW (1997) Effects of fire size and pattern on early succession in Yellowstone National Park. *Ecological Monographs* **67,** 411–433.

Turner MG, Baker WL, Peterson CJ and Peet RK (1998) Factors influencing succession: lessons from large, infrequent natural disturbances. *Ecosystems* **1,** 511–523.

Turner MG, Romme WH and Tinker DB (2003) Surprises and lessons from the 1988 Yellowstone fires. *Frontiers in Ecology and Environment* **1,** 351–358.

Tyler MJ (1994) *Australian Frogs: A Natural History*. Revised edn. Reed Books, Sydney.

Tyndale-Biscoe H (2005) *Life of Marsupials*. CSIRO Publishing, Melbourne.

Tyndale-Biscoe CH and Smith RF (1969) Studies of the marsupial glider, *Schoinobates volans* Kerr III. Response to habitat destruction. *Journal of Animal Ecology* **38,** 651–659.

Ulbricht R, Hinrichs A and Ruslim Y (1999) Technical guidelines for salvage felling in rehabilitation areas after forest fires. *Report 1 of the Sustainable Forest Management Project*. Samarinda, Indonesia.

Urban D, O'Neill RV and Shugart HH (1987) Landscape ecology: a hierarchical perspective can help scientists understand spatial patterns. *BioScience* **37,** 119–127.

Van der Meer PJ and Dignan P (2007) Regeneration after 8 years in artificial canopy gaps in mountain ash (*Eucalyptus regnans* F. Muell.) forest in south-eastern Australia. *Forest Ecology and Management* **244,** 102–111.

van der Rhee R and Loyn RH (2002) The influence of time since fire and distance from fire boundary on the distribution of arboreal marsupials in the *Eucalyptus*-dominated forest in the Central Highlands of Victoria. *Wildlife Research* **29,** 151–158.

Vanderwel MC, Malcolm JR and Smith SM (2006) An integrated model for snag and downed woody debris decay class transitions. *Forest Ecology and Management* **234,** 48–59.

van Pelt R (2007) *Identifying Mature and Old Forests in Western Washington*. Department of Natural Resources, Olympia, Washington State.

Van Wagner CE (1976) The line intersect method in forest fuel sampling. *Forest Science* **14,** 20–26.

Veblen TT, Hadley KS, Nel EM, Kitsberger T, Reid M and Villalba R (1994) Disturbance regime and disturbance interactions in a Rocky Mountain subalpine forest. *Journal of Ecology* **82,** 125–135.

Veblen TT, Donoso C, Kitzberger T and Rebertus AJ (1996) Ecology of southern Chilean and Argentinean *Nothofagus* forests. In *The Ecology and Biogeography of Nothofagus Forests*. (Eds TT Veblen, RS Hill and J Read) pp. 293–353. Yale University Press, New Haven, Connecticut.

Vertessey RA and Watson FG (2001) Factors determining relations between stand age and catchment water balance in mountain ash forests. *Forest Ecology and Management* **143,** 13–26.

Victorian Government (1992) *Flora and Fauna Guarantee Strategy: Conservation of Victoria's Biodiversity*. Department of Conservation and Environment, Melbourne.

Viggers KL and Lindenmayer DB (1995) The use of tiletamine hydrochloride and zolazepam hydrochloride for sedation of the mountain brushtail possum, *Trichosurus caninus* Ogilby (Phalangeridae, Marsupialia). *Australian Veterinary Journal* **72**(6), 215–216.

Viggers KL and Lindenmayer DB (2000) A population study of the mountain brushtail possum (*Trichosurus caninus*) in the central highlands of Victoria. *Australian Journal of Zoology* **48**(2), 201–216.

Viggers KL and Lindenmayer DB (2004) A review of the biology of the short-eared possum *Trichosurus caninus* and the mountain brushtail possum *Trichosurus cunninghami*. In *The Biology of Australian Possums and Gliders*. (Eds RL Goldingay and SM Jackson) pp. 490–505. Surrey Beatty & Sons, Sydney.

von Hartman L (1971) Population dynamics. In *Avian Biology. Vol. 1*. (Eds DS Farner and JR King), pp. 391–459. Academic Press, London.

Wace N (1977) Assessment of dispersal of plant species: the car-borne flora of Canberra. *Ecological Society of Australia* **10**, 166–186.

Wadleigh L and Jenkins MJ (1996) Fire frequency and the vegetative mosaic of a spruce fir forest in northern Utah. *Great Basin Naturalist* **56**, 28–37.

Wallace LL (Ed.) (2004) *After the Fires: The Ecology of Change in Yellowstone National Park*. Yale University Press, New Haven, Connecticut.

Walters CJ (1986) *Adaptive Management of Renewable Resources*. Macmillan, New York.

Ward SJ (1990a) Life history of the feathertail glider, *Acrobates pygmaeus* (Acrobatidae: Marsupialia) in south-eastern Australia. *Australian Journal of Zoology* **38**, 503–517.

Ward SM (1990b) Life history of the eastern pygmy possum, *Cercatetus nanus* (Burramyidae: Marsupialia) in south-eastern Australia. *Australian Journal of Zoology* **38**, 287–304.

Ward P (2004) The father of mass extinctions. *Conservation in Practice* **5**(3), 12–19.

Wardell-Johnson G and Horowitz P (1996) Conserving biodiversity and the recognition of heterogeneity in ancient landscapes: a case study from south-western Australia. *Forest Ecology and Management* **85**, 219–238.

Warren WG and Olsen PF (1964) A line intersect technique for assessing logging waste. *Forest Science* **10**, 267–276.

Watson DM (2001) Mistletoe: a keystone resource in forests and woodlands worldwide. *Annual Review of Ecology and Systematics* **32**, 219–249.

Watson FG, Vertessy RA, McMahon TA, Rhodes BG and Watson IS (1999) The hydrologic impacts of forestry on the Maroondah catchments. Cooperative Research Centre for Catchment Hydrology/ Melbourne Water, Melbourne.

Wayne AF (2005) The ecology of the koomal (*Trichosurus vulpecula hypoleucus*) and ngwayir (*Pseudocheirus occidentalis*) in jarrah forests of south-western Australia. PhD thesis. Australian National University, Canberra.

Webb G (1995) Habitat use and activity patterns in some south-eastern Australian skinks. In *Biology of Australasian Frogs and Reptiles*. (Eds G Grigg, R Shine and H Ehmann) pp. 23–30. Royal Zoological Society of NSW, Sydney.

Welsh AH, Cunningham RB, Donnelly CF and Lindenmayer DB (1996) Modelling the abundance of rare species: statistical models for counts with extra zeros. *Ecological Modelling* **88**(1–3), 297–308.

Welsh AH, Lindenmayer DB, Donnelly CF and Ruckstuhl A (1998) Use of nest trees by the mountain brushtail possum (*Trichosurus caninus*) (Phalangeridae : Marsupialia). IV. Transitions between den trees. *Wildlife Research* **25**(6), 611–625.

Welsh AH, Cunningham RB and Chambers RL (2000) Methodology for estimating the abundance of rare animals: seabird nesting on North East Herald Cay. *Biometrics* **56**, 22–30.

West Arnhem Fire Management Agreement (2009) The West Arnhem Fire Management Agreement.

Westerling AL, Hidalgo HG, Cayan DR and Swetnam TW (2006) Warming and earlier spring increase western US forest wildfire activity. *Science* **313**, 940–943.

Whelan RJ (1995) *The Ecology of Fire*. Cambridge University Press, Cambridge.

Whelan R, Rodgerson L, Dickman CR and Sutherland EF (2002) Critical lifecycles of plants and animals: developing a process-based understanding of population changes in fire-prone landscapes. In *Flammable Australia: The Fire Regimes and Biodiversity of a Continent*. (Eds RA Bradstock, JE Williams and AM Gill) pp. 94–124. Cambridge University Press, Melbourne.

White PS and Pickett STA (1985) Natural disturbance and patch dynamics: an introduction. In *The Ecology of Natural Disturbance and Patch Dynamics*. (Eds STA Pickett and PS White) pp. 3–13. Academic Press, Orlando, Florida.

Wilkinson DA, Grigg GC and Beard LA (1998) Shelter selection and home range of echidnas, *Tachyglossus aculeatus*, in the highlands of south-east Queensland. *Wildlife Research* **25**, 219–232.

Wilson S and Swan G (2007) *A Complete Guide to the Reptiles of Australia*. New Holland, Sydney.

Wintle BA and Lindenmayer DB (2008) Adaptive risk management for certifiably sustainable forestry. *Forest Ecology and Management*. Doi: 10.1016/j.foreco.2008.06.042.

Woinarski JCZ (1999) Fire and Australian birds: a review. In *Australia's Biodiversity. Responses to Fire: Biodiversity.* (Eds AM Gill, JCZ Woinarski and A York) Technical Paper No. 1, pp. 55–112. Environment Australia, Canberra.

Woinarski JC and Cullen JM (1984) Distribution of invertebrates on foliage in forests in south-eastern Australia. *Australian Journal of Ecology* **9**, 207–232.

Woldendorp G and Kennan R (2005) Coarse woody debris in Australian forest ecosystems: a review. *Australian Ecology* **30**, 834–843.

Woodgate P, Peel W, Ritman KT, Coram JE, Brady A, Rule AJ and Banks JC (1994) *A Study of the Old-growth Forests of East Gippsland.* Department of Conservation and Natural Resources, Melbourne.

Woodgate PW, Peel BD, Coram JE, Farrell SJ, Ritman KT and Lewis A (1996) Old-growth forest studies in Victoria, Australia. Concepts and principles. *Forest Ecology and Management* **85**, 79–84.

Woodward FL and Williams BG (1987) Climate and plant distribution at global and local scales. *Vegetatio* **69**, 189–197.

Yamamoto S-I (1992) The gap theory in forest dynamics. *Botanical Magazine (Tokyo)* **105**, 375–383.

Index

Acacia spp. 120
Agile Antechinus (*Antechinus agilis*) 19–20, 39–40, 117, 119, 142–3, 145, 172, 237
ALEX 152, 153, 156, 158, 160, 165, 209
Alpine Ash (*Eucalyptus delegatensis*) 9, 12–13, 28, 47, 49–52, 58, 63–4, 75, 77, 90, 118, 144, 145, 172–3, 176, 208, 241, 248
animal distribution and abundance 133–49
    2009 wildfires 146
    arboreal marsupials 134–42
        broad-scale climatic factors: BIOCLIM analyses 135–9
        landscape factors 139
        microhabitat-level factors 141–2
        patch-level factors 139–40
        spatial prediction 141
        stand-level factors 140
        tree level 141
    background datasets 134
    birds 143–5
    forest pattern 148–9
    importance of trees with hollows 82–3
    knowledge gaps 147–8
    lessons learned 146–7
    patterns 133, 148
    small terrestrial mammals 142–3
animal communities, composition of 167–80
    background datasets 168
    birds 172–6
        assemblage, factors influencing composition 173–4
        assemblage, relative resilience 175–6
        assembly rules 174
        body size patterns 176
        rank abundance distributions (RADs) 174–5
        species richness 172–3
    co-occurrence patterns and the search for an indicator species 169–70
    forest pattern 177
    knowledge gaps 177
    lessons learned 176
    resource partitioning among arboreal marsupials 170–3
    species richness 168–9, 172–3
animal occurrence 131
    in shrub layers 119–20
    in understorey trees 115–19
    see also population viability analysis (PVA)
animals, partitioning of tree hollow resources
    characteristics of occupied trees 83–4
    co-occupancy 101–4
    factors influencing occupancy 84–102
    long-term tree occupancy 101
    shifting baselines and levels of tree occupancy 101
Antarctic Beech (*Nothofagus moorei*) 61
arboreal marsupials 14–17
    comparisons of extinction risks 162
    distribution and abundance 134–42
        broad-scale climatic factors: BIOCLIM analyses 135–9
        landscape factors 139
        microhabitat-levels factors 141–2
        patch-level factors 139–40
        spatial prediction 141
        stand-level factors 140
        tree level 141
    field surveys 29–38
    and multiple tree hollows (den-swapping) 102–5
    occurrence in understorey trees 115–17
    populations, viability and fire 188–90
    PVA modelling of 152–3, 160–3
    and rainforests 65
    resource partitioning among 170–3
        body size 171
        habitat and nest tree requirements 171–2
        home range 171
        life history attributes 171
    species richness 168–9
    wildlife corridors 225
artificial hollows 106–8
ash-type eucalypt forest 47–69
    background datasets 48
    bioclimatic domains of species 52
    bioclimatic envelope 49, 50, 51
    in a changing climate 261–2
    climate envelopes for each species 48–9
    climatic and environmental analyses 52
    distribution patterns 52–3
    environmental factors 55
    forest pattern 58–9
    frequency curves for various species 49–52
    knowledge gaps 58
    lessons learned 58
    natural disturbances 58
    overstorey 47, 48
    successional trajectories 53–5
    old growth 55–8
    spatial composition 55
    understanding distribution of understorey 48

Austral Mulberry (*Hedycarya angustifolia*) 14, 115

bark streamers 75
bats 20–1
Baw Baw Frog (*Philoria frosti*) 26, 27, 28, 99, 222, 223, 240
BIOCLIM analyses 11, 48–9, 50–2, 93, 135–9, 201
biological legacies, loss of 208–9
biodiversity conservation 215, 218, 220–1
    mitigating logging impacts 229–44
    monitoring 245–54
    reserves 217–27
biodiversity management practices 267
biodiversity research 266–7
birds 21–4
    composition of communities 172–6
        assemblage, factors influencing composition 173–4
        assemblage, relative resilience 175–6
        assembly rules 174
        body size patterns 176
        rank abundance distributions (RADs) 174–5
        species richness 172–3
    distribution and abundance 143–5
    dusk counts 41
    field surveys 40–1
    occurrence in understorey trees 117
    point interval counts 40–1
    and rainforests 65–6, 67
Black Rock Skink (*Egernia saxatilis*) 25, 26
Black Wallaby (*Wallabia bicolor*) 20
Blackwood (*Acacia melanoxylon*) 13
Broad-toothed Rat (*Mastcomys fuscus*) 17
Brown Thornbill (*Acanthiza pusilla*) 67, 144
Brown-headed Honeyeater (*Melithreptus brevirostris*) 65, 67
Bush Rat (*Rattus fuscipes*) 17, 19, 20, 39–40, 64, 65, 76, 119, 125, 142–3, 145, 172, 237
bushfires *see* wildfires 2009

Cambarville 10, 30, 33, 37–8, 63, 75, 101, 102, 103–4, 110, 138, 252
carbon budgets and disturbance regimes 190
carbon economy and management of montane ash forests 262–4
carbon storage values 265–6
caveats 7, 258
Central Highlands of Victoria 3, 7
    climate 11
    fauna 14–28
    flora 11–14, 61–3
    geology 11
    location 9–11
    log diameter 127
    logging 193–4
    monitoring program 245–54

Mountain Ash forest 193
    reserve system 217
    soils 11
    structure and composition of vegetation 29
Chocolate Wattled Bat (*Chalinolobus morio*) 20, 22
clearfelling 7
    conventional operations 194–6
    impacts on ground cover and coarse woody debris 199–200
    impacts on multi-aged stands 198–9
    impacts on overstorey trees 197–8
    impacts on understorey vegetation 199
    landscape-level impacts 200–2
    stand-level impacts 197–200
    studies on the impacts 196–7
    see also logging
climate 11
climate change 261–2, 263–4
Common Brushtail Possum (*Trichosurus vulpecula*) 16
Common Eastern Froglet (*Crinia signifera*) 26, 27
Common Ringtail Possum (*Pseudocheirus peregrinus*) 14, 15–17, 18, 31, 32, 36, 83, 92, 102, 169, 170, 171
Common Wombat (*Vombatus ursinus*) 19, 20, 125
composition of the forest 45
Coventry's Skink (*Niveoscincus conventryi*) 25, 26
Crescent Honeyeater (*Phylidonyris pyrrhoptera*) 67, 117, 144
Crested Shrike-tit (*Falcunculus frontatus*) 21, 67
Crimson Rosella (*Platycercus elegans*) 21, 66, 67, 144, 145
Cunningham's Skink (*Egernia cunninghami*) 25, 26

den-swapping and arboreal marsupials 102–5
    possible reasons for 104–5
disturbance regimes 179
    human disturbance: logging 193–204
    important features of natural disturbances 182–3
    salvage logging effects 205–14
    wildfire 183–8
    see also logging; natural disturbance regimes: fire
Dogwood (*Cassinia aculeata*) 13, 115, 126
dusk counts 41
Dusky Antechinus (*Antechinus swainsonii*) 19–20, 39–40, 117, 142, 143, 145, 172, 237
Dusty Daisy Bush (*Olearia phlogopappa*) 13, 114

Eastern Broad-nosed Bat (*Scotorepens orion*) 20, 23
Eastern Falsistrelle (*Falsistrellus tasmaniensis*) 20, 23
Eastern Grey Kangaroo (*Macropus giganteus*) 19, 20
Eastern Pygmy Possum (*Cercartetus nanus*) 14, 15, 16, 17, 18, 30, 169
Eastern Small-eyed Snake (*Cryptophyis nigrescens*) 25, 26
Eastern Spinebill (*Acanthorhynchus tenuirostris*) 67, 144
Eastern Whipbird (*Psophodes olivaceus*) 67, 125
Eastern Yellow Robin (*Eopsaltria australis*) 21, 66, 67, 144

Echidna (*Tachyglossus aculeatus*) 19, 20
ecological processes 6–7, 268–9
ecological reserves 218, 221, 229
ecologically sustainable forest management 229, 234, 239, 243, 247, 258, 260–1, 269
ecosystem-oriented research 177, 268

fauna 14–28
    distribution and abundance (importance of trees with hollows) 82–3
    old-growth montane ash forest 57
    see also arboreal marsupials; animal communities, composition of; animal distribution and abundance; animal occurrence; animals, partitioning of tree hollow resources; terrestrial native mammals
Feathertail Glider (*Acrobates pygmaeus*) 14, 15, 16, 17, 18, 29, 30, 31, 34–5, 36, 83, 102, 169, 170, 171
field survey methods
    arboreal marsupials 29–38
    birds 40–1
    knowledge gaps 44
    small terrestrial mammals 38–40
    summary 42–4
    vegetation measures 41–2
fire *see* natural disturbance regimes: fire
fish 26–8
Flame Robin (*Petroica phoenicea*) 21, 67, 173
flora 11–14
forest cover 45
forest management 215
    ecologically sustainable 229, 234, 239, 243, 247, 258, 260–1, 269
    mitigating logging impacts 229–44
    monitoring 245–54
    reserves 217–27
forest patterns 6–7, 58–9, 68–9, 110, 122–3, 129–30, 148–9, 165–6, 177, 191–2, 203, 214, 226, 244, 254, 268–9
forest structure 71
forest thinning 267
Forest Wattle (*Acacia frigiscens*) 13, 114, 115, 117, 118, 141
frogs 26, 27

Gang-gang Cockatoo (*Callopcephalon fimbriatum*) 21, 67
genetic tools in field surveys 37–8, 39–40
geology 11
Golden Whistler (*Pachycephala pectoralis*) 67, 144
Gould's Long-eared Bat (*Nyctophilus gouldii*) 20, 23
Gould's Wattled Bat (*Chalinolobus gouldii*) 20, 22
Greater Glider (*Petauroides volans*) 14, 16, 17, 18, 29, 31, 32–3, 35, 57, 59, 65, 74, 82–4, 95, 101, 102, 105, 135, 139, 140, 141, 147, 152, 153, 160–1, 162, 163, 164, 165, 170, 171, 172, 188–9, 200, 219–20, 223, 224, 225, 249–50, 253, 261
Grey Fantail (*Rhipidura fuliginosa*) 67, 144

Grey Shrike-thrush (*Colluricincla harmonica*) 65, 67, 144, 145
ground cover, clearfelling impacts on 199–200

habitat islands 235–6, 238
hairtubing in field surveys 34–5, 39
Hazel Pomaderris (*Pomaderris aspera*) 13, 15
Highland Copperhead (*Austrelaps ramsayi*) 25, 26
Highlands Forest Skink (*Anepischtos maccoyi*) 25, 26
Horsefield's Bronze Cuckoo (*Chrysococcyx basilis*) 66
human disturbance *see* logging

invertebrates 28

King Parrot (*Alisterus scapularis*) 21
knowledge gaps and field surveys 44

landscape-level impacts of clearfelling 200–2
Large Forest Bat (*Vespadelus darlingtoni*) 20, 22
Laughing Kookaburra (*Dacelo novaeguineae*) 66, 67
Leadbeater's Possum (*Gymnobelideus leadbeateri*) 3, 4, 5, 14, 15
    meso-scale reserves 223
    PVA modelling of 153–60
        conservation, management recommendations 159–60
        group sizes for reintroduction 154–5
        importance of ensembles of habitat patches 156–8
        interpatch dispersal 155–6
        mitigating logging effects 158–9
        viability of single populations 154
        wildfire effects 158
Lesser Long-eared Bat (*Nyctophilus geoffroyi*) 20, 23
Little Forest Bat (*Vespadelus vulturnus*) 20, 22
logged sites and accelerated fall of trees with hollows 81–2
logging 193–203
    Central Highlands Victoria 193–4
    comparing logged and unlogged sites 196
    conventional clearfelling operations 194–6
    forest pattern 203
    knowledge gaps 202–3
    impacts on overstorey trees 197–8
    landscape-level impacts of clearfelling 200–2
    lessons learned 202
    mitigating impacts 158–9, 229–44
        background datasets 230
        cutting experiment 235–8
        forest pattern 244
        human and natural disturbances 230–2
        impacts of salvage logging 240–2
        knowledge gaps 243–4
        legacy retention in salvage logged areas 241
        lessons learned 243
        management at stand level 232–3

mechanical disturbance 241–2
post-harvest seedbed preparation 242
Silvicultural Systems Project 233–4
timing of salvage logging 242
Variable Retention Harvest System (VRHS) 234–5, 237, 238–40, 243
varied salvage intensity 240–1
montane ash forests 196
salvage logged area, legacy retention in 241
salvage logging effects 205–14
conversion of multi-aged stands 209–10
differences between conventional and 205–6
forest pattern 214
impacts of 240–2, 208–12
impacts on plants 210
impacts on stand structure 208
knowledge gaps 214
lessons learned 211–14
loss of biological legacies 208–9
montane ash forests 206–8
potential impacts 208–12
simulations on wildlife persistence 209
wildfires 2009 211
stand-level impacts of clearfelling 197–200
studies on the impacts of clearfelling 196–7
unloggable areas 222
see also clearfelling
logs 125–30
background datasets 126
biological legacies in montane ash forests 129–30
diameter and overall volume 127
forest pattern 129–30
key structural features of montane ash forests 125–6
knowledge gaps 129
lessons learned 129
and moss cover 128
Long-nosed Bandicoot (*Parameles nasuta*) 19, 20, 125

mammals *see* terrestrial native mammals
marsupials *see* arboreal marsupials
Masked Owl (*Tyto novaehollandiae*) 24
meso-scale reserves 221–4, 229
monitoring programs 245–54
background datasets 246
Central Highlands of Victoria 247–9
features of good 246–7
forest pattern 254
how to keep monitoring going 251–2
knowledge gaps 254
lessons learned 253–4
and other research studies 252–3
results 249–51
why they often fail 246
montane ash forests 9, 12
biology and ecology 7

carbon economy and management 262–4
carbon storage values 265–6
composition of the understorey and shrub layers 114–15
climate change 261–2
fauna of old-growth 57
logging 196
logs as biological legacies in 129–30
logs as key structural features 125–6
reserves 218–21
salvage logging effects 206–8
self-thinning 53
shifting baselines and levels of tree occupancy 101
moss cover and logs 128
Mountain Ash (*Eucalyptus regnans*) 9, 12–13, 28, 47, 49–55, 56–8, 63–4, 65–6, 67, 77–8, 90, 107, 114, 115, 118, 126, 127, 128, 145, 172–3, 176, 186, 188, 193, 211, 224, 235, 238–9, 263, 265
Mountain Brushtail Possum (*Trichosurus cunninghami*) 7, 14, 16, 17, 18, 29, 30, 31, 32, 33, 34, 37–8, 39, 62, 63, 65, 68, 75, 82, 83, 92, 93, 101–5, 110, 116–17, 119, 122, 125, 135, 136, 138, 139–40, 141–2, 148, 152, 153, 161–3, 165, 169, 170, 172, 199, 200–1, 219, 225, 249–50, 252
Mountain Correa (*Correa lawrenciana*) 13, 115
Mountain Hickory Wattle (*Acacia obliquinervia*) 13, 54, 114, 115, 117, 120
multi-aged stands
clearfelling impacts on 198–9
salvage logging effects 209–10
Musk Daisy Bush (*Olearia argophylla*) 13, 114, 126
Myrtle Beech (*Nothofagus cunninghamii*) 13, 15, 61–4, 65, 66, 69, 114, 115, 125, 143, 173, 188, 199

natural disturbance regimes: fire 181–92, 230–2
background datasets 182
carbon budgets and disturbance regimes 190
forest pattern 191–2
important features of natural disturbance 182–3
intermediate-intensity wildfires 186–7
intervals 187–8
knowledge gaps 191
lessons learned 191
post-fire ecological recovery 264–6
prescribed burning 190
viability of arboreal marsupial populations 188–90
wildfire 183–8
wildfires 2009 42, 58, 108, 164
nest boxes 35–7
effectiveness 107
occupancy 107
studies 106–7
wildfires 2009 108

old-growth forest 56–8, 189, 222
Olive Whistler (*Pachycephala olivacea*) 21

O'Shannassy Water Catchment 9, 55, 58, 77–8, 90, 91, 156, 157, 158, 164, 189, 190, 200, 218, 219, 220, 227, 262, 265
overstorey trees, impacts of clearfelling on 197–8
overstorey trees with hollows 73–111
    2009 wildfires 108
    arboreal marsupials and multiple tree hollows 102–5
    artificial hollows 106–7
    animal distribution and abundance 82–3
    background datasets 74–5
    bark streamers 75
    different kinds of hollow trees 75–7
    forest pattern 110–11
    knowledge gaps 109–10
    lessons learned 108–9
    nest box studies 106–7
    partitioning by animals 83–102
    recruitment 78
    tree decay and collapse 78–82
    where do hollow trees occur in the landscape? 77–8

Peron's Tree Frog (*Littoria peronii*) 26, 27
Pink Robin (*Petroica rodinogaster*) 65, 66, 67, 68
plot-level measures 41–2
Pobblebonk (*Limnodynastes dumerilii*) 26, 27
population viability analysis (PVA) 151–66
    2009 wildfires 164
    arboreal marsupials, comparisons of extinction risks 162
    arboreal marsupials, modelling of 152–3, 160–3
    background 151–2
    background datasets 153
    forest pattern 165–6
    Greater Glider 160–1
    knowledge gaps 165
    Leadbeater's Possum, modelling of 153–60
        conservation, management recommendations 159–60
        group sizes for reintroduction 154–5
        importance of ensembles of habitat patches 156–8
        interpatch dispersal 155–6
        mitigating logging effects 158–9
        viability of single populations 154
        wildfire effects 158
    lessons learned 164–5
    model testing 163–4
    Mountain Brushtail Possum 161–2
Powelltown State Forest 10, 106, 138, 207, 244
Powerful Owl (*Ninox strenua*) 24
prescribed burning 190

radio-tracking 37
rainforest 61–9
    and arboreal marsupials 65
    background datasets 62–3
    and birds 65–6, 67
    cool temperate 64–6
    forest pattern 68–9
    knowledge gaps 68
    lessons learned 67
    Myrtle Beech 63–4
    and small terrestrial mammals 64–5
RAMAS 152, 153, 154
rank abundance distributions (RADs) 174–5
Red-bellied Black Snake (*Pseudechis porphyriacus*) 25, 26
Red-browed Treecreeper (*Climacteris erythrops*) 21
Red-necked Wallaby (*Macropus rufogriseus*) 19, 20
reptiles 24–6
research projects, maintaining long-term 259–60
reserves 217–27
    biodiversity conservation 217–27
    ecological 218, 221, 229
    forest management 217–27
    forest pattern 226
    knowledge gaps 226
    lessons learned 225–6
    limitations of large ecological 221
    meso-scale 221–4, 229
    montane ash 218–21
    why large ecological reserves are important 218
    wilderness and biodiversity conservation 221
    Yarra Ranges National Park 218–21
Resource Assessment Commission (RAC) 55, 56
Rough Tree Fern (*Cyathea australis*) 13, 15, 114

salvage logging effects 205–14
    conversion of multi-aged stands 209–10
    differences between conventional and 205–6
    forest pattern 214
    impacts of 240–2, 208–12
    impacts on plants 210
    impacts on stand structure 208
    knowledge gaps 214
    lessons learned 211–14
    loss of biological legacies 208–9
    montane ash forests 206–8
    potential impacts 208–12
    simulations on wildlife persistence 209
    wildfires 2009 211
Sambar Deer (*Cervus unicolor*) 21, 267
Shining Gum (*Eucalyptus nitens*) 9, 12–13, 14, 47, 49–52, 58, 63–4, 77, 78, 90, 144, 172–3, 248
shrub layers 13
    animal occurrence 119–20
    and relationship with the collapse of overstory trees with hollows 121
Silver Wattle (*Acacia dealbata*) 13, 15, 114, 115, 116, 118, 120, 141
Silvereye (*Zosterops lateralis*) 67, 144
silvicultural systems 7, 107, 109, 164, 231–2, 233–5, 238, 243, 268

Soft Tree Fern (*Dicksonia antarctica*) 13, 114
soils 11
Sooty Owl (*Tyto tenebricosa*) 4, 24, 57, 59, 200, 219, 224
Southern Forest Bat (*Vespadelus regulus*) 20, 22
Southern Sassafras (*Atherosperma moschatum*) 13, 114
Southern Toadlet (*Pseudophryne semiornata*) 26, 27
Southern Water Skink (*Eulamprus tympanum*) 25, 26
Spencer's Skink (*Pseudomoia spenceri*) 25, 26
spotlighting 32–3
Spotted Pardalote (*Pardalotus punctatus*) 65, 67
Spotted-tailed Quoll (*Dasyurus maculatus*) 19, 20
stagwatching 29–32, 43, 75, 83, 248, 258
stand-level impacts of clearfelling 197–200
stand-level measures 41–2
statistical science, critical roles of 258
Stinkwood (*Zieria arborescens*) 14, 115
Striated Pardalote (*Pardalotus striatus*) 57, 67, 144, 145
Striated Thornbill (*Acanthiza lineata*) 67, 144, 145
Sugar Glider (*Petaurus breviceps*) 14, 16, 17, 18, 29, 31, 32–3, 35, 36, 82, 83, 102, 105, 118, 135, 140, 141, 170, 171, 172, 225, 249
Superb Lyrebird (*Menura novaehollandiae*) 21, 28, 67, 125
Swamp Rat (*Rattus lutreolus*) 19, 20
Swamp Wallaby (*Wallabia bicolor*) 19

terrestrial feral animals 21
terrestrial native mammals 17–20
    distribution and abundance 142–3
        microhabitat-level factors 142–3
        patch-level factors 142
    occurrence in understorey and shrub layers 115, 117
    and rainforests 64–5
thinning, forest 267
Tiger Snake (*Notechis scutatus*) 24, 25, 26
Toolangi State Forest 10, 106, 253, 244
trapping and field surveys 33–4, 38–9
tree decay and collapse 78–82
tree fern layer 120–1
Tree Geebung (*Persoonia arborea*) 13, 115
tree hollows
    accelerated tree fall on logged sites 81–2
    animal distribution and abundance 82–3
    artificial 106–7
    characteristics of occupied trees 83–4
    co-occupancy 101–4
    datasets 74–5
    den-swapping 102–5
    different kinds of hollow trees in montane ash forest landscapes 75–7
    factors influencing occupancy 84–102
    long-term tree occupancy 101
    multiple use by arboreal marsupials 102–5
    partitioning of resources by animals 83–102
    recruitment 78
    shifting baselines and levels of tree occupancy 101
    tree decay and collapse 78–82
    where do they occur in the landscape? 77–8
    wildfires 2009 108
tree-level measures 41
Tussock Skink (*Pseudomoia entrecasteauxii*) 25, 26

understorey trees 13, 113–23
    animal occurrence 115–19
    arboreal marsupials 115–17
    background datasets 115
    birds 117
    broad composition in montane ash forests 114–15
    disturbance 120–1
    forest pattern 122–3
    knowledge gaps 122
    layer and relationship with the collapse of overstorey trees with hollows 121
    lessons learned 122
    small terrestrial mammals 117
    variation in nutrient content of wattle tree species 118–19
understorey vegetation, clearfelling impacts on 199

Variable Retention Harvest System (VRHS) 234–5, 237, 238–40, 243
vegetation measures (field surveys) 41–2, 43
vertebrate biota and rainforests 64–6
Victorian Christmas Bush (*Prostanthera lasianthos*) 14, 115
volunteers, critical role of 258–9
VORTEX 152, 153, 154, 155, 161, 162

wattle tree species, variation in nutrient content of 118–19
Whistling Tree Frog (*Littoria ewingii*) 26, 27
White-browed Scrubwren (*Sericornis frontalis*) 67, 144, 145
White-striped Freetail Bat (*Austronomus australis*) 20, 22
White-throated treecreeper (*Cormobates leucophaeus*) 67, 117, 144, 145
White's Skink (*Liopholis whitii*) 25, 26
wildlife corridors 101, 160, 218, 222, 223, 225, 226
wildfire (natural disturbance) 183–8
    intermediate-intensity 186–7
    see also fire
wildfires 2009 42, 58, 108, 164, 211

Yarra Ranges National Park 217, 218–21
Yellow-bellied Glider (*Petaurus australis*) 14, 16, 17, 18, 32, 56, 57, 59, 83, 102, 105, 135, 139, 146, 170, 171, 200, 219, 224, 231, 249, 263
Yellow-tailed Black Cockatoo (*Calyptorhynchus funereus*) 21, 263, 264